Biology *of* Spiders

Second Edition

RAINER F. FOELIX

New York Oxford
OXFORD UNIVERSITY PRESS
GEORG THIEME VERLAG
1996

Oxford University Press

Oxford New York
Athens Auckland Bangkok Bogota Bombay
Buenos Aires Calcutta Cape Town Dar es Salaam
Delhi Florence Hong Kong Istanbul Karachi
Kuala Lumpur Madras Madrid Melbourne
Mexico City Nairobi Paris Singapore
Taipei Tokyo Toronto

and associated companies in
Berlin Ibadan

Published by Oxford University Press, Inc.
198 Madison Avenue, New York, New York 10016

Oxford is a registered trademark of Oxford University Press

Library of Congress Cataloging-in-Publication Data
Foelix, Rainer F., 1943–
[Biologie der Spinnen. English]
Biology of spiders / Rainer F. Foelix. — 2nd ed.
p. cm.
Includes bibliographical references and index.
ISBN 0-19-509593-6; ISBN 0-19-509594-4 (pbk.)
1. Spiders. I. Title.
QL458.4.F6313 1996
595.4′4—dc20 95-47791

5 7 9 8 6 4

Printed in the United States of America
on acid free paper

Preface

It certainly came as a surprise that the first edition of *Biology of Spiders* (1982) was out of print within a few years. It has taken me several more years to prepare an updated second edition, first in German (1992) and now in English. For a single author, it has been a real challenge to cover the entire literature on arachnology, given the many thousands of publications that have appeared over the past ten years. The constant dilemma I was faced with was what to include from this wealth of information and what to omit. My main goal has always been to write a *readable* book on the biology of spiders and not a comprehensive review of arachnology. This approach seemed justified, judging from the reactions to the first edition. Some critical remarks were, nevertheless, taken to heart and I hope that the improvements can be recognized in the new edition. There will still be a few topics—especially those I am not directly familiar with—that may not be covered as thoroughly as some of my colleagues wished. I can only ask their indulgence for any shortcomings, but also to voice their criticisms.

I would like to thank all my colleagues who have given me their critiques and suggestions on the previous book and have thus helped to improve this new edition. Particular thanks go to Friedrich Barth, Elisabeth Bauchhenss, Jon Coddington, Heinrich Homann, Herb Levi, Yael Lubin, Rüdiger Paul, Hans Peters, Jerome Rovner, Paul Selden, Ernst-August Seyfarth, Gabriele Uhl, Fritz Vollrath, and Jürg Wunderlich. I am also grateful to many other arachnologists who have generously provided me with their exceptional photographs or diagrams. My warmest thanks to my friend and colleague Peter Kaufmann for his unfailing support on the scanning electron microscope and on computer matters, and also to Daniel Palestrina for his advice to improve many illustrations.

This second edition is dedicated to the late Dr. H. Homann (1894–1993), who has been a fatherly friend and my mentor in arachnology. His scientific work as well as his modest and kind personality will remain to be an inspiration to me.

Aarau
Spring 1996

R. F. F.

Contents

1

An Introduction to Spiders

Spiders are distributed all over the world and have conquered all ecological environments, with perhaps the exception of the air and the open sea. Most spiders are relatively small (2–10 mm body length), yet some large "tarantulas" may reach a body length of 80–90 mm. Male spiders are almost always smaller and have a shorter life span than females.

All spiders are carnivorous. Many are specialized as snare builders (web spiders), whereas others hunt their victims (ground spiders or wandering spiders). Insects constitute the major source of prey for spiders, but certain other arthropods are often consumed as well.

A spider's body consists of two main parts: an anterior portion, the *prosoma* (or cephalothorax), and a posterior part, the *opisthosoma* (or abdomen). These are connected by a narrow stalk, the *pedicel* (Fig. 1). The prosoma's functions are mainly for locomotion, for food uptake, and for nervous integration (as the site of the central nervous system). In contrast, the opisthosoma fulfills chiefly vegetative tasks: digestion, circulation, respiration, excretion, reproduction, and silk production.

The prosoma is covered by a dorsal and a ventral plate, the *carapace* and the *sternum*, respectively. It serves as the place of attachment for six pairs of extremities: one pair of biting chelicerae and one pair of leglike pedipalps are situated in front of four pairs of walking legs. In mature male spiders the pedipalps are modified into copulatory organs—a quite extraordinary feature, not found in any other arthropod. The opisthosoma is usually unsegmented, except in some spiders considered to have evolved from ancient species (Mesothelae). In contrast to the firm prosoma, the abdomen is rather soft and sacklike; it carries the spinnerets on its posterior end.

A Sketch of Spider Systematics

At present taxonomists recognize about 34,000 spider species which they group into approximately 100 families. How this diversity should be arranged into a "natural" system of classification is still very much a matter of controversy. This is best illustrated by the fact that about 20 different spider classifications have been proposed since 1900.

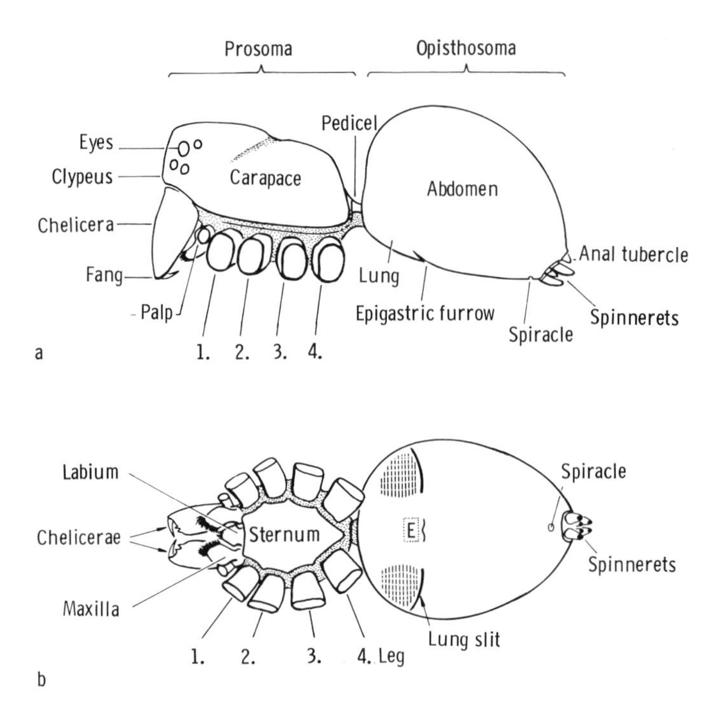

FIG. 1. External appearance of a spider's body: (*a*) lateral view; (*b*) ventral view, E = epigynum.

The order of spiders (Araneae) is usually divided into three suborders, the Mesothelae, the Mygalomorphae, and the Araneomorphae. Until recently the Mygalomorphae were referred to as Orthognatha because of the nearly parallel alignment of their chelicerae, while the Araneomorphae correspond to the former Labidognatha, which possess vertical chelicerae opposing each other (Fig. 2). The Mesothelae represent the phylogenetically oldest spiders because they exhibit a clearly segmented abdomen, as well as several other "primitive" characters. The Mygalomorphae comprise all the "tarantulas;" their chelicerae lie parallel to each other and their spinnerets are often reduced.

More than 90% of all spiders belong to the Araneomorphae (Labidognatha). Their classification into higher taxa is still problematical. Formerly one separated the Cribellatae from the Ecribellatae, based on the presence of a spinning plate (*cribellum*) situated in front of the spinnerets as the distinguishing character of the Cribellatae. All Araneomorphae without such a cribellum were grouped together as Ecribellatae. Nowadays it is generally assumed that all spiders were originally cribellate, and that the ecribellate spiders evolved later by a reduction or loss of the cribellum. However, several aspects remain unclear—for example, possible parallel evolutions (convergences) among cribellate and ecribellate spiders. This "Cribellate problem" will be discussed in more detail in chapter 10.

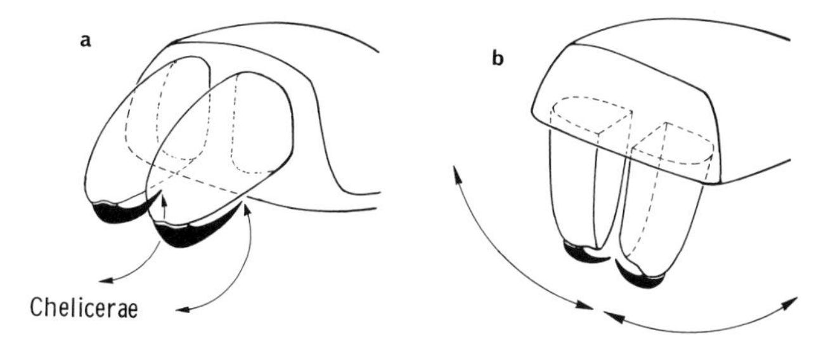

FIG. 2. Movement of the chelicerae in (*a*) orthognath and (*b*) labidognath spiders. (After Kaestner, 1969.)

Among the Ecribellatae, some spider families with simple genital structures —the so-called Haplogynae—were separated from those with complex genital structures, the Entelegynae. This classifiction dates back to Eugène Simon's "Histoire Naturelle des Araignées" (1892–1903). Over the past years, however, several arachnologists have voiced the opinion that the Haplogynae are not really a homogenous group (Brignoli, 1975; Lehtinen, 1975; Platnick, 1975). Despite their arguments, there are still some families that represent "classical" haplogyne spiders—for instance, the Scytotidae, the Pholcidae, and the Dysderidae.

Following Simon's classification, the Entelegynae were further divided into Dionycha and Trionycha, depending on whether the walking legs have two or three tarsal claws. Although this subdivision also became questionable, there is again some justification to maintain at least some "classical" Dionycha, such as the Salticidae, the Clubionidae, and the Thomisidae.

Since the following text will often refer to certain spider families, the main families and their systematic position will be listed here (for more details of modern spider systematics, see Fig. 215).

Order Araneae

1. Suborder Mesothelae	(1 family)
Fam. Liphistiidae	(40 species)
2. Suborder Mygalomorphae (Orthognatha)	(15 families)
Fam. Atypidae	(30 species)
Ctenizidae	(400 species)
Dipluridae	(250 species)
Theraphosidae	(800 species)
3. Suborder Araneomorphae (Labidognatha)	(90 families)
Fam. Dysderidae	(250 species)
Pholcidae	(350 species)
Scytotidae	(150 species)
Amaurobiidae	(350 species)
Dictynidae	(500 species)
Eresidae	(100 species)

Clubionidae	(1,500 species)
Gnaphosidae	(2,200 species)
Salticidae	(4,400 species)
Thomisidae	(2,000 species)
Lycosidae	(2,200 species)
Pisauridae	(600 species)
Oxyopidae	(500 species)
Agelenidae	(800 species)
Araneidae	(2,600 species)
Linyphiidae	(3,700 species)
Theridiidae	(2,200 species)
Uloboridae	(200 species)

In order to familiarize the uninitiated reader with this seemingly abstract system, the following natural history of some spider families will serve as an introduction.

Funnel-web Spiders (Agelenidae)

Funnel-web spiders are familiar to most of us. In European houses, for example, we find *Tegenaria* usually in the bathroom, often trapped in the tub, where it cannot scale the smooth walls. Aside from its considerable size (10 mm body length), *Tegenaria* is quite conspicuous because of its long, hairy legs (12–18 mm) and the two long spinnerets protruding from its abdomen (Fig. 3). Outdoors we can readily find the somewhat smaller *Agelena* in short grass or low bushes. The sheet webs of agelenids usually cover vegetation, or bridge the corners of buildings. The flat web narrows like a funnel on one end, forming a small silken tube. This retreat is open on both ends, and most of the time the spider sits there in ambush, its outstretched front legs poised to receive vibrations from

Fig. 3. (*a*) Juvenile house spider (*Tegenaria*) sitting at the entrance of her retreat. (*b*) Typical agelenid sheet web, covered with early morning dew. (Photo: Paas.)

the web. When an insect blunders onto the web, the spider quickly darts out from its hideout, bites the victim, and carries it back. The actual feeding process always takes place inside the retreat. During the return to the tube the spider shows remarkably good orientation. For this reason funnel-web spiders have been a favorite subject for sensory physiologists (see chapter 4).

The water spider *Argyroneta aquatica* was long considered to be a member of the agelenid family, but is now placed into its own family (Argyronetidae). It is the only spider that lives constantly under water. Rather than build a web, she attaches an air bubble to a water plant and uses it as a residence. She hunts mostly fly larvae or small crustaceans, which she catches as she swims about freely under water. To eat the prey the spider must return to her diving bell. The abdomen of the water spider is always encased in a shiny air bubble, and this silvery reflection has earned her the scientific name *Argyroneta* (Greek, *argyros* = silver). From time to time the air bag is replenished at the water surface. Thus the respiration of a water spider does not differ in principle from that of her land-living relatives.

Orb-web Spiders (Araneidae)

The most impressive web design belongs to the orb weavers. The orb web of the common garden spider certainly represents the best-known type of all webs (Figs 108, 110). The spider either sits right in the center of the web or hides in a retreat outside it. Insects flying into the web become stuck to the sticky threads long enough for the spider to rush out from the hub to bite or wrap its victim.

Araneids are among the most successful spider families, as the enormous diversity of their species (2,600) testifies. Thus it comes as no surprise to find that there are hundreds of structural variations on the orb-web design as well. Some examples will be given in chapter 5. The body structures of araneids may also vary considerably; most notable are the tropical orb weavers, which can be very colorful and exotically shaped (Fig. 4).

An orb web is typical not only of the Araneidae but also two other spider families, the Tetragnathidae and the Uloboridae. Uloborids build an orb web that is very similar to the webs of the araneids, but that differs from them in one important aspect: the catching threads are not studded with glue droplets but are decorated instead with an extremely fine mesh of cribellate silk ("hackle band;" Fig. 104).

Wolf Spiders (Lycosidae)

Wolf spiders are vagabonds that lie in ambush or freely hunt their prey. They are best recognized by their characteristic eye arrangement of four uniformly small eyes in the anterior row of eyes and two large median eyes in the posterior row (Figs 5, 9). About 2,200 different species occur all over the world, and they may vary quite a bit in size. Smaller wolf spiders (4–10 mm body length) roam freely among stones or low vegetation; only the larger representatives (*Arctosa, Trochosa, Alopecosa*; 10–20 mm) dig burrows. Certain species live close to the

FIG. 4. (*a*) The best known of all orb weavers, the garden spider *Araneus diadematus*. (Photo: Grocki and Foelix.) (*b*) This *Gasteracantha versicolor* from Madagascar bears long spines on its abdomen, as exotic araneids often do. (Photo: Emerit.) (*c*) A North American orb weaver, *Micrathena gracilis*, hanging on a horizontal thread. Note the position of the spinnerets (arrow).

water and can even walk on its surface (Fig. 133). Members of the aptly named genus *Pirata* hunt insects on the water surface, or even dive after tadpoles or small fish (Gettmann, 1978). A few species of wolf spiders (*Aulonia, Hippasa, Sossipus*), thought to be more primitive varieties, actually build webs reminiscent of the sheet webs of agelenids (Brady, 1962; Job, 1968, 1974).

The most famous wolf spider is certainly the Mediterranean tarantula (the name being derived from the Italian town of Taranto). True tarantulas (*Lycosa, Hogna*) can reach an impressive 30 mm of body length, but they are not related to the big tropical "tarantulas," the mygalomorphs, also known as bird spiders. Although tarantulas have long had a reputation as dangerous spiders, the ancient fear of their poisonous bite has been proven to be quite wrong. Probably any bites alleged to be from a tarantula were in fact inflicted by black widow spiders. Tarantulas live in silk-lined burrows in the soil. Some species even construct a sort of lid to close the tube, creating a burrow quite similar to that of the trapdoor spiders (Ctenizidae). At night tarantulas leave their burrows to

FIG. 5. Portrait of the famous Mediterranean tarantula (*Lycosa tarentula*). (Photo: Orta-Ocaña.)

prowl in search of insects. However, wolf spiders generally do not actively run down their prey, as their name might suggest, but sit quietly and wait until a victim happens to come by (Ford, 1978; Stratton, 1985).

Wolf spiders react mainly to vibrations caused by the wing beat or by the characteristic walking pattern of the prey. Visual cues also play a role in detecting prey, but the eyes of wolf spiders perceive only a coarse image, and thus only objects very close by can serve as visual stimuli. This becomes apparent during courtship, when the dark palps or front legs of the male are waved in a species-specific manner to attract the attention of the female.

Female wolf spiders are well known for their brood care. After laying their eggs, they attach the egg case to their spinnerets and carry it around wherever they go. Some weeks later, just before the young spiderlings are ready to leave the cocoon, the mother rips the cocoon wall so that the young can emerge. As soon as the spiderlings have crawled out, they clamber onto their mother's back (Fig. 176). Since they may number more than one hundred, they huddle there in several layers. They ride on their mother's abdomen for about a week, then gradually disperse and take in food for the first time.

Another group of spiders, the Ctenidae, is often considered as a separate family, but can also be classified as a subfamily of the Lycosidae (Homann, 1971). The most notorious ctenid spider is the extremely poisonous and aggressive *Phoneutria* (Fig. 39d). A less ferocious ctenid spider, which will be mentioned in many of the following chapters, is *Cupiennius salei* from Central America (Fig. 137).

FIG. 6. (*a*) The crab spider *Misumenoides formosipes* is eye-catching because of its bright yellow abdomen. Note the strong front legs, which are used to seize prey. (*b*) Portrait of the the crab spider *Xysticus*. (Photo: Beinbrech.)

Crab Spiders (Thomisidae)

Crab spiders lie quietly in ambush and do not build webs (Fig. 6). They sit motionless on leaves or in blossoms where, with attentively outstretched legs, they await landing insects. Their small eyes can produce sharp images only at very short distances, yet they perceive motions as far as 20 cm away (Homann, 1934). If prey comes within reach (0.5–1 cm), it is seized by the spider's strong front legs and then paralyzed by its poisonous bite. Even quite large insects, such as butterflies or bumblebees, are successfully attacked (Fig. 195b). The victim is sucked out through the tiny bite holes. Since its exoskeleton remains intact, the victim appears practically unharmed when the spider has finished its meal (Foelix, 1996).

Crab spiders may be very colorful—they are often white or bright yellow, and some are green. To some degree adult females can adapt their coloration to the background on which they sit. Even the less colorful species are usually well camouflaged and are hard to detect among the vegetation. The name crab spider comes from their ability to walk sideways very adroitly. The family

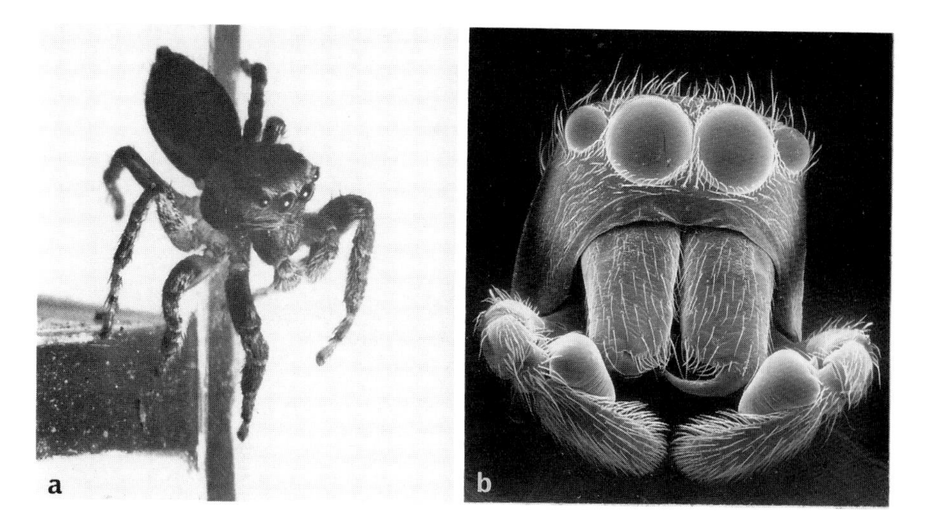

FIG. 7. (*a*) Jumping spider about to leap off an edge. The front legs are raised; only the hind legs provide the thrust for the jump. (*b*) Portrait of the jumping spider *Heliophanes*. (Photo: Chu and Foelix.)

Philodromidae was formerly grouped together with the Thomisidae, although they bear only a superficial resemblance to crab spiders (Homann, 1975). Most notably, their legs are all of equal length, a feature typical of wandering spiders.

Jumping Spiders (Salticidae)

At least for an arachnologist the "jumpers" are among the most attractive, if not congenial, spiders. They are all rather small (3–10 mm) with short, stout legs and a square prosoma. Most conspicuous are the anterior eyes, which occupy the entire front of the carapace (Fig. 7). Jumping spiders react very definitely to visual stimuli such as passing insects or the approaching finger of an observer: first they turn to face the object, then they walk closer. They can distinguish different shapes at a distance of less than 10 cm; this limit can be observed while the spider stalks prey, and also during courtship. Males often possess conspicuously marked legs, which they use for display toward the females. The hunting behavior of jumping spiders reminds one of cats, for they stalk and pursue their prey until it is close enough for a final pounce. Long jumps up to 16 cm can be seen when a jumping spider flees. Before it jumps, the spider always attaches a safety thread to the ground, so that if it falls off an edge it will simply be held back by the thread and can quickly climb back the few centimeters to the point of takeoff.

Jumping spiders are most active during the day. They prefer sunshine; in cloudy or rainy weather they withdraw inside small silken nests. These shelters not only protect them from the environment but also permit them to molt safely, to build egg cases, and to hibernate.

2

Functional Anatomy

Prosoma

The dorsal plate of the prosoma is called the carapace. It bears a distinct indentation along its midline (Fig. 8). This indented area extends on the inside of the carapace as a solid cuticular ridge, which serves as the attachment site for the dorsal muscles of the sucking stomach. Several furrows radiate from the dorsal groove: two shallow, diverging lines separate an anterior cephalic part from a posterior, thoracic part. Even fainter subdivisions may be recognized laterally, pointing from the center of the carapace toward the coxae of the legs. It seems doubtful whether these furrows correspond to the original segmentation of the prosoma, although embryological studies indicate that the prosoma is formed by six fused segments.

The "head" part of the prosoma bears the eyes and the chelicerae. Most spiders have eight eyes, which are arranged in specific patterns in the various families. Usually the eyes lie in two rows (sometimes in three), and accordingly they are referred to as anterior lateral eyes (ALE), anterior median eyes (AME), posterior lateral eyes (PLE), and posterior median eyes (PME) (Fig. 9). The relative position of the eyes is very important for the systematic classification of spiders. Just by looking at the arrangement and relative size of the different eyes one can often immediately determine the family of a particular spider. For example, all jumping spiders (Salticidae) possess one row of large frontal eyes, with the AME being the largest; wolf spiders have four uniformly small frontal eyes, but rather large PLE and PME (Figs 5, 9). The area between the anterior row of eyes and the edge of the carapace is called the *clypeus*.

Often the eyes stand close together in pairs, sometimes clearly raised above the surrounding cuticle, as in the Theraphosidae and Filistatidae. In their most extreme arrangement, we find them perched on the ends of eye stalks, as in some midget spiders (Micryphantidae; Fig. 10). Often the males of this family have prosomal glands, which open through tiny pores onto the head region. Their secretion is sucked up by the females during courtship and apparently serves a special function in sexual behavior (Blest and Taylor, 1977; Schaible et al., 1986). Some spiders have fewer than the usual eight eyes. The so-called six-eyed spiders (Dysderidae), the spitting spiders (Scytodidae), and some daddy-longlegs spiders (Pholcidae), have only six. In some species we find a reduction to four (*Tetrablemma*, Theridiidae) or even to two eyes (*Nops*,

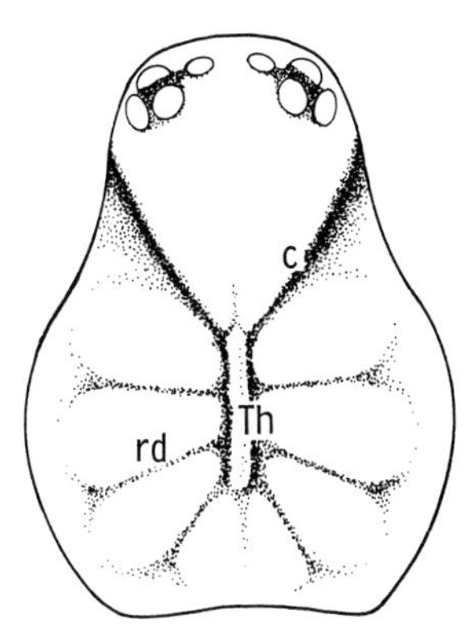

FIG. 8. Carapace of the "primitive" spider *Hypochilus*, dorsal view. Two grooves (*c*) extend from the thoracic furrow (Th), separating the carapace into a "head" and a "thoracic" portion. The latter is further subdivided by radial furrows (rd).

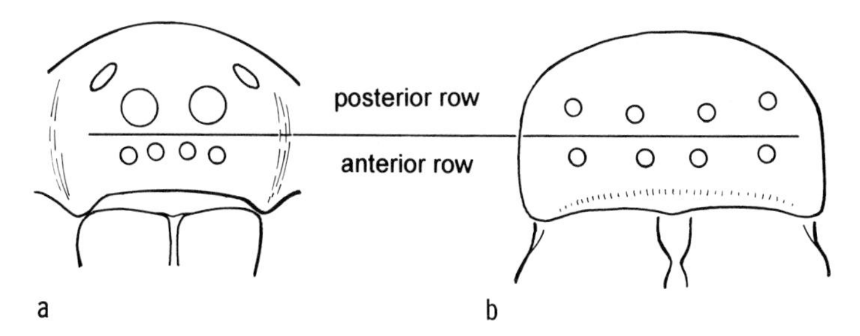

FIG. 9. Eye pattern in (*a*) the wolf spider *Lycosa* and (*b*) the orb weaver *Tetragnatha*. The wolf spider has uniformly small eyes in the anterior row, the orb weaver has relatively small eyes in both rows of eyes. [(*a*) After Kaston, 1972; (*b*) After Bristowe, 1958.]

Caponiidae), and some cave-dwelling spiders have lost their eyes altogether (Millot, 1949).

Sternum

An undivided sternal plate (sternum) lies on the ventral side of the prosoma (Fig. 1). Developmentally it is derived from four fused *sternites*; the partition lines can sometimes just be discerned in very young spiders. Anteriorly, a small medial plate, the *labium*, is attached to the sternum. In most spiders, the labium and sternum are hinged together by a cuticular membrane (Fig. 11). Both the sternum and carapace are rather stiff parts of the prosomal exoskeleton. They are, however, connected by a soft pliable area, the *pleurae* (the stippled area of Fig. 1), which enables them to move in relation to each other.

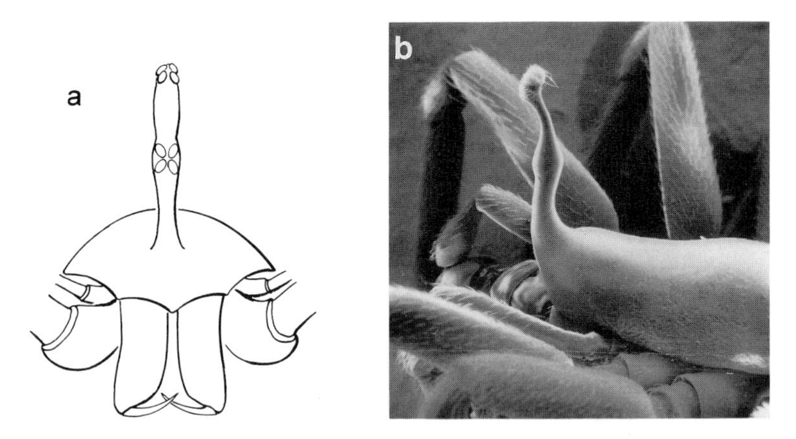

FIG. 10. Aberrant position of the eyes in a male midget spider, *Walckenaeria acuminata*. (*a*) Front view of the prosoma with eye stalk; (*b*) lateral view, 80 ×. [(*a*) After Crome, 1957; (*b*) Photo: Foelix and Kaufmann.]

Chelicerae

The chelicerae are the first appendages of the prosoma. In the spider embryo they lie behind the mouth opening, but during subsequent development they migrate to an anterior position, as do the antennae of other arthropods. Each chelicera consists of two parts, a stout basal part and a movable articulated fang. The inner edge of the fang is finely serrated and is apparently used to clip silk threads (H. M. Peters, 1982). Normally the fang rests in a groove of the

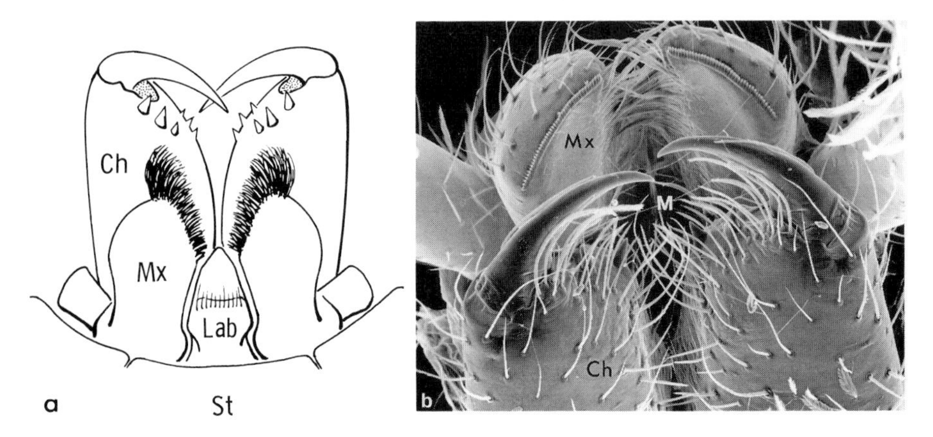

FIG. 11. (*a*) Mouth parts of a wolf spider (*Lycosa*) as seen ventrally. Ch = chelicerae, Lab = labium, Mx = maxilla (endite), St = sternum. (After Kaston, 1972.)
(*b*) Mouthparts of a jumping spider (*Portia*), frontal view. The mouth opening (M) lies between the two maxillae (Mx) and behind the chelicerae (Ch). 215 ×.

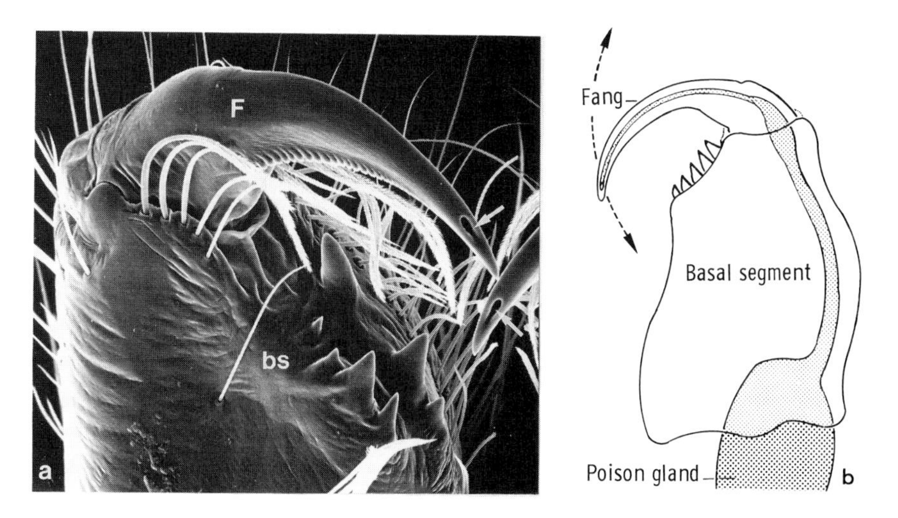

Fig. 12. (*a*) The movable cheliceral fang (F) normally rests between the cheliceral teeth of the basal segment (bs). Note the serrated edge of the fang and the opening of the poison gland (arrow) (*Portia*, 250 ×). (*b*) Schematic drawing of a chelicera. Arrows indicate possible movements of the fang. (After Millot, 1930, 1949.)

basal segment like the blade of a pocket knife. When the spider bites, the fangs move out of their groove and penetrate the prey. At the same time, poison is injected through a tiny opening at the tip of the fang (Fig. 12).

Both sides of the cheliceral groove are often armed with cuticular teeth. These act as a buttress for the movable fang. Spiders whose chelicerae are equipped with such teeth mash their prey into an unrecognizable mass. Spiders without such teeth (e.g., thomisids) can only suck out their victims through the small bite holes (Fig. 36a) (Homann, 1985; Foelix, 1996). The number and size of the cheliceral teeth are important diagnostic characteristics for the taxonomist.

The chelicerae are used not only for subduing prey or for defense, but also serve as "pliers" for all kinds of grasping; therefore they have been referred to as the spider's "hands." The versatility of the chelicerae is best demonstrated when we look at their main tasks in various spider families. Trapdoor spiders (Ctenizidae) use their chelicerae to dig burrows, nursery-web spiders (Pisauridae) for carrying egg cocoons, and orb weavers (Araneidae) to transport small prey. In the long-jawed orb weavers (Tetragnathidae) and the cribellate Dictynidae, the males and females interlock their chelicerae during mating.

Some spiders, such as the Archaeidae from Madagascar, possess extremely large chelicerae (Fig. 13a), and they literally spear their prey with these enormous tools. Often male spiders have larger chelicerae than the females (Fig. 13b). The common zebra spider, *Salticus scenicus* (Fig. 140), is a typical example of such sexual dimorphism. In another jumping spider (*Myrmarachne*), the male's chelicerae are five times larger than in the female, but they lack a fang

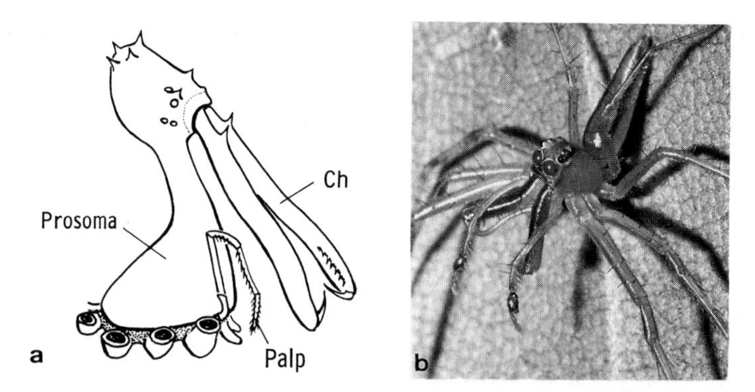

Fig. 13. Extreme development of the chelicerae. (*a*) In *Archaea workmani.* (After Legendre, 1972.) (*b*) In the male lynx spider *Lyssomanes viridis.* (Photo: Hill.)

duct and therefore cannot envenom prey (Pollard, 1994). Instead, they skewer their victims on their elongated cheliceral fangs. Finally, some money spiders (Linyphiidae) possess chelicerae with stridulatory organs for producing sounds (see chapter 9).

It has already been noted that Orthognatha and Labidognatha move their chelicerae in quite different manners (Fig. 2). The opposed chelicerae of Labidognatha have supposedly the advantage of a much larger span, so that even rather large prey could be overpowered (Kaestner, 1952). However, direct observation and experiments do not support this theory. Many orthognath (mygalomorph) spiders subdue quite large prey, sometimes even twice their own size (Nentwig and Wissel, 1986). This is due to other factors than the cheliceral span, e.g., an aggressive wrapping behavior.

There is another commonly held belief that has been challenged recently, namely the strict division into "primitive" orthognath and "higher" labidognath spiders (Kraus and Kraus, 1993). Some ancient spiders, e.g., Mesothelae or Hypochilidae, have their chelicerae neither parallel nor opposing each other, but in an intermediate position termed *plagiognath* (Fig. 14). It is conceivable that this condition actually represents the older arrangement from which orthognathy and labidognathy was derived.

Pedipalps and Mouth Parts

The second pair of appendages are the *pedipalps.* Their segmentation corresponds to that of the legs, except that one segment, the *metatarsus,* is lacking (Fig. 16). Despite their general resemblance to legs, the palps are usually not used for locomotion. Instead, they often play a role during prey catching, when they constantly touch and manipulate the prey.

The most notable modification of the palps is found in male spiders. Male palps act as copulatory devices. The coxae of the pedipalps represent another specialization, for these have been transformed into chewing mouth parts

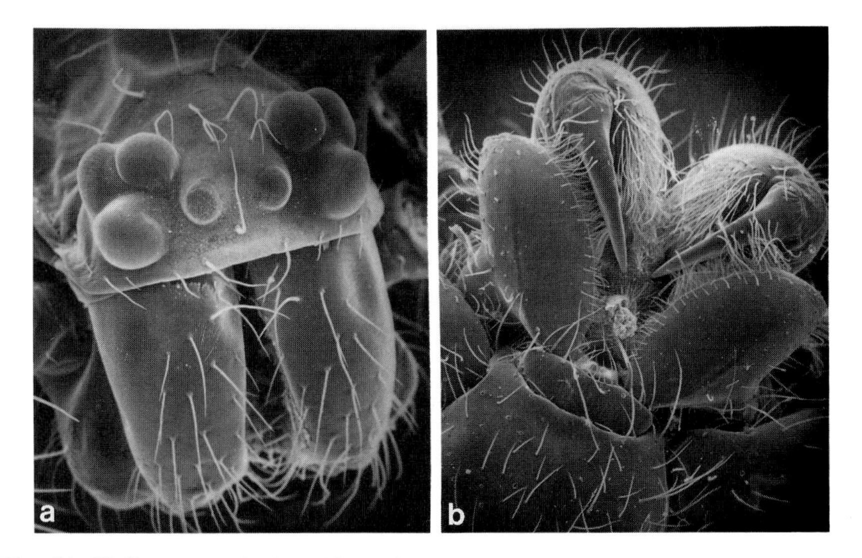

Fig. 14. Chelicerae can be found in an intermediary position between orthognath and labidognath in some "primitive" spiders, e.g., in *Hypochilus*. (*a*) Frontal view, (*b*) ventral view, 50 ×. (Photos: Gnatzy and Foelix.)

(*maxillae* or *endites*). In "primitive" spiders, such as Mesothelae and tarantulas, the maxillae are only slightly modified, whereas in the Labidognatha they are broadened laterally. In most Labidognatha the anterior rim of each maxilla is clearly serrated; this rim, the *serrula* (Fig. 15), is used as a saw for cutting into the prey. The inner sides of the maxillae are fringed by a dense cover of hair that acts as a filter while the spider is sucking in the liquefied food.

The mouth opening is bordered laterally by the maxillae, in front by the rostrum, and in the back by the labium. These four mouth parts form the mouth proper, which leads into a flattened *pharynx*. The pharynx consists of a movable, hinged front (rostrum) and a backwall (labium) and is lined by cuticular platelets. These contain very fine grooves covered by small teeth, which together function as a microfilter. The pharyngeal lumen can be widened by the action of several muscle bands (Fig. 27). Thus the pharynx acts as a suction pump.

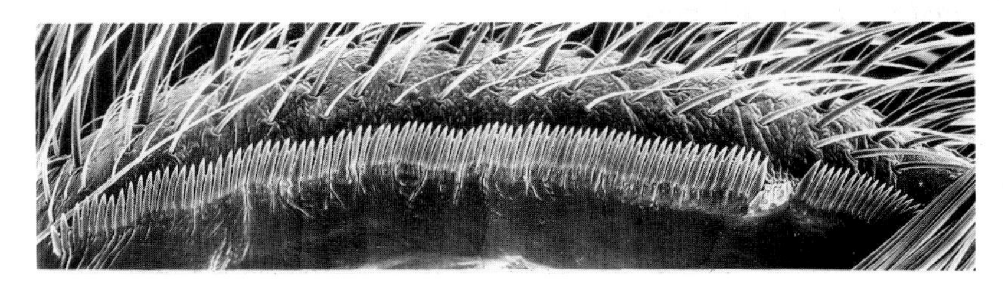

Fig. 15. The serrula, a sawlike cuticular ridge on the palpal coxae (*Cupiennius*), 220 ×.

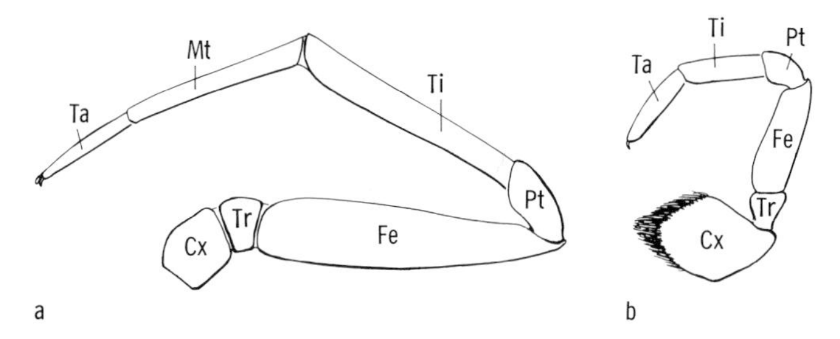

FIG. 16. Segmentation of (*a*) a walking leg and (*b*) a pedipalp. Ta = tarsus, Mt = metatarsus, Ti = tibia, Pt = patella, Fe = femur, Tr = trochanter, Cx = coxa. (After Kaston, 1972.)

Walking Legs

Four pairs of legs fan out radially from the pliable connection between carapace and sternum. Each leg has seven segments: a *coxa* and a *trochanter*, which are both short; a long *femur* and a kneelike *patella*; a slender *tibia* and *metatarsus*; and, finally, a *tarsus* with two or three claws (Fig. 16). Usually the front legs (1 and 2) are relatively long, and the first pair of legs in particular is often used as feelers to probe the environment. The sensory capacity of the legs stems from a variety of sensory hairs that densely cover the distal leg segments.

Tarsal Claws The tip of the tarsus bears two bent claws, which are generally serrated like a comb (Figs 17, 18). Some spiders (Trionycha) have an additional middle claw. All claws arise from a single cuticular platelet. They can be lifted or lowered as a unit by the action of two antagonistic muscles, the *musculus levator praetarsi* and the *musculus depressor praetarsi* (Fig. 24). The middle claw is quite important for web spiders, because they use it to catch hold of the silk threads of their webs—in fact, only the middle hook and not the large main claws grasps the thread. The thread is pushed by the middle claw against serrated bristles situated opposite the claws (Fig. 18b). There it is held, snagged in the little notches of these bristles. To release the thread, the middle claw is lifted by the *musculus levator praetarsi*; the inherent elasticity of the thread simply causes it to spring back out of the clasp of the claw (Wilson, 1962b).

Scopulae Many hunting spiders possess dense tufts of hair, the *scopulae*, directly under the claws (Fig. 19a). In some spiders, especially in "tarantulas" (mygalomorphs), the entire ventral side of the tarsus and metatarsus may be covered by such scopula hairs.

All spiders that have scopulae on their feet can easily walk on smooth vertical walls, or even on window panes. Under experimental conditions *Cupiennius* could hold about ten times its own body weight, when sitting on a vertical glass plate (Seyfarth, 1985; pers. comm.). This remarkable ability can be explained if we look at the fine structure of the scopula hairs. The ventral surface of each hair splits into thousands of fine cuticular extensions ("end feet"), giving the hair the appearance of a hand broom (*scopa* means broom in Latin).

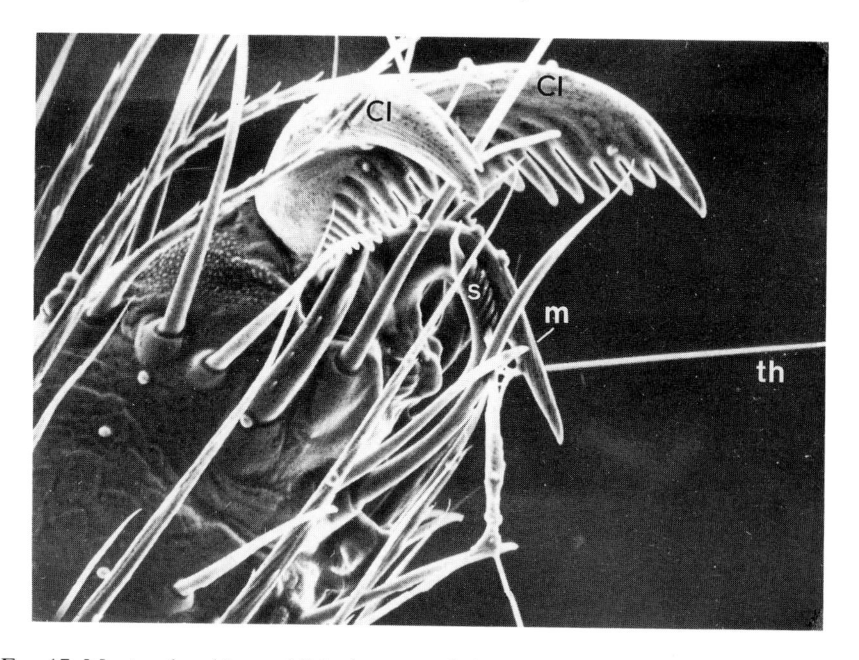

FIG. 17. Most web spiders exhibit three tarsal claws: two combed main claws (Cl) and a smooth middle hook (m). The silk thread (th) is grasped only with the middle hook and pushed against serrated bristles (s; cf. FIG. 18) (*Araneus diadematus*, 620 ×).

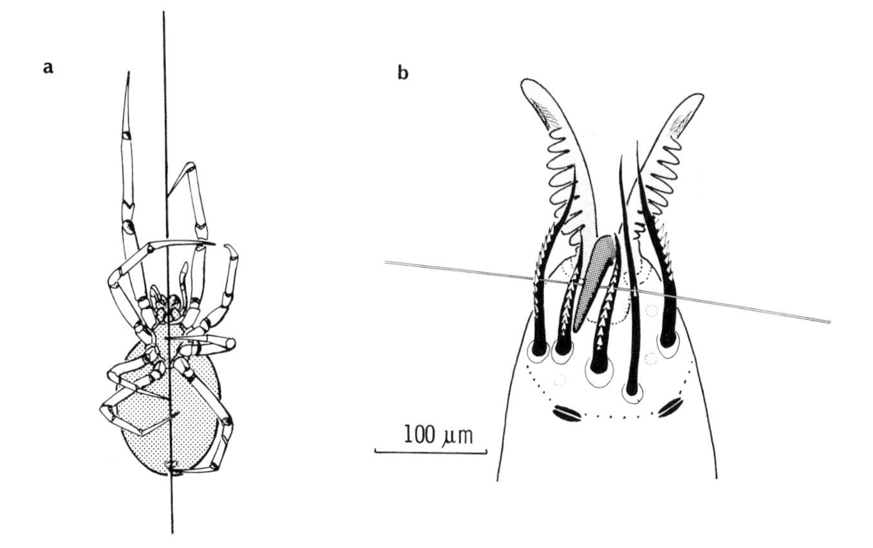

FIG. 18. (*a*) The orb weaver *Zygiella* climbing a vertical thread. (After Frank, 1957.) (*b*) Grasping a thread: interaction of the middle hook and serrated bristles. (After Foelix, 1970a.)

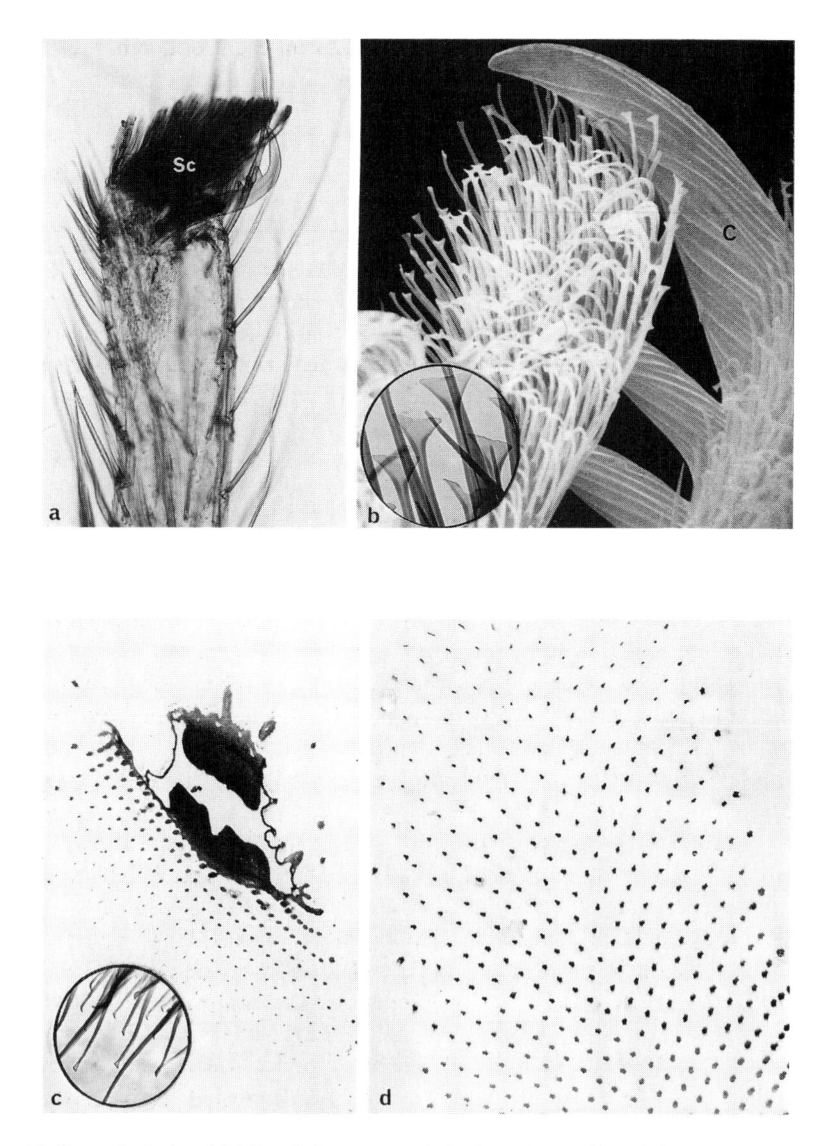

Fig. 19. Scopula hairs. (*a*) Tip of the tarsus of the jumping spider *Salticus scenicus*. The claw tuft is formed by many scopula hairs (Sc), 160 ×. (*b*) Part of a claw (C) and a scopula hair of a *Philodromus aureolus*, 3,600 ×. (Photo: Gnatzy and Foelix.) At higher magnification (inset) the triangular "end feet" of the scopula hair are clearly visible, 6,200 ×. (*c*) Cross-section of a scopula hair from the wolf spider *Lycosa punctulata*. The ventral side bears hundreds of cuticular extensions, but typical "end feet" are lacking, 2,800 ×; inset: 13,000 ×. (*d*) Horizontal section of a scopula hair (of *Lycosa*) illustrating the "footprint" that a single scopula hair produces when in contact with the substrate, 4,800 ×. (Photos: Chu-Wang and Foelix.)

A crab spider with only about 30 scopula hairs on each foot, can nevertheless achieve 160,000 contact points with the substrate, because each scopula hair has 500–1,000 end feet (Foelix and Chu-Wang, 1975). Thus, the actual contact between the foot and the substrate is mediated by the thousands of microscopic "end feet" (Fig. 19b–d). The commonly held belief that the scopulae function like suction cups is erroneous. Electrostatic forces are not involved either. The spider's surefooted grip is achieved merely by the forces of physical adhesion. Adhesion is enhanced by the capillary forces of an extremely thin water film on the substrate (Homann, 1957; Roscoe and Walker, 1991). If the water film is absent, as in Teflon foil, then even scopulae-equipped spiders begin to slide or fall off, although normally they can easily walk upside down on glass plates. It has been calculated that a spider weighing 3 g would have 70 ponds (or 0.7 N) of capillary force available, provided that all possible contact points are established (Homann, 1957). This explains why a spider can still hold fast when some legs are lifted during walking. Incidentally, the same phenomenon of highly efficient adhesion forces is known in some reptiles, such as in the skinks, which can walk surefootedly on walls and ceilings (Hiller, 1968). One problem that is not quite resolved is how such a grip is loosened again. It has been suggested that a spider can, by degrees, detach its scopula hairs hydraulically by gradually increasing the hemolymph pressure (Rovner, 1978).

In their natural surroundings, spiders with scopulae certainly have the advantage of being able to climb securely on overhanging rocks or leaves. Thus certain environments that normally would be inaccessible become available. The contact with a smooth substrate is always made by the distal claw tufts, not by the hairs covering the ventral sides of tarsi and metatarsi. A close inspection shows that the adhesive side of those scopulate hairs normally does not point toward the substrate but actually faces the leg surface. However, their adhesive function comes into play during prey capture, i.e., when a struggling victim needs to be held down securely (Rovner and Knost, 1974; Rovner, 1978). Just before seizing a prey, the scopulate hairs can be erected hydraulically, thereby exposing their adhesive sides (Foelix et al., 1984).

Leg Joints and Musculature The movability of the seven leg segments is determined not only by the extensive musculature but also by the different types of joints. Most joints are *dicondylous*, that is, they can move only in one plane (vertical movements; Fig. 20). In contrast, the coxa–trochanter joint represents a kind of "saddle joint" and thus can be moved forward and backward. This lateral movability is absolutely necessary for locomotion, and is most important for legs 2 and 3 (see chapter 6). The joint between patella and tibia also moves laterally, and assists in positioning the leg tip precisely (Frank, 1957). A so-called passive joint lies between the tarsus and metatarsus; although a hinge is present, the muscles are not involved in the movement between the two segments (Figs 21, 24).

There is little variation in the leg musculature of different spider families. For instance, 33 different leg muscles have been described for the orb weaver *Zygiella* (Frank, 1957); the "tarantula" *Eurypelma* (synonym *Dugesiella*), with

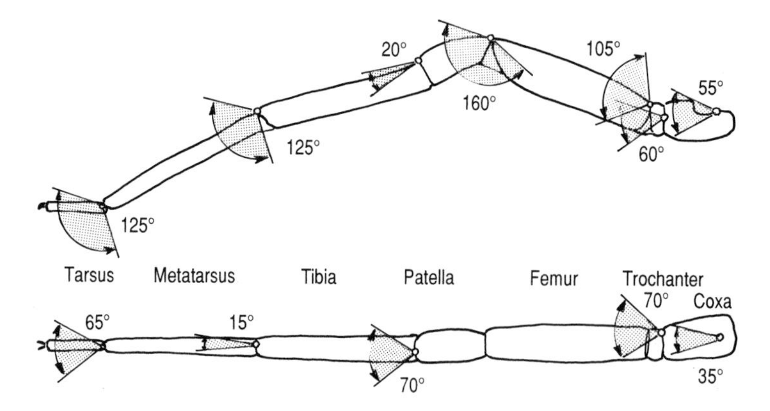

FIG. 20. Range of movements for the leg joints in the wandering spider *Cupiennius*. The angles that each joint can move were measured on a live animal. Top: Lateral view of leg 1. Bottom: Dorsal view of leg 1. (After Seyfarth, unpublished.)

30, has just about as many (Dillon, 1952; Ruhland, 1976). Cursory spiders, which need to hold down their struggling prey, tend to have better developed muscles than web spiders, which subdue their prey by wrapping (Clarke, 1986). Most joints are equipped with several muscles (Fig. 21) that either bend the joint (the *flexors*) or stretch it (the *extensors*). Two remarkable exceptions are the femur–patella joint and the tibia–metatarsal joint, both of which lack extensors (Petrunkevitch, 1909). How these joints can be stretched at all puzzled arachnologists for a long time. Later it was shown that the extension is caused by a hydraulic mechanism, that is, an increase in the hemolymph pressure (Ellis, 1944; Parry and Brown, 1959a,b; R. S. Wilson, 1970; Wilson and Bullock, 1973; Anderson and Prestwich, 1975).

The hydraulically mediated extension of these joints can be elicited by gingerly squeezing a leg with forceps. In contrast, no such movement occurs if the leg is squeezed after its tip has been cut off. An important question is where the "pressure pump" is located within the living spider. It turns out that a contraction of entosternal muscles (Fig. 34b; Shultz, 1991), which traverse the carapace vertically, leads to a reduction in the volume of the prosoma and thus to increased pressure there (Fig. 22). This hemolymph pressure has been measured directly in several spiders (Homann, 1949; Parry and Brown, 1959a,b). In a "tarantula," pressures of 40–60 mm of mercury (Hg) were measured during rest and 100 mm Hg during walking. At maximal activity (struggling), values of 480 mm Hg were observed (Stewart and Martin, 1974). These values are generally comparable to human systolic and diastolic blood pressures, which for a 20-year old are normally 120 and 80 mm Hg, respectively.

Histologically, all spider leg muscles are typical striated skeletal muscles. Each cell consists of many myofibrils, with light I-bands (isotropic) and darker A-bands (anisotropic). The Z-disks, which define the functional unit of a myofibril (the *sarcomere*) can be seen clearly under the light microscope. The ultrastructure of the contractile elements (*actin* and *myosin*) is very similar to

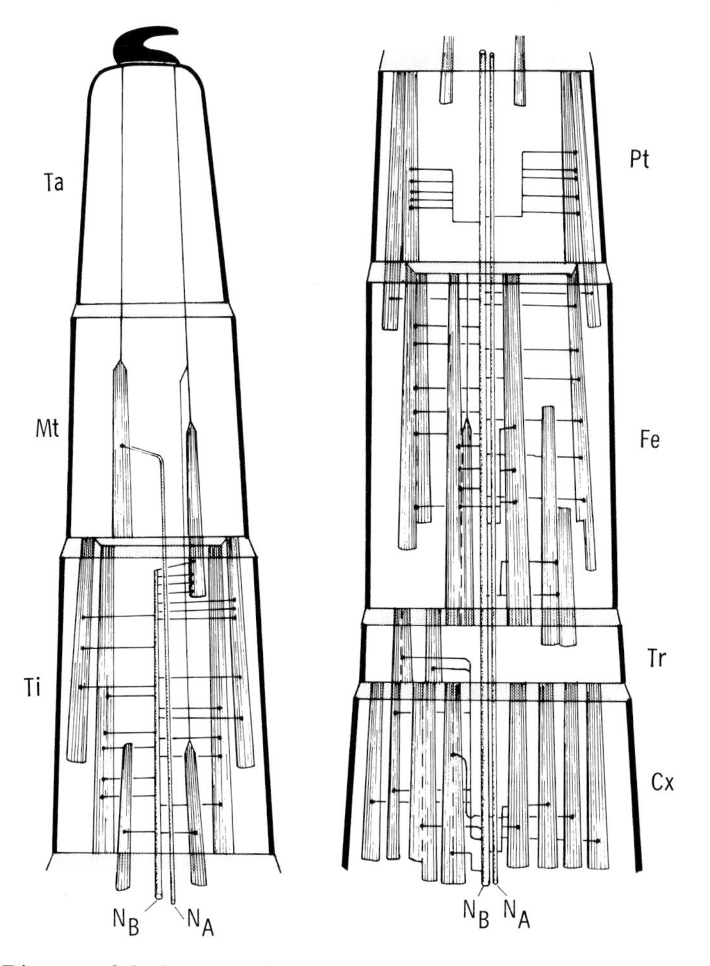

Fig. 21. Diagram of the leg musculature and its innervation in the "tarantula" *Dugesiella*; N_A, N_B = motor nerves A and B. (After Ruhland and Rathmayer, 1978.)

that of skeletal muscles of other arthropods (Fig. 23; Zebe and Rathmayer, 1968; Sherman and Luff, 1971; Fourtner, 1973). However, spider muscles contain only few mitochondria (Linzen and Gallowitz, 1975), and since mitochondria ultimately provide the energy for the cell, it is not surprising that spider muscles fatigue rapidly. It has been known for many years that, although spiders can achieve high levels of momentary activity, they usually become exhausted after a few seconds of exertion (Bristowe and Millot, 1933). During activity, mainly anaerobic energy sources (phosphate, glycogen) are used, whereas fat is burned during rest (Prestwich, 1988a,b; Paul, 1990; Eschrich and Paul, 1991). After a period of high activity, a long period of recovery follows with an elevated aerobic metabolism. D-Lactate accumulates as an end product in the muscle tissue, then spreads into the hemolymph, causing a drop in pH. This metabolic acidosis contributes to the rapid exhaustion in spiders.

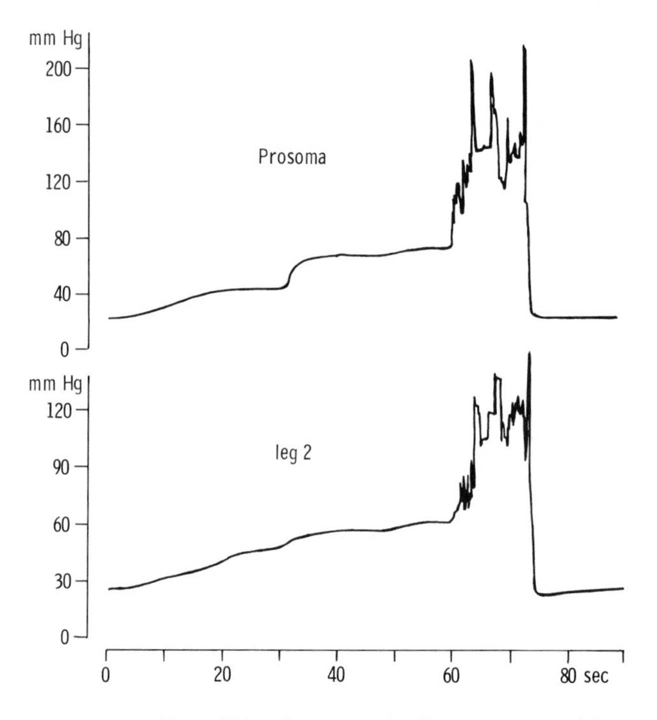

Fig. 22. Synchronous recording of blood pressure in the prosoma and leg 2 of the "tarantula" *Dugesiella.* In the resting state the blood pressure is about 30 mm Hg. The onset of activity is marked by a concomitant increase of blood pressure in the prosoma and in the leg. (After Stewart and Martin, 1974.)

Some muscles are elongated at one end into long "tendons." These tendons consist of fine cuticular tubes covered by a thin epithelium (*hypodermis*). Every time a spider molts, the cuticular part of the tendon is discarded along with the shed skin (*exuvium*). A typical example for such tendons is provided by the claw muscles located in the metatarsus and tibia. Their long tendons operate the claws as reins manipulate a bridle (Fig. 24). It would be beyond the scope of this book to list all the different muscles in a spider's body. I shall give just one example to show how complex the muscle arrangement can be: the pedicel alone is provided by 36 muscles, 17 paired and 2 unpaired (Dierkes and Barth, 1995). Readers with a deeper interest in spider myology are referred to the detailed works of Palmgren (1978, 1981).

Innervation The innervation of spider muscles is polyneuronal, that is, each muscle cell is supplied by several (usually three) motor nerve fibers (Rathmayer, 1965b). The nerve endings are distributed over the entire length of a muscle fiber where they form numerous neuromuscular contacts (synapses; Fig. 23b).

Each spider leg is traversed by three main nerves (A, B, C). Nerve A is the smallest and contains sensory as well as motor fibers (Rathmayer, 1965a). The many thousands of nerve fibers (axons) that form the large nerve C (Fig. 25)

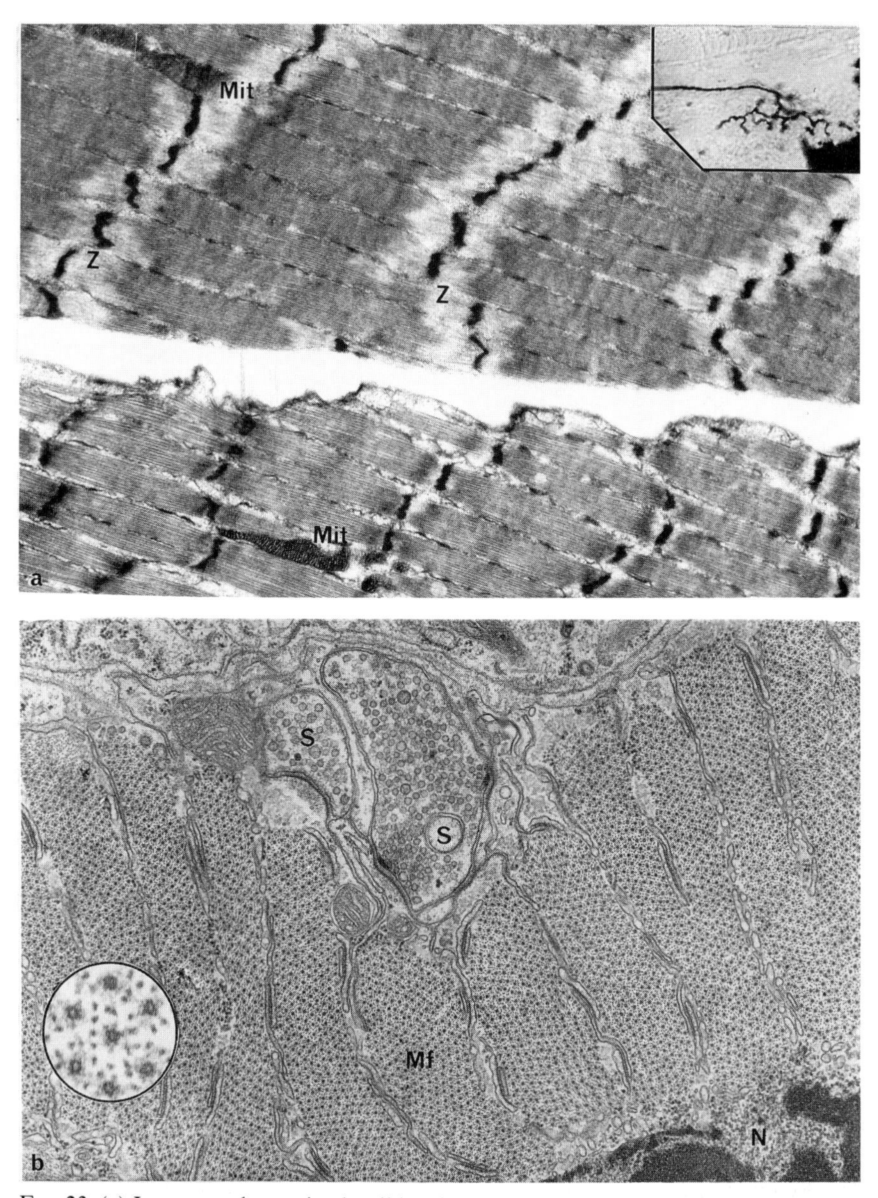

FIG. 23. (*a*) Leg musculature in the tibia of *Pholcus phalangioides*, longitudinal section. The muscle fiber pictured in the upper half is in the relaxed state. The micrograph clearly shows dark A-bands and light I-bands with distinct Z-lines. The lower muscle fiber is contracted, hence the light I-bands have almost disappeared. Mitochondria (Mit) are rare. 7,000 ×. Inset: motor nerve endings branch on the surface of the muscle and form multiple synapses (Golgi preparation). 340 ×. (*b*) Leg musculature in the patella of *Zygiella x-notata*, cross-section. The muscle cell is subdivided into myofibrils (Mf), which radiate from the nucleus (N). Each myofibril consists of a hexagonal array of myosin filaments and of less regularly distributed thin actin filaments (inset, 93,000 ×). Two neuromuscular contacts (S) are seen on the surface of the muscle cell. 14,000 ×.

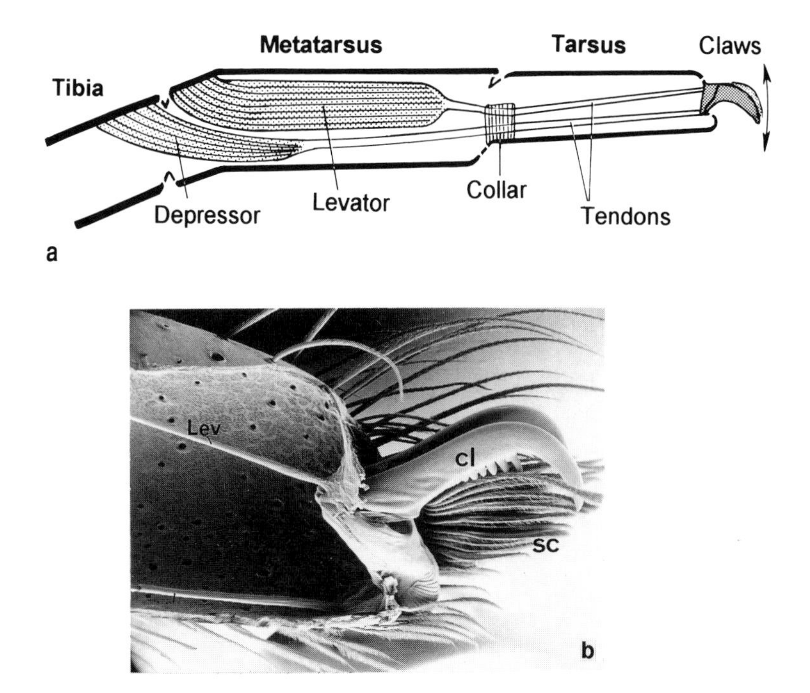

FIG. 24. Musculature of the tarsal claws. (*a*) The m. levator praetarsi (Levator) lifts the claws; the m. depressor praetarsi (Depressor) lowers them. The long tendons of both muscles pass through a collar near the metatarsal joint. (*b*) Exuvium of the tarsus of *Phidippus*, opened longitudinally. Note the attachment of the cuticular tendons (Lev, Depr); cl = claws, sc = scopula. [(*a*) After Sherman and Luff, 1971; (*b*) from Hill, 1977a. Copyright © Linnaean Society of London.)

arise from the numerous receptors on the spider leg and is thus purely sensory. Nerve B consists of rather few but large motor fibers. Physiologically, spiders have both fast and slow contracting muscle fibers, and it has been reported (Brenner, 1972) that slow muscle fibers have an inhibitory innervation similar to that of insect and crustacean muscles. Recent histochemical and electrophysiological investigations revealed four different types of muscle fibers: chiefly A fibers, which contract rapidly but also fatigue quickly; B and C fibers, which are somewhat slower but contract for a longer time; and D fibers, which are rich in glycogen and can remain contracted permanently (Maier et al., 1987; Paul et al., 1991).

Leg Receptors All leg segments are covered with various sensory hairs, discussed here only briefly (for details see chapter 4). Most of these *sensilla* are movable, articulated setae, or bristles, which function as mechanoreceptors (touch, vibration). Each leg receptor is associated with several primary sensory cells, and consequently, sensory nerves are built up by thousands of separate sensory fibers (Fig. 25; Foelix et al., 1980).

A particularly interesting type of sensillum is the *trichobothrium*, a very thin

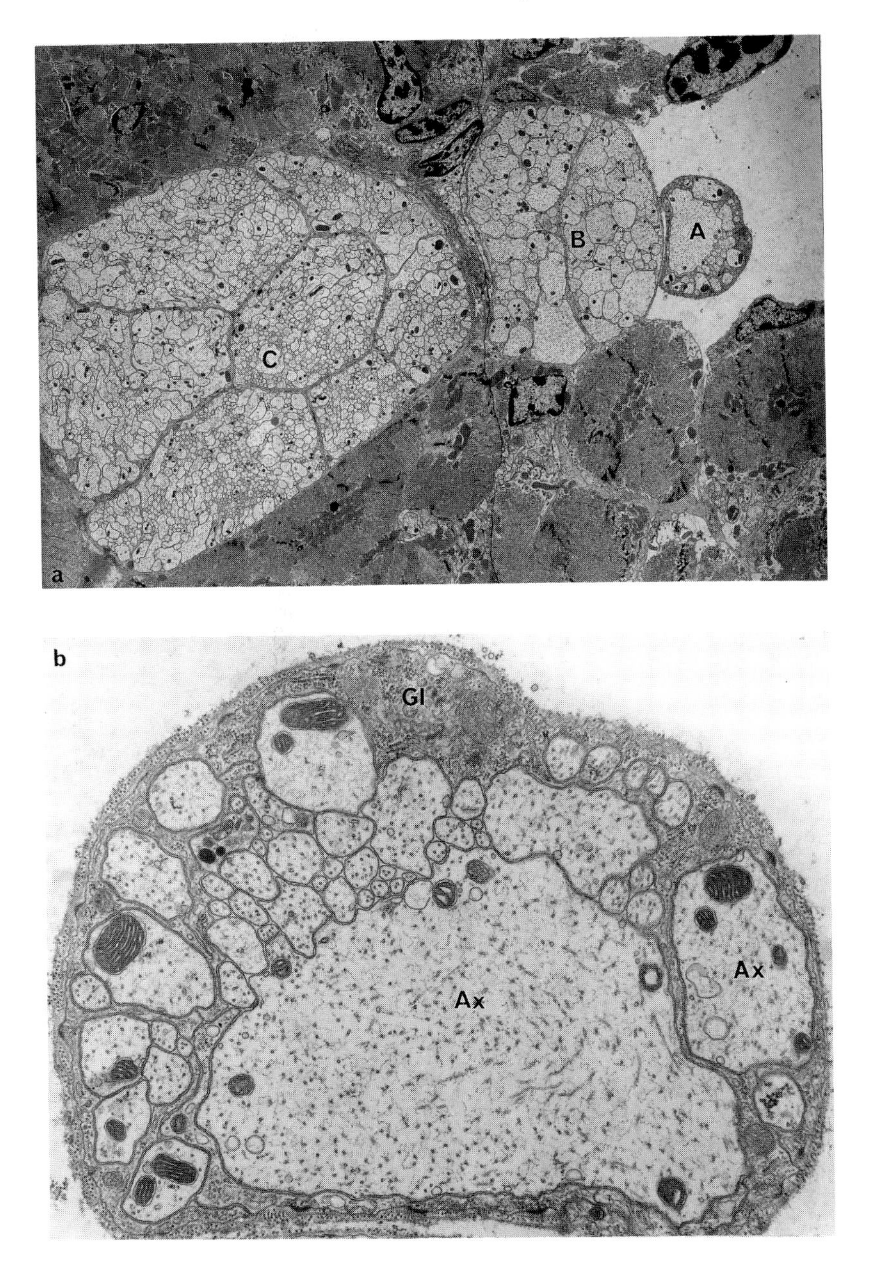

Fig. 25. (*a*) Cross-section of the three leg nerves (A, B, C) in the femur of *Zygiella*. Nerve C is purely sensory and contains about 5,000 small fibers. Nerve B is a motor nerve consisting of about 120 large fibers. 2,400 ×. (*b*) The small nerve A is made up of two motor axons (Ax) and about 50 sensory axons. Peripherally the nerve fibers are enveloped by a glial cell (Gl). 14,500 ×.

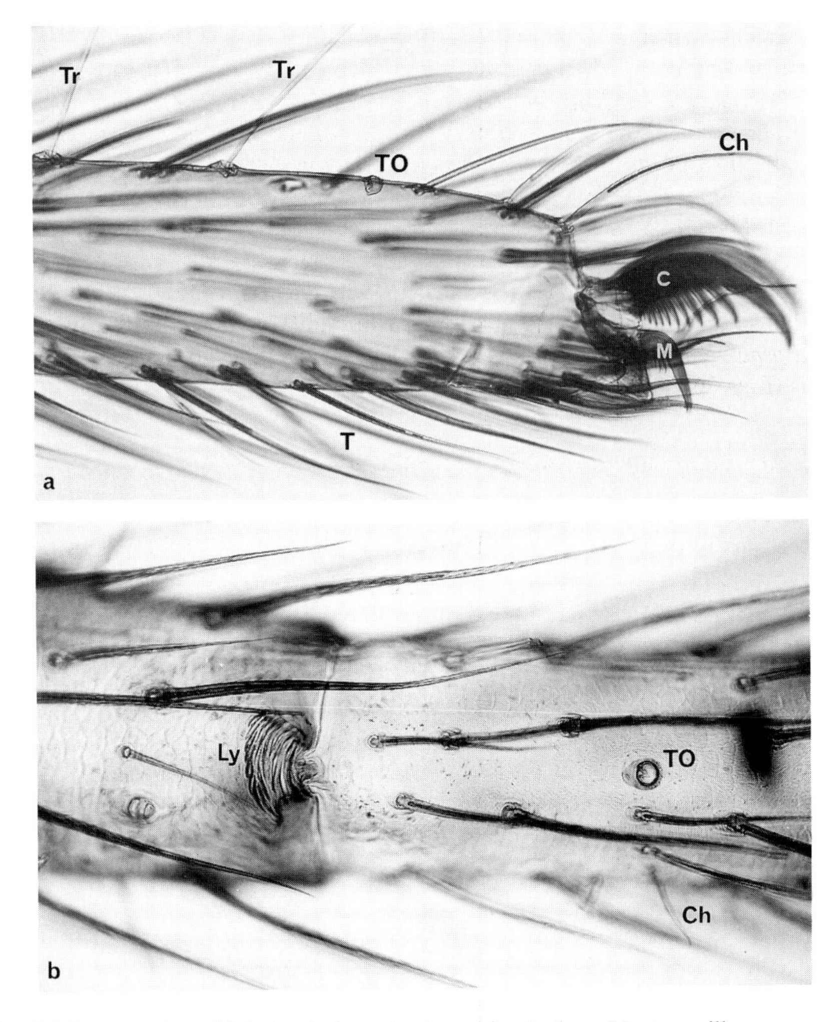

FIG. 26. Leg receptors. (*a*) Lateral view ot a tarsus in *Agelena*. Most sensilla are simple tactile hairs (T), but a few slender trichobothria (Tr) and chemosensitive hairs (Ch) can also be seen. TO = tarsal organ, C = main claws, M = middle hook. 250 ×. (*b*) Dorsal view of the tarsus–metatarsus joint in *Araneus*. The metatarsal lyriform organ (Ly) lies close to the joint; the tarsal organ (TO) superficially resembles an empty hair socket. 230 ×.

hair set almost at a right angle to the leg axis (Figs 26, 62). Originally the trichobothria were thought to represent hearing organs (Dahl, 1883), but later they were considered "touch-at-a-distance" receptors, because they react to air currents and low-frequency air vibrations (Görner and Andrews, 1969).

Less conspicuous mechanoreceptors are the *slit sensilla*, which occur on all leg segments, singly or in groups (*lyriform organs*; Figs 26, 63). The slits of the lyriform organs are usually located near the leg joints, where they measure the strain in the surrounding cuticle (Barth, 1972a,b). Inside the joints lie groups of proprioceptors, which gather information on the position of a particular leg joint (Rathmayer, 1967; Rathmayer and Koopmann, 1970).

In addition to the mechanosensitive sensilla, we also find *chemosensitive hairs* on all distal leg segments (Foelix, 1970b; Foelix and Chu-Wang, 1973b). These hairs are characterized by an open tip in which several nerve fibers are exposed directly to the environment. It has been known for a long time that spiders can test the chemical quality of a substrate merely by probing it with the tips of their legs (Bristowe and Locket, 1926; Kaston, 1936; Bristowe, 1941). Another chemoreceptor, more precisely a humidity receptor, is the so-called *tarsal organ*, which usually forms a spherical pit on the dorsal surface of each tarsus (Figs 26, 68).

Internal Organs of the Prosoma: An Overview

Inside the prosoma lie the central nervous system (CNS), part of the intestinal tract, a pair of poison glands, and an extensive musculature (Palmgren, 1978) for the extremities, the pharynx, and the sucking stomach. Figure 27 provides a simplified picture of the spatial relationships of the various organs. A special feature is the narrow esophagus, which traverses the CNS horizontally, dividing it into a supra- and a subesophageal ganglion. Another peculiarity of spiders is the branching of the midgut. These branches (*diverticula*) extend into the entire prosoma and may even enter the leg coxae (Fig. 42). Similar gut diverticula fill the abdomen, but there they are further subdivided into many lobules. A more detailed description of the prosomal organ systems follows in chapters 3 and 4.

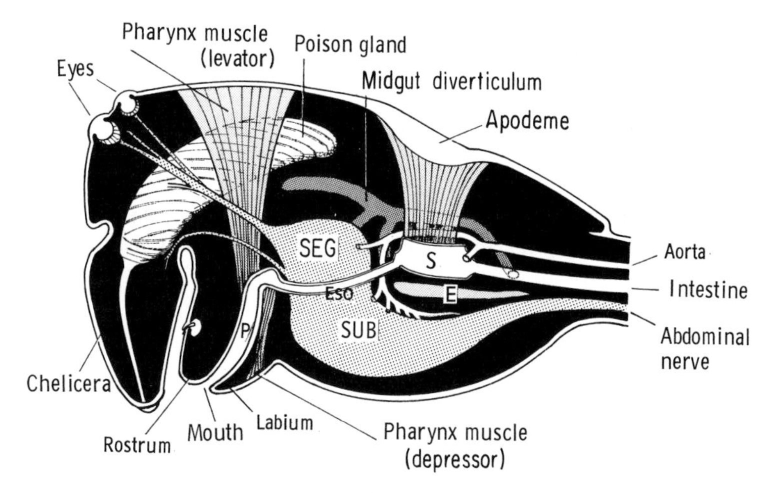

FIG. 27. Longitudinal section of the prosoma. E = endosternite, Eso = esophagus, P = pharynx, SEG = supraesophageal ganglion, S = sucking stomach, SUB = subesophageal ganglion.

Opisthosoma

An Overview of the Internal Organs

The midgut enters the opisthosoma through the narrow stalk of the pedicel. Many lobed diverticula branch off the main tract before it widens into the large *stercoral pocket* (Figs 28, 30). From there a short hindgut connects to the anus. Embedded between the many branched gut diverticula are: the heart and the abdominal arteries; the excretory organs (Malpighian tubules), which empty into the stercoral pocket; the respiratory organs (*book lungs* and *tubular tracheae*); the reproductive organs (*ovaries* and *testes*); and the various *spinning glands*, which are connected to the spinnerets via long ducts.

Segmentation

In most spiders the opisthosoma is soft and expansible. It usually shows no trace of segmentation. Only the Mesothelae, believed to represent an ancient form from which present-day spiders are derived, possess a clearly segmented abdomen. A close look at the opisthosoma in this suborder provides an idea of its basic organization for all spiders.

Twelve abdominal segments can be distinguished in all. Each consists of a dorsal *tergite* and a ventral sternite. Both parts are connected by the pliable pleurae (Fig. 1), which are not segmented, but form a continuous band on each side (Fig. 29). The first abdominal segment is merely a small stalk (pedicel) that unites the prosoma and the opisthosoma. The next segments (2–6, or body

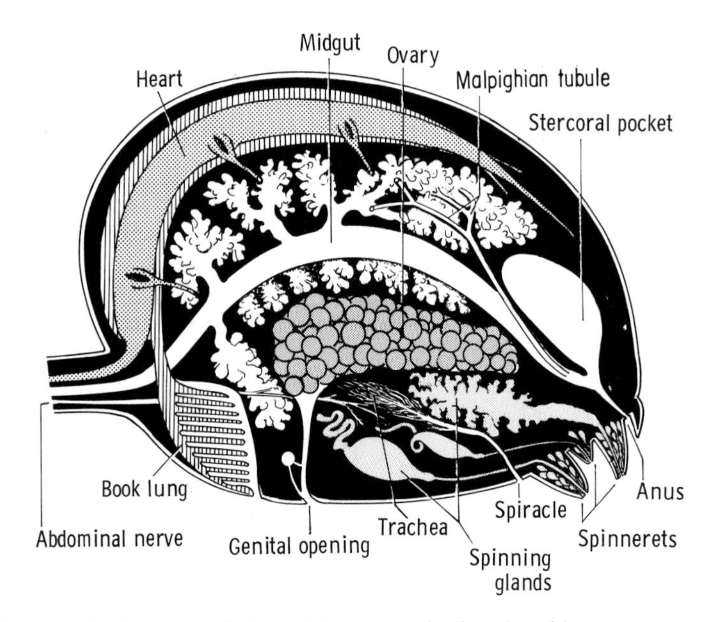

Fig. 28. Longitudinal section of the opisthosoma of a female spider.

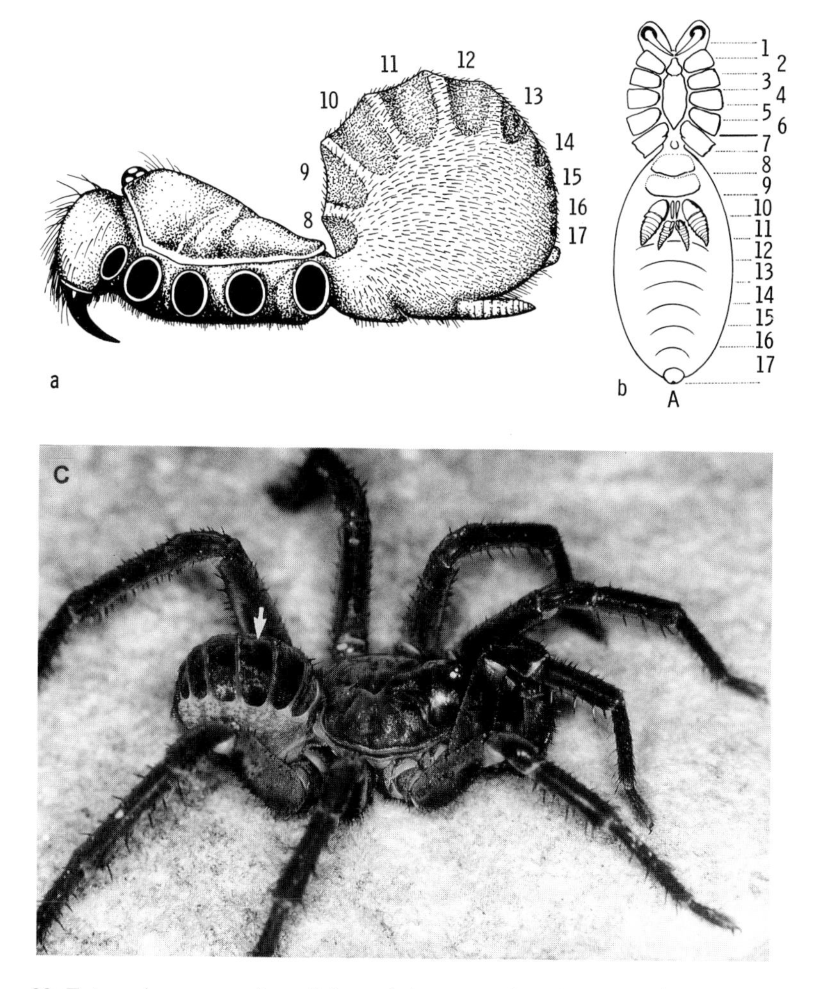

FIG. 29. External segmentation of the opisthosoma of *Liphistius*. (After Millot, 1949.) (*a*) Lateral view: the platelike tergites are marked with the numbers of the respective body segments. (*b*) Ventral view: two pairs of book lungs are illustrated on segments 8 and 9; segments 10 and 11 bear two pairs of spinnerets each. A = anus. Numbers in parentheses refer to the abdominal segments. (*c*) The abdominal tergites (arrow) are clearly seen in this *Liphistius* from Thailand. (Photo: Knoflach and Schwendinger.)

segments 8–12) are the largest, and in these the tergites are most easily recognizable. The posterior segments (13–17) become gradually smaller and terminate in the *anal tubercle* (segment 17).

Looking at the ventral side, the second sternite is by far the largest. Its posterior margin forms the entrance to the respiratory organs (book lungs) and to the reproductive organs (*epigastric furrow*). In the Mesothelae a second pair of book lungs lies underneath the sternite of the third segment. This second pair is, however, lacking in most other spiders. The fourth and fifth sternites

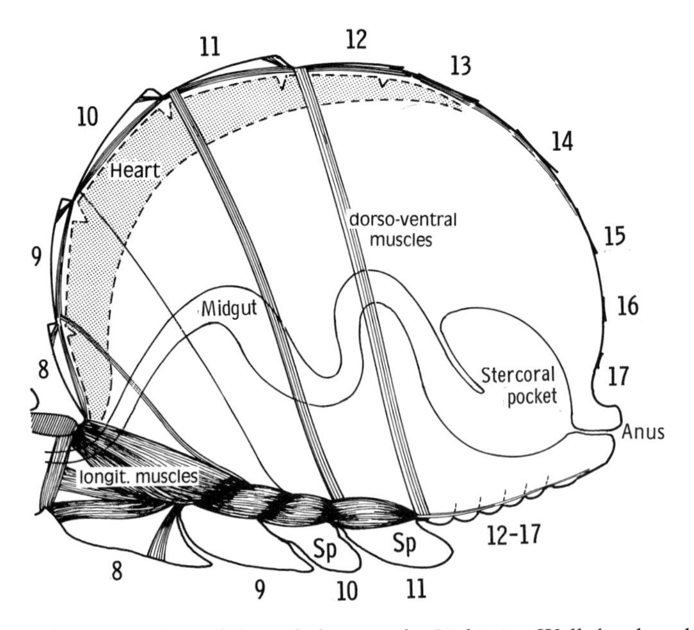

FIG 30. Internal segmentation of the opisthosoma in *Liphistius*. Well-developed longitudinal muscles connect the segments ventrally, whereas the dorsal longitudinal muscles are inconspicuous. Dorsoventral muscles connect tergite and sternite of the same segment. Sp = spinnerets. (After Millot, 1949.)

each carry a pair of spinnerets. The next sternites (6–12) are not as clearly marked and are recognizable only as a series of ventral furrows. As can be seen from Figure 29, the spinnerets of the Mesothelae lie approximately in the midregion of the abdomen, but in all other spiders they are always situated caudally. The terminal position of the spinnerets is the result of a longitudinal expansion of the third sternite during phylogeny such that the spinnerets became displaced posteriorly. Immediately in front of the spinnerets, at the borderline between the third and the fourth sternite, another respiratory opening (*spiracle*) can be seen in many spiders. This is the entrance to the tubular tracheae.

The segmentation of the opisthosoma is visible not only externally but also in the internal organization of the abdomen. This becomes obvious when one considers the abdominal musculature (Fig. 30). On the one hand, a segmentally arranged longitudinal musculature traverses the abdomen dorsally from tergite to tergite and ventrally from sternite to sternite. On the other hand, some muscles run intrasegmentally (that is, within a single segment) in a dorsoventral direction, connecting the corresponding tergites and sternites. The heart is also segmentally organized. In spiders with a transparent cuticle the heart tube is visible from the outside along the dorsal midline of the opisthosoma. On either side of this line are several small indentations of the cuticle (*apodemes*) that serve as insertion sites for the dorsoventral musculature. The serial arrangement of apodemes is further evidence of segmentation.

Exoskeleton

As members of the arthropods, spiders possess a hard external body shell, the *exoskeleton*. The exoskeleton is made of a stiff material called the *cuticle*. The function of the cuticle is manifold: it is the building material of the entire body surface, the joint membranes, tendons, apodemes, sensory hairs, and even the lining of the respiratory and reproductive organs. Apart from its purely structural function, the cuticle also serves to protect the spider and prevents desiccation of its body (Nemenz, 1954, 1955). A spider cuticle is quite similar in composition to an insect cuticle (Barth, 1969, 1970), and accordingly, various layers of epi-, exo-, meso-, and endocuticle can be distinguished (Fig. 31). In contrast to most insects, however, spiders possess a mesocuticle even in the adult stage.

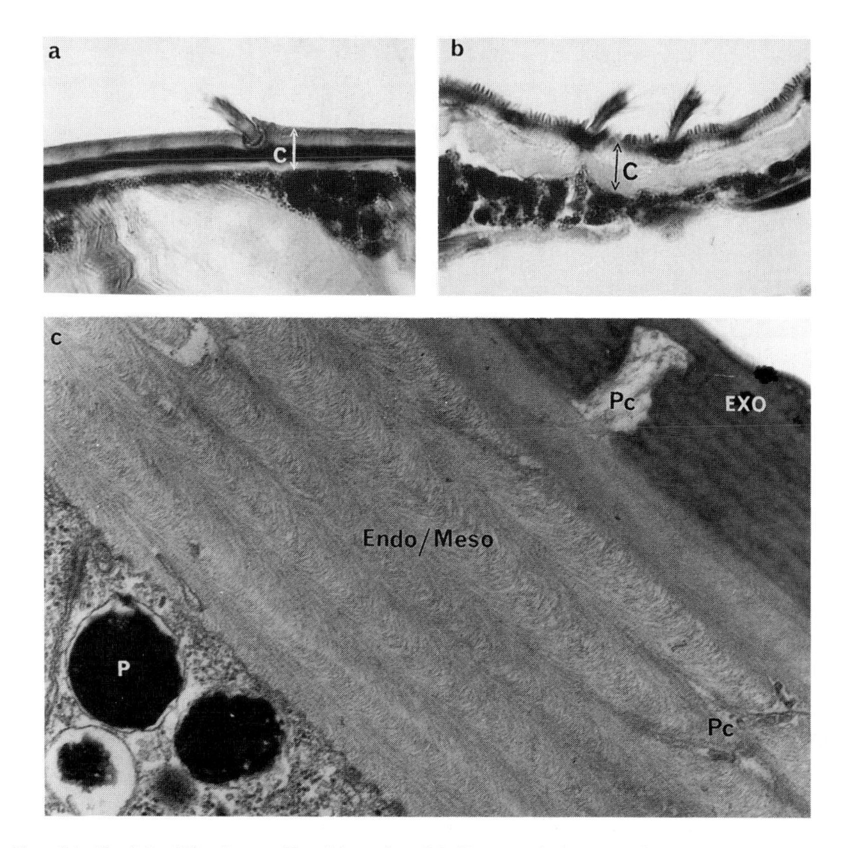

Fig. 31. Cuticle (C) of a wolf spider after Mallory staining. In the prosoma (*a*) three layers (exo-, meso- and endocuticle) are apparent, whereas the opisthosoma (*b*) is made up mainly of endocuticle. 650 ×. (*c*) Electron microscopical view of the leg cuticle in *Zygiella*. Endo- and mesocuticle (Endo/Meso) exhibit the typical lamellar structure. EXO = exocuticle, P = pigment granules of hypodermis cells, Pc = pore canals. 31,500 ×.

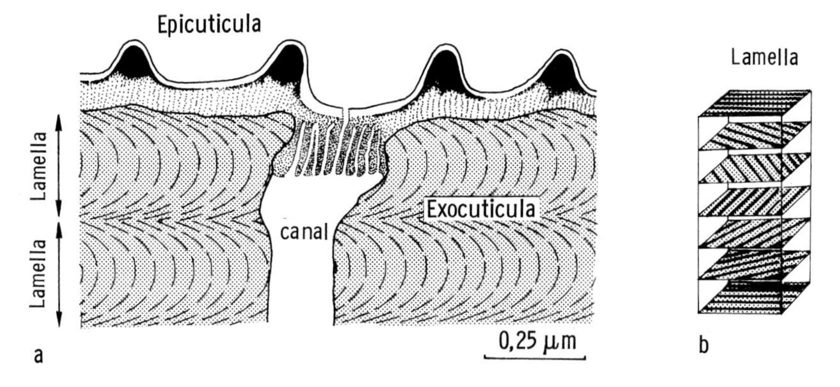

FIG. 32. Epicuticle and part of the exocuticle of *Cupiennius*. (After Barth, 1969.) (*a*) The lamellae of the exocuticle are traversed by a pore canal that splits into delicate channels shortly before reaching the epicuticle. (*b*) Each cuticular lamella consists of many parallel layers of differently oriented microfibers; in sections this gives the impression of arcuate structures as pictured in (*a*). (After Barth, 1973a.)

The epicuticle is extremely thin, and because it forms the outermost layer, it determines the permeability properties of the entire cuticle. Mechanically abraded epicuticle is replaced by material supplied by the pore canals, which traverse the cuticle layers vertically and terminate at the surface. The exocuticle, lying just beneath the epicuticle, is much thicker. It consists of many stacked lamellae (Figs 31, 32), which are structures also characteristic of the meso- and endocuticle.

Each lamella is composed of microfibers that are ostensibly arranged in a paraboloid fashion. After careful analysis, however, it becomes apparent that each lamella is made up of many thin layers of microfibers (Barth, 1973a); these microfibers are oriented in the same direction within any given layer but change their direction by increments from layer to layer so that a complete rotation (180°) of the fiber direction takes place within one lamella (Fig. 32b).

From a technological viewpoint the cuticle is thus a kind of laminated composite material. Its basic component is a protein into which the microfibers, made of chitin, a polysaccharide, are incorporated. This construction confers great strength and, at the same time, high elasticity to the cuticle. Most likely, it is the exocuticle that is toughest, and hence this layer is most responsible for the overall qualities of the cuticle. This becomes evident in the soft cuticle of the opisthosoma or of the joint membranes, where an exocuticle is lacking altogether (Fig. 33). In contrast, the stiff sensory hairs consist almost entirely of exocuticle, as do the shed skins (*exuvia*) after a molt. Although endo- and mesocuticle do not differ in their structure, they react differently to certain dyes (for example, Mallory stain; Fig. 31a). Since the endocuticle is the last cuticular layer to be secreted by the epithelial cells, it seems that the mesocuticle is simply a more sclerotized form of cuticle. The pore canals mentioned earlier constitute the direct connection between epithelial cells and the cuticular surface. They

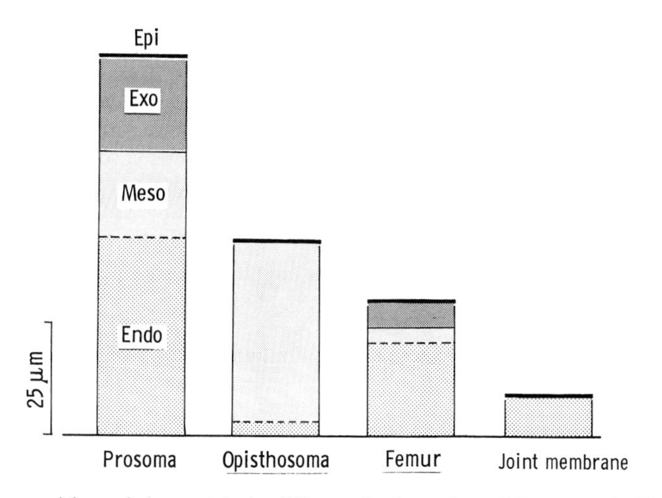

FIG. 33. Composition of the cuticle in different body regions (*Cupiennius*). The hard exocuticle is well developed in the prosoma and the legs (femur), but is lacking in the opisthosoma and the joint membranes. (After Barth, 1973a.)

measure 1–2 μm in width and spiral through the cuticular lamellae. Only near the epicuticle do they begin to fan out into finer canals (Figs 31c, 32).

In addition to the exoskeleton, the underlying epithelium (hypodermis), which secretes the cuticle, also belongs to the integument. It has often been claimed, from light microscopical observations, that the hypodermis is syncytial, because cell borders are usually unrecognizable. All electron microscopic investigations, however, have shown convincingly that the epithelial cells are

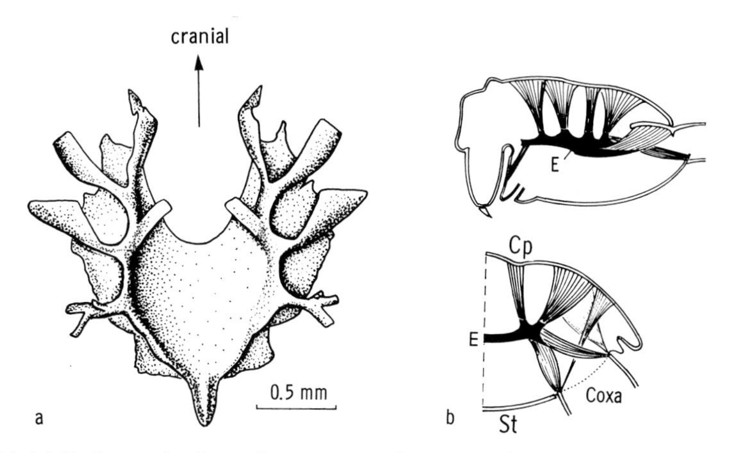

FIG. 34. (*a*) Endosternite from the prosoma of *Araneus*, dorsal aspect. The lateral processes serve as insertion sites for muscles. (After Gerhardt and Kaestner, 1938.) (*b*) Endosternite (E) with attached muscles: (top) longitudinal section; (bottom) corresponding cross-section of one-half of the prosoma. Cp = carapace, St = sternum. (After Whitehead and Rempel, 1959.)

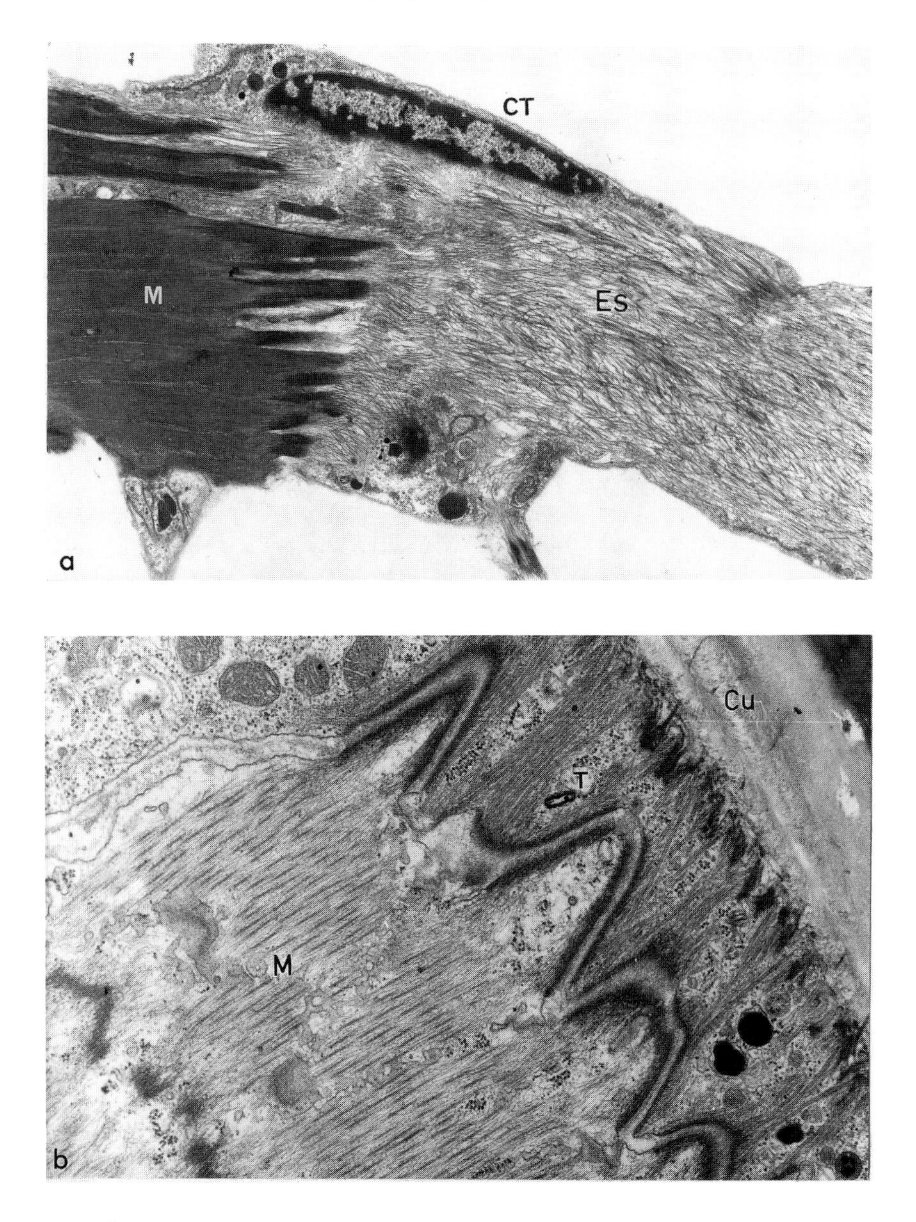

Fig. 35. (*a*) Endosternite of the house spider *Tegenaria*, longitudinal section. The endosternite (Es) consists of a fibrous matrix which is secreted by connective tissue cells (CT). The muscle cells (M) interdigitate directly with the endosternite. 4,400 ×. (*b*) The attachment of a muscle cell (M) to the leg cuticle (Cu) is always mediated by an intercalated tendon cell (T). Note the conspicuous desmosomal connections between tendon cell and muscle cell. The tendon cells are tightly packed with microtubules, which probably transmit tension from the muscle to the cuticle (Cu). 21,000 ×. (Photo: Choms and Foelix.)

separated by cell membranes. The epithelial cells normally form a single layer, yet other cells, for example, sensory and gland cells, may be interspersed among them. Each epithelial cell may contain inclusions, such as pigment granules or guanine crystals.

The different pigments that give spiders their typical coloration (yellow, orange, red, brown, and black) belong predominantly to the *ommochromes*, rarely to the black melanin (Seligy, 1972; Holl, 1987a). Some spiders are distinctly green (e.g., *Araniella cucurbitina, Micromata virescens*), which is due to certain bile pigments (*biliverdin*; Holl, 1987b). White is produced by a reflection from guanine crystals.

Endoskeleton

It is a little-known fact that spiders also possess an internal skeleton. Parts of the exoskeleton invaginate into the body. Such invaginations are termed apodemes or *entapophyses*, and they serve as attachment points for muscles. A typical example is the tergal apodeme (Fig. 27), which projects from the thoracic groove of the carapace into the prosoma and provides the attachment site for the dorsal muscles of the sucking stomach. Besides this ectodermal endoskeleton, spiders also have mesodermal skeletal elements, called *entosterna*. The mesodermal endoskeleton is quite different from the ectodermal exoskeleton. Chitin fibers, which are typical for the exoskeleton, are completely lacking in the endoskeleton (Cutler and Richards, 1974).

Histologically, the entosterna resemble the cartilage of vertebrates. Connective tissue cells secrete a homogeneous substance into which collagen fibers are incorporated (Baschwitz, 1973). The largest of these entosterna is the *endosternite*, a bowl-like platelet that lies in the middle of the prosoma (Figs 34, 35). The many lateral projections of the endosternite serve for the attachment of the lateral stomach muscles and also for muscles of the limbs. Additional but smaller entosterna lie between various muscle strands in the opisthosoma (Fig. 30). It is noteworthy that the muscle cells are inserted *directly* into the entosterna. In contrast, the muscles attached to the exoskeleton always possess a specialized intercalated epidermal cell, the tendon cell (Smith et al., 1969; Baschwitz, 1974). During molting the muscles do not really detach from their insertion points; they always maintain their connection with tendon cells.

3

Metabolism

Spiders have developed an unusual mode of food intake: digestion is initiated outside the body. After the prey has been subdued by a poisonous bite or wrapped with silk, the spider vomits some digestive fluid from the intestinal tract onto the victim. After a few seconds a drop of the predigested liquid prey is sucked in and this process is then repeated many times.

Feeding behavior differs markedly among the various spider families and depends on whether cheliceral teeth are present or not. Theridiids and thomisids, which have few or no cheliceral teeth, inflict only a small wound on their prey. The digestive fluid is pumped in and out through this hole, and the dissolving tissue is gradually sucked out. After the meal the prey remains as an empty shell that appears externally unharmed (Fig. 36a). Spiders with cheliceral teeth mash up their prey so that it can hardly be identified afterwards. For instance, when an orb weaver has finished with a fly, it leaves only a small ball of conglomerated cuticular parts (Fig. 36b). Similarly, after a large "tarantula" has captured a frog, the combined action of the digestive fluid and the continuous mashing of the chelicerae soon produces an unrecognizable mass of tissue with only a few protruding bones (Gerhardt and Kaestner, 1938).

Although spider poison may contain some proteolytic enzymes (Kaiser and Raab, 1967), it plays an insignificant role in actual digestion. Little is known about the glands situated inside the maxillae and the rostrum. Probably their secretions serve mainly as saliva to soften the food.

Poison Glands and Toxicity

Spiders belong to the *actively poisonous* animals, i.e., they use their poison offensively to paralyze or to kill their prey. The quick immobilization is certainly the primary function of the poison—the lethal effect is only secondary (Friedel, 1987; Friedel and Nentwig, 1989).

All spiders except for the family Uloboridae possess a pair of poison glands located in the prosoma (Figs 27, 37). Each poison gland consists of a long cylindrical part and an adjoining duct, which terminates at the tip of the cheliceral fang (Fig. 12). A conspicuous muscle layer spirals around the body of the gland and serves to expel the venom rapidly as the muscle contracts. The muscle fibers are cross-striated and have their own motor nerves, which

FIG. 36. (*a*) Bite marks of the cheliceral fangs (arrows) on the head of a blister beetle (Meloidae), inflicted by social spiders (*Stegodyphus dumicola*). The prey's exoskeleton remains completely intact; the tiny bite holes only become visible under a microscope. 110 ×. (Photo: Foelix and Kaufmann). (*b*) Remnants of a blowfly after a garden spider (*Araneus diadematus*) had been feeding on it. Only few cuticular pieces such as some bristles or the corneal lenses of the compound eye can still be identified. 150 ×. (Photo: Peters and Hüttemann).

facilitates the rapid triggering of venom release. The glandular epithelium itself is also innervated (Järlfors et al., 1971); possibly this nervous supply stimulates or regulates the synthesis of poison.

In the large "tarantulas" (mygalomorphs) the poison glands are quite small and lie inside the chelicerae. In contrast, most labidognath spiders have relatively large poison glands that may extend out of the chelicerae and reach far into the cephalothorax (Fig. 37). In some extreme cases (*Filistata*, for example) the glands may be even larger and subdivided into lobules. The most curious specialization is found in the spitting spider *Scytodes* (Fig. 38), in which the gland consists of an anterior part that produces the venom and of a posterior part that produces a glue-like substance (Kovoor and Zylberberg, 1972).

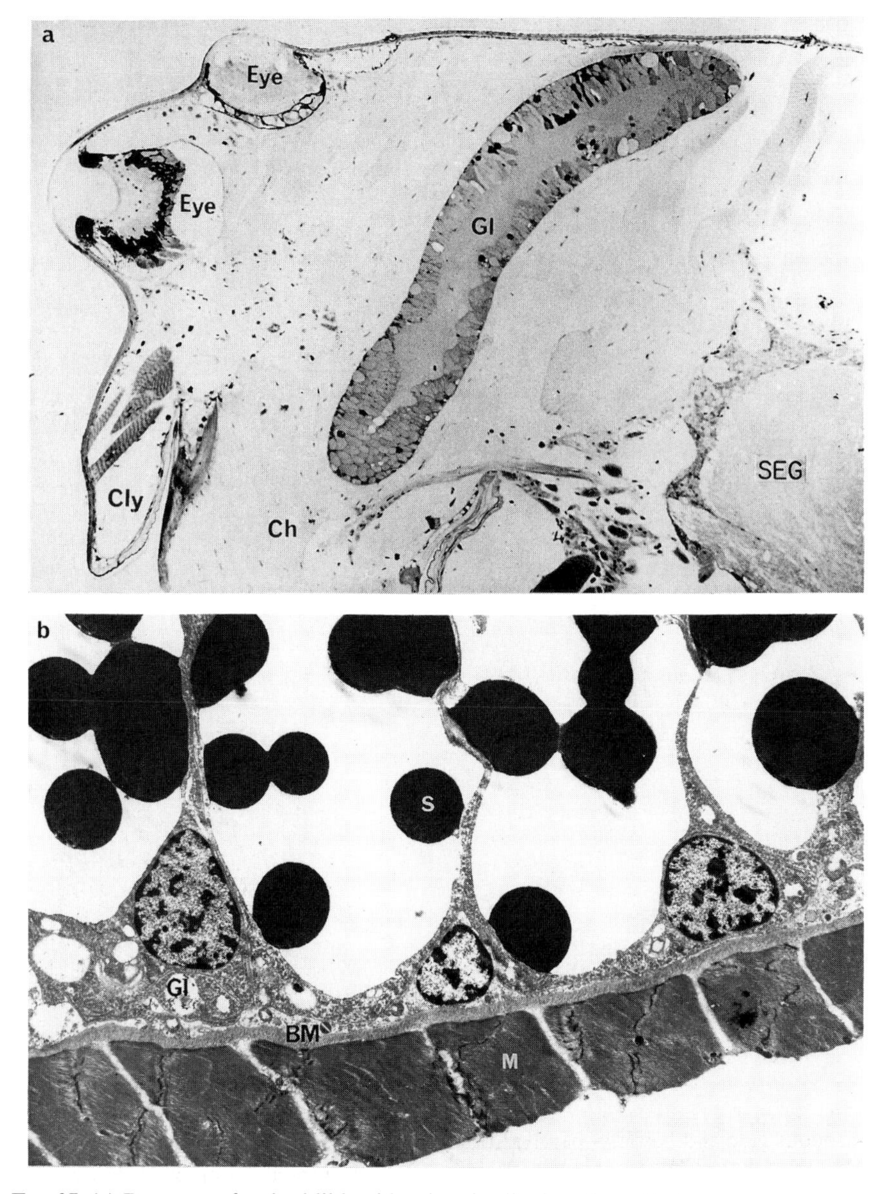

FIG. 37. (*a*) Prosoma of a theridiid spider, longitudinal section. The large poison glands (Gl) extend up to the carapace. Ch = chelicerae, Cly = clypeus, SEG = supraesophageal ganglion. 100×. (*b*) The wall of a poison gland (*Tegenaria*) consists of a (spiraled) muscle layer (M) which is connected via a basal lamina (BM) to the glandular cells (Gl). Secretory droplets (S) containing poison accumulate toward the center of the lumen. 2,800×.

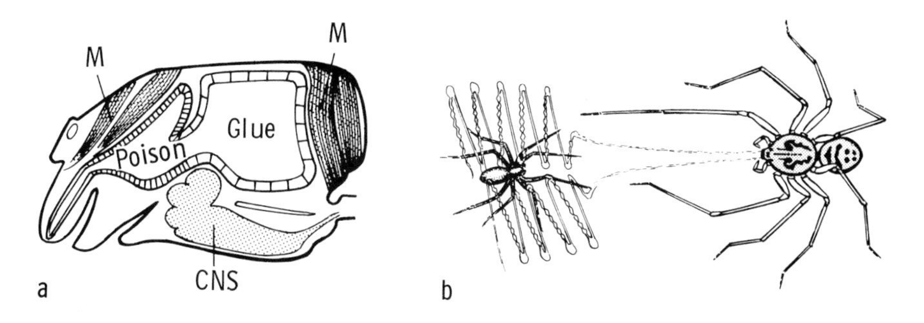

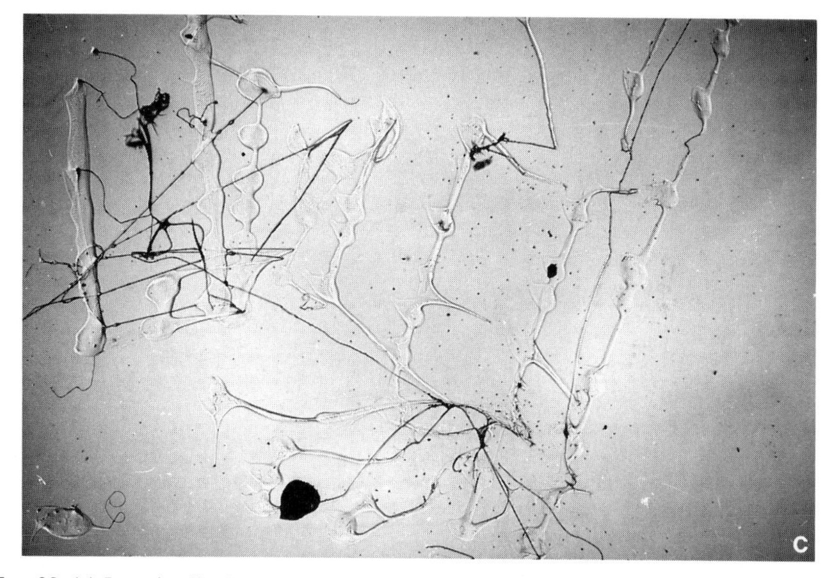

FIG. 38. (*a*) Longitudinal section of the prosoma of the spitting spider *Scytodes*. The enormous poison glands consist of an anterior portion that produces venom and a posterior part that synthesizes a gluey substance. Rapid contraction of the prosomal muscles (M) squirts out a mixture of poison and glue from the chelicerae. CNS = central nervous system. (After Millot, 1949.) (*b*) Prey capture method of *Scytodes*. Prey passing within a distance of 1–2 cm is quickly (140 ms) squirted with poisonous glue and becomes paralyzed and stuck to the ground. (After Bristowe, 1958.) (*c*) Spitting pattern of a *Scytodes* on a glass slide (preparation by H. Bürgis). 35 ×.

Scytodes catches its prey by quickly spitting glue onto it, thereby fixing it to the ground before it is poisoned (Monterosso, 1928; Dabelow, 1958; Bürgis, 1980, 1990).

Chemically, spider venom is heterogeneous in that it may contain many different substances. It is a mixture mostly of large, neurotoxic polypeptides (molecular weight 5,000–13,000) and smaller biogenic amines and amino acids; proteolytic enzymes may also be present (Kaiser and Raab, 1967; Habermehl, 1975; Bachmann, 1976). Since practically all spiders have poison glands, they

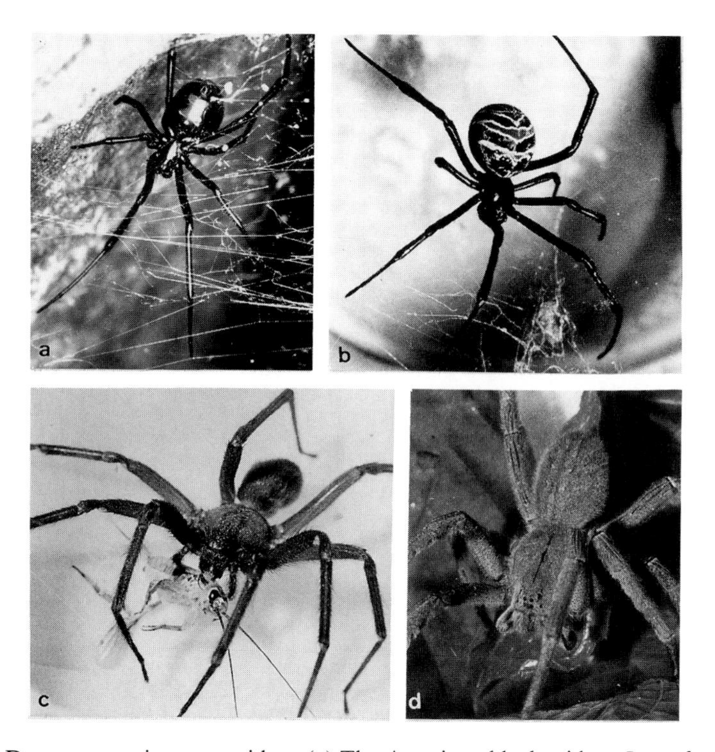

FIG. 39. Dangerous poisonous spiders. (*a*) The American black widow *Latrodectus mactans* with the typical (red) hourglass mark on the ventral side of the abdomen. (Photo: Cooke.) (*b*) Another *Latrodectus* species from South America, dorsal aspect. The black abdomen has contrasting white and red chevron markings. (*c*) A brown recluse spider (*Loxosceles*) capturing a cricket. (*d*) The ctenid *Phoneutria fera* a highly aggressive spider from South America. (Photo: Foelix and Flatt.)

are all potentially poisonous—at least with regard to their normal prey. However, only about 20–30 of the 34,000 species of spiders are dangerously poisonous to man (Schmidt, 1973; Maretić, 1975).

The prime example is certainly the black widow spider, *Latrodectus mactans*, from the family Theridiidae (Fig. 39). The bite itself is not particularly painful and often is not even noticed (Maretić, 1983, 1987). The first real pain is felt after 10–60 minutes in the regions of the lymph nodes, from where it spreads to the muscles. Strong muscle cramps develop and the abdominal muscles become very rigid (this is an important diagnostic feature!). Another typical symptom is a contorted facial expression, called *facies latrodectismi*, which refers to a flushed, sweat-covered face, swollen eyelids, inflammed lips and contracted masseter muscles. If the breathing muscles of the thorax become affected, this can eventually lead to death. Besides the strong muscle pain, the black widow spider venom (BWSV) also elicits psychological symptoms, which range from anxiety feelings to actual fear of death. Apparently the toxin can pass the blood–brain barrier and directly attack the central nervous system.

Without any treatment the symptoms will last for about 5 days and a complete recovery may take weeks. About 50 years ago, lethality was 5% in the USA (Thorp and Woodson, 1945), but is now less than 1% (Zahl, 1971). The best treatment against a bite from a black widow is a combination of calcium gluconate and antivenin (e.g., Lyovac; McCrone and Netzloff, 1965) injected intravenously. Calcium causes the pain to subside quickly and the antidote binds to the toxin. The patient feels relieved within 10–20 minutes and will completely recover in a few hours.

The poison (BWSV) is a neurotoxin that affects the neuromuscular endplates (Figs 23, 80b) but also synapses in the CNS. The synaptic vesicles become completely depleted, causing a permanent blockage of the synapse (Clark et al., 1972; Griffiths and Smyth, 1973; Tzeng and Siekevitz, 1978; Wanke et al., 1986). One component of the poison (α-Latrotoxin) binds to a presynaptic receptor of cholinergic synapses (Meldolesi et al., 1986). This is in contrast to toxins in orb-web spiders, which act on synapses that use glutamate as neurotransmitter (Kawai et al., 1982; Michaelis et al., 1984). Over the past years a number of spider toxins have been used extensively in neurobiological research because they block specific ion channels (e.g., for Ca^{2+}) of the cell membrane (Adams et al., 1989; Jackson and Parks, 1989).

As a measure for the toxicity of a given substance, the lethal dose 50 (LD_{50}) is usually given. This term refers to the dosage of venom that will kill 50% of injected experimental animals. For the venom of *Latrodectus mactans*, the LD_{50} for a mouse equals 0.0009 mg per gram of body weight. As can be seen in Table 1, however, susceptibility to the venom varies greatly among different animals (Maroli et al., 1973).

Some animals, e.g., horses, cows, and sheep, are more sensitive to a black widow spider's bite than man and significant losses of cattle have been reported from such bites. Other animals, such as rats, rabbits, dogs and goats, seem to be much hardier (Maretić and Habermehl, 1985).

"Tarantulas" (mygalomorphs), despite their large size, are much less dangerous than is commonly thought. A tarantula bite, it is true, is deadly for mice or insects, but for a man it is hardly worse than a wasp sting (Schanbacher

TABLE 1. LD_{50} of Venom from *Lactrodectus mactans* for Various Animals. (After Maroli et al., 1973.)

	Lethal Dose 50	
Animal	*mg/animal*	*mg/g body weight*
Frog	2.18	0.145
Chicken	0.19	0.002
Mouse	0.013	0.0009
Cockroach	0.015	0.0027
Fly	0.000013	0.0006

et al., 1973). The poison consists of various polypeptides (molecular weight 5,000–20,000) and is most likely a neurotoxin (Perret, 1973; Lee et al., 1974). In a detailed analysis of a tarantula poison (*Eurypelma*), the following low-molecular-weight components were found: ATP, ADP, AMP, glutamic and aspartic acid, GABA and glucose; a high-molecular-weight protein (molecular weight 40,000) was identified as the enzyme hyaluronidase (Savel-Niemann, 1989). It might be added that American tarantulas are rather docile and bite only if provoked, whereas African (*Pterinochilus*) and asiatic species (*Poecilotheria*) are a bit more aggressive and can inflict painful bites.

The European tarantulas (Lycosidae) have never deserved their bad reputation, which has stuck to them since the Middle Ages (Crome, 1956b). The symptoms of tarantism were probably related to many other causes, ranging from other spider bites, notably those of the black widow, to epilepsy. However, some tropical lycosid spiders may well be dangerous, for example, *Lycosa erythrognatha* of Brazil (Lucas, 1988). After sustaining a bite, the skin surrounding the injured area becomes necrotic, and secondary bacterial infections are likely to occur (Bücherl, 1956).

Another well-known spider whose bite is much feared is the American brown recluse *Loxosceles reclusa* (Sicariidae; Fig. 39c). The bite causes local swelling as well as necrosis of the skin. The responsible enzyme is a sphingomyelinase that can cause a deep wound (Kurpiewski et al., 1981). The poison is also hemolytic (Bernheimer et al., 1985), and this may lead to kidney failure and sometimes to death.

In Australia, the funnel-web spider *Atrax robustus* (Dipluridae or Hexathelidae) is dangerous. Effects of the bite are severe pains, shivering, muscle cramps, loss of eyesight, and paralysis of the breathing center (Gage and Spence, 1977; Spence et al., 1977). Whereas the poison (Robustoxin; Sheumack et al., 1985) has severe effects on primates, dogs, cats, and rabbits are almost immune to it. In humans, only about a dozen fatal cases are known from *Atrax* bites (Gray, 1981).

In Europe, there are two other spiders one should handle cautiously, although neither is really dangerous—*Cheiracanthium* (Clubionidae) and the water spider *Argyroneta* (Argyronetidae). Their bites are painful, but the symptoms (itching, shivering, vomiting, slight fever) disappear within 2–3 days (Habermehl, 1974; Wolf, 1988). Most of the smaller spiders, however, cannot even pierce the human skin. In those rare cases where the skin is broken (as may happen from our garden spider), the effects, local swelling, blisters, and raised body temperature, are negligible (Maretić, 1971).

Most poisonous of all are certain ctenid spiders (Fig. 39d) of South America (Bücherl, 1953, 1971; Schenberg and Pereira Lima, 1971). In contrast to most other dangerous spiders mentioned above, these ctenid spiders are extremely aggressive. Their bite is very painful and can cause a shock. The observed effects are sweating, accelerated heart beat, and feeling cold and tense. The poison is neurotoxic by activating the Na^+ channels of nerve and muscle membranes (Entwistle et al., 1982; Fontana and Vital Brazil, 1985). One representative, *Phoneutria*, injects about 8 mg (dry weight) of venom per bite. This amount would theoretically be sufficient to kill 300 mice (LD_{50}=0.013 mg/animal). Nevertheless, the bite is rarely deadly for humans, probably because only small amounts of toxin are injected during a defensive bite.

Finally, it should be stressed that, at least statistically, spider bites are much less dangerous to man than the poisonous stings of bees, wasps, and hornets. This is because we encounter those hymenopterans more often than spiders, and in addition, often in swarms (Maretić, 1982).

From a biological view point, the poison of spiders is primarily designed to paralyze their prey, i.e., mainly insects; defensive bites against large animals (including man) are only secondary. Some spiders possess strong and quickly acting venom, others have much less effective toxins. Among the first category are spiders (e.g., Thomisidae and Mimetidae) that attack potentially dangerous prey such as bees, bumblebees, or other spiders. Many web spiders, which carefully wrap their prey before biting it, belong to the second category (Fig. 142; Friedel, 1987).

Internal Digestion

The liquefied food is sucked through the narrow mouth opening by the action of the muscles of the pharynx and the stomach. An initial filtering effect is brought about by the many bristles bordering the mouth opening. A second filtration occurs in the flattened pharynx, as the food passes the rostral cuticular lining. This *palate plate* possesses a median furrow and thousands of laterally situated cuticular platelets arranged like shingles (Fig. 40). Only very small particles ($< 1\ \mu m$) can pass through this filter, whereas all larger ones get caught. This can be nicely demonstrated by feeding a spider a suspension of India ink in a neutral red solution. The entire gut content will turn red but will contain only a very few grains of India ink; the ink particles become lodged in the lamellae of the palate plate (Bartels, 1930). Obviously, the spider needs to clean this filter device, and this is done with an antiperistaltic stream of digestive fluid. When a carmine dye solution was fed to a spider, it took only 30 minutes for the dye particles to appear at the mouth parts in the form of a solidified pellet. Shortly thereafter the pellet was brushed off with the palps (Zimmermann, 1934).

In the lateral walls of the pharynx lie cells that are ostensibly connected with the sense of taste (Millot, 1936). However, this conclusion is based on histological observations and still needs confirmation (see chapter 4). The fact that spiders readily accept a drop of water but quickly reject a weak quinine solution proves that they are capable of tasting. However, the seat of taste reception need not be located only beyond the mouth, but could also be in the contact chemoreceptors (taste hairs) on the maxillae and the palps.

The sucking stomach serves as the main pump for food intake. In contrast to the narrow esophagus, the stomach is of large diameter. In cross-section it typically exhibits the shape of a collapsed square (Figs 40, 41). This shape is produced by the pliable cuticular walls, which are closely apposed in the resting

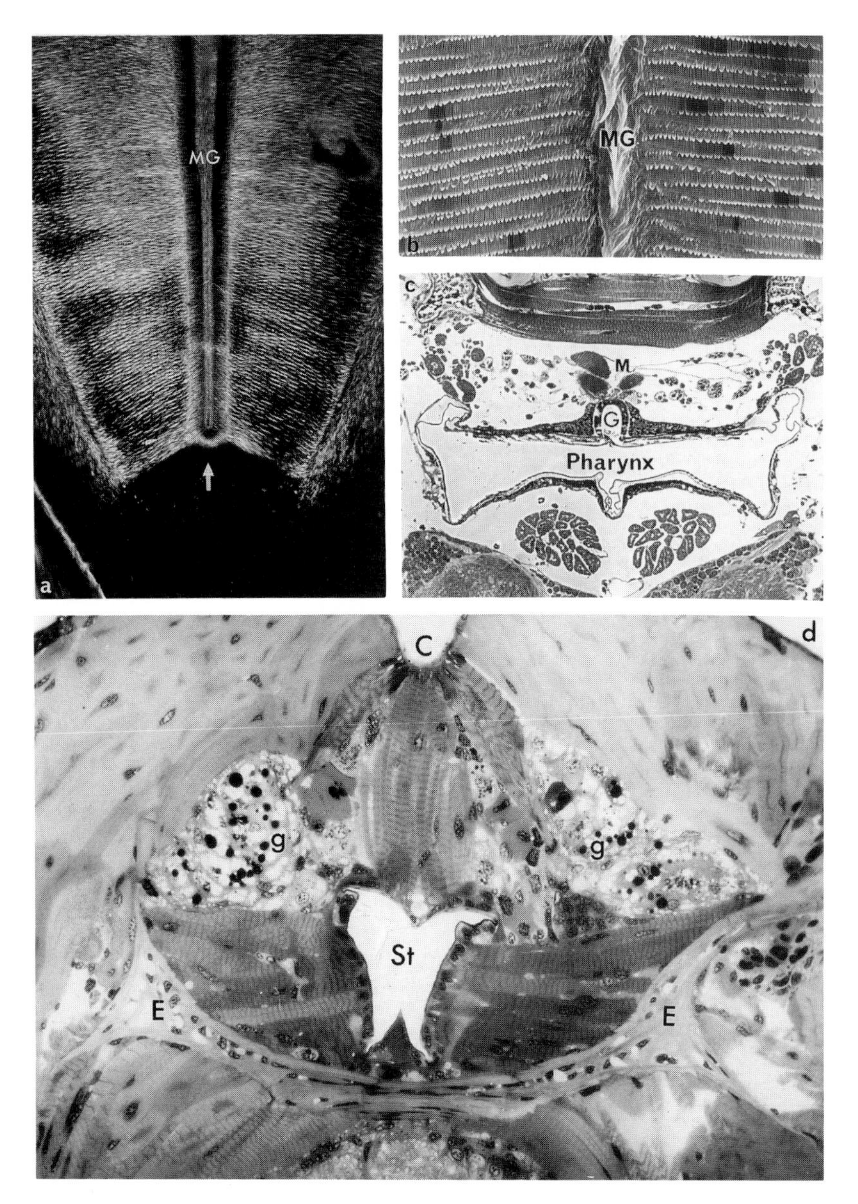

FIG. 40. Rostral plate in *Cupiennius*. (*a*) Inside view near the mouth opening. The liquefied food is sucked into the mouth in the direction of the arrow and is filtered at the lateral ridges. MG = median groove. 80 ×. (*b*) Each lateral ridge consists of cuticular platelets arranged like shingles. 200 ×. (*c*) Horizontal section through the pharynx of *Zygiella*. G = median groove with inserting muscles (M); these muscles expand the pharynx. 150 ×. (*d*) Cross-section of the sucking stomach (St) in a theridiid spider. Note the striated muscles connecting the stomach wall dorsally to the carapace (C) and laterally to the endosternite (E). g, Midgut diverticula. 280 ×.

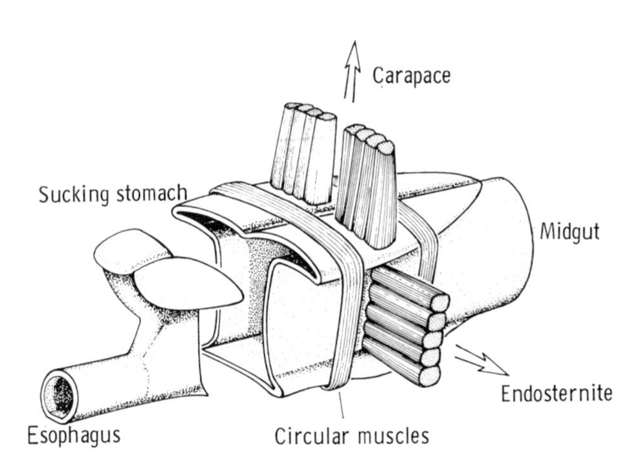

FIG. 41. Diagram of the sucking stomach. Dilation muscles connect dorsally to the carapace and laterally to the endosternite. Smaller circular muscles can constrict the lumen of the sucking stomach. (After Kaestner, 1969.)

state. Several strong muscle bands connect the upper stomach wall with the dorsal apodeme and the lateral walls with the endosternite. When all these muscles contract, the stomach lumen increases greatly and the stomach thus functions as a suction pump. Several thin annular muscles, which lie between the expanding muscles, act antagonistically; that is, they act to diminish the stomach lumen (Schimkewitsch, 1884). A precise coordination of both sets of muscles causes wavelike contractions of the sucking stomach. Thus despite an extremely small mouth opening, the food is taken in rather quickly.

Regulatory valves, which are present at the entrance and exit of the sucking stomach, ensure that the food passes in a caudal direction as the annular muscles contract (Legendre, 1961a).

The midgut has its origin directly behind the sucking stomach. Its proximal part is still situated in the prosoma, while its median and posterior portions are in the opisthosoma. The proximal part gives off two lateral extensions, each of which splits into several fingerlike processes (Figs 27, 42). The arrangement and shape of these diverticula may vary considerably among different spider families; in most cases the diverticula extend into the coxae of the walking legs.

The branches of the midgut fill out most of the opisthosoma and surround many organs, such as the spinning glands and the reproductive tracts. Gut diverticula are also present in the prosoma and may be numerous there. For instance, in jumping spiders the midgut extensions even penetrate between the eyes (Fig. 70b). Such an extensively developed intestinal system explains in part why spiders can survive for a long time without eating. In one experiment with black widows, adult spiders were not fed at all, yet continued to live for 200 days (Kaston, 1970). Well-nourished "tarantulas" often do not feed for several months.

The entire pharynx, as well as the sucking stomach, is covered by a thin cuticular lining. The midgut lacks such a lining and is therefore the only part

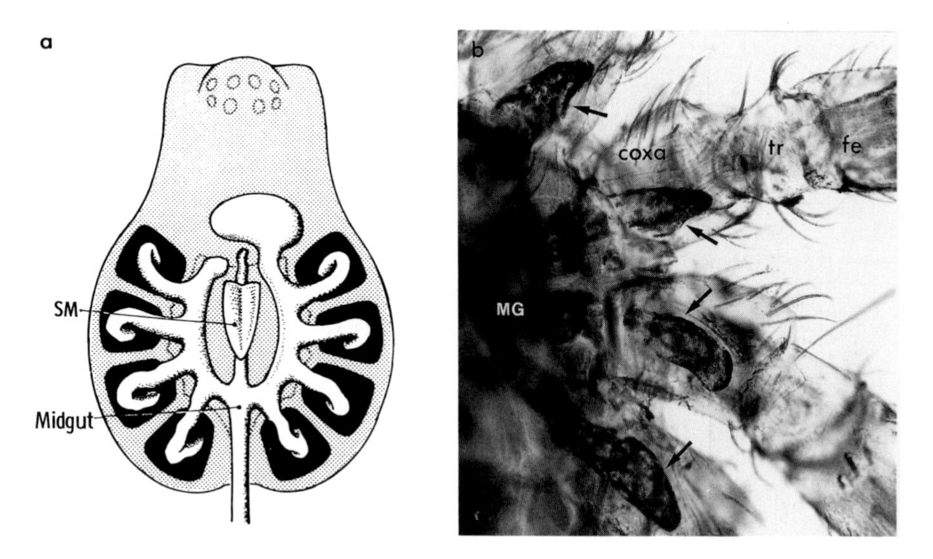

FIG. 42. Midgut. (*a*) Behind the sucking stomach (SM) branches of the midgut extend through the entire prosoma and even enter the legs. *Tegenaria*, dorsal view. (After Plateau, 1877, and Zimmermann, 1934). (*b*) Whole mount of a young wolf spider showing midgut branches (MG, arrows) ending blindly in the leg coxae. tr = trochanter, fe = femur. 45 ×. (Preparation by H. Moor.)

suitable for digestion, that is, the absorption of nutrients. It has been known for a long time that spiders possess two different kinds of cells in the intestinal epithelium; *secretory cells* and *resorptive cells* (Bertkau, 1885; Millot, 1926; Fig. 43). The secretory cells contain digestive enzymes, which can be seen as dark cytoplasmic granules under the microscope. Most of the intestinal cells are of the resorptive type. They can be recognized by their many inclusions (the "food vacuoles"). The resorptive cells process the nutrients further and then pass them on to the underlying interstitial tissue or into the hemolymph.

After prey has been captured, the following cellular events take place in the intestinal epithelium (Nawabi, 1974). Within a few minutes the secretory cells empty their enzymatic granules into the gut lumen. After about an hour, while the prey may still be in the process of being sucked out, the first food vacuoles appear inside the resorptive cells. Initially the food vacuoles are seen only at the cell's apex, but soon they become distributed throughout the entire cell. Eight hours after food intake has begun, most of the secretory cells have become void of enzymatic granules. At the same time, the first excretory products can be detected basally in the resorptive cells. After 24 hours the secretory cells have resynthesized their enzymatic granules and the resorptive cells contain only a few food vacuoles. Excretory products are often concentrated apically in the resorptive cells, or they may be deposited as crystals in the underlying interstitial tissue. This tissue also serves as a reservoir for accumulated glycogen and lipids.

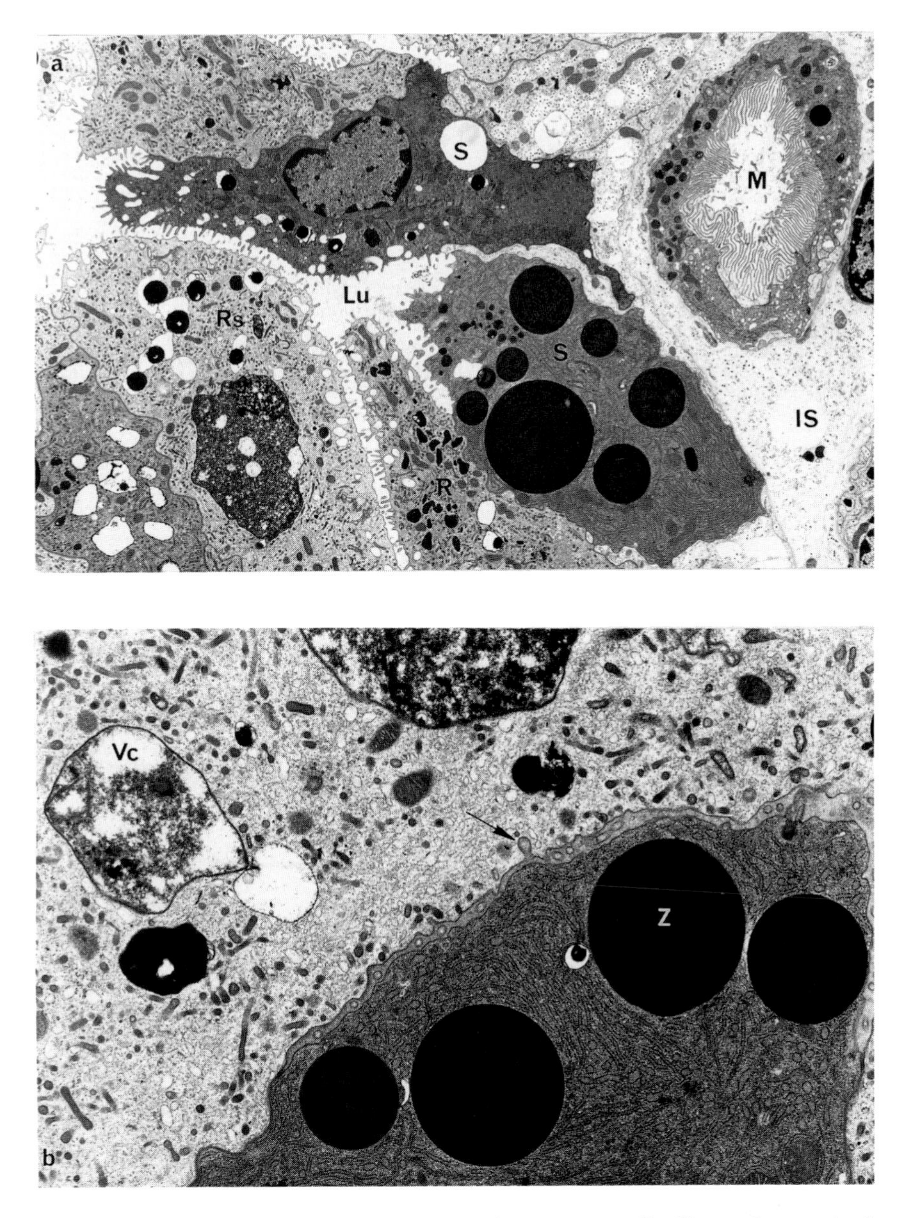

FIG. 43. Midgut diverticula of *Tegenaria*. (*a*) The secretory cells (S) are characterized by many densely stained enzyme granules, whereas the resorptive cells (Rs) possess irregular food vacuoles. Lu = gut lumen, M = Malpighian tubule, IS = interstitial tissue. 3,000 ×. (*b*) Secretory cell with zymogen granules (Z) and extensive endoplasmic reticulum; resorptive cell with tubular indentations (pinocytosis, arrow) and large food vacuoles (Vc). 9,000 ×.

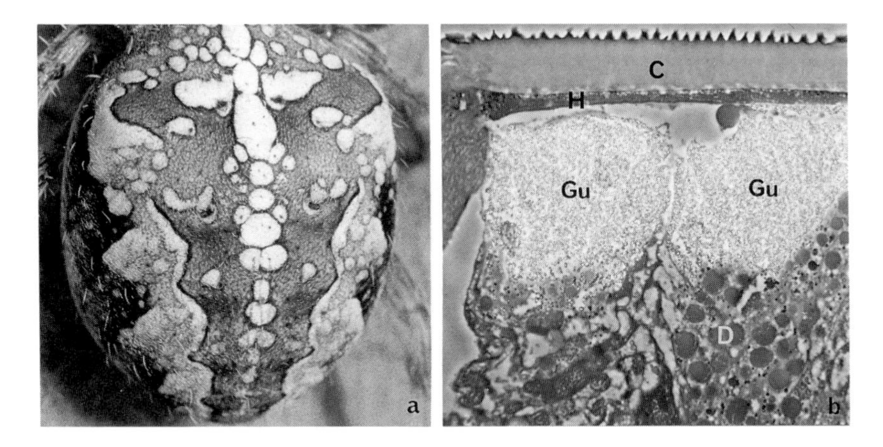

Fig. 44. (*a*) Abdomen of the garden spider *Araneus diadematus*, dorsal view. The markings in the form of a white cross are caused by guanine cells, which shine through the transparent cuticle. (Photo: Foelix and Grocki.) (*b*) Histological section of the abdomen showing guanine crystals deposited in specialized guanine cells (Gu) at the periphery of the gut diverticula (D). C = cuticle, H = hypodermis. 400 ×.

In addition to the secretory and resorptive cells, the intestinal epithelium contains two other cell types, *basal cells*, which eventually replace old secretory and resorptive cells, and *guanocytes*, in which guanine crystals become deposited (Millot, 1926; Seitz, 1972c, 1975). The guanocytes can be thought of as specialized resorptive cells, which take up metabolites (e.g., purines) and store them temporarily in the form of crystals. Guanocytes may form a contiguous cell layer directly beneath the hypodermis (Fig. 44b). Since the guanine inclusions totally reflect the incoming light they appear white and can be seen through transparent parts of the cuticle. The white cross pattern on the abdomen of the common garden spider *Araneus diadematus* (Fig. 44a) is a well-known example of such an accumulation of guanine. The gut cells also contain mineral-rich particles (*spherites*), which may store certain ions (e.g., calcium), or help in excretion (e.g., guanine) or in detoxification (e.g., lead) (Ludwig and Alberti, 1988).

Whereas the shape of the midgut appears rather confusing owing to its elaborate branching, the next section of the digestive system is structurally much simpler. The posterior end of the midgut widens to form the stercoral pocket, or cloacal chamber. It consists mainly of a large, blind sack into which the Malpighian tubules (see below) exit, and it connects to the anus via the short hindgut. The excrement stored in the stercoral pocket is periodically passed to the outside at the anus.

Excretion

The fact that the intestinal cells themselves can accumulate guanine and thereby perform an excretory function has already been mentioned. The main excretory

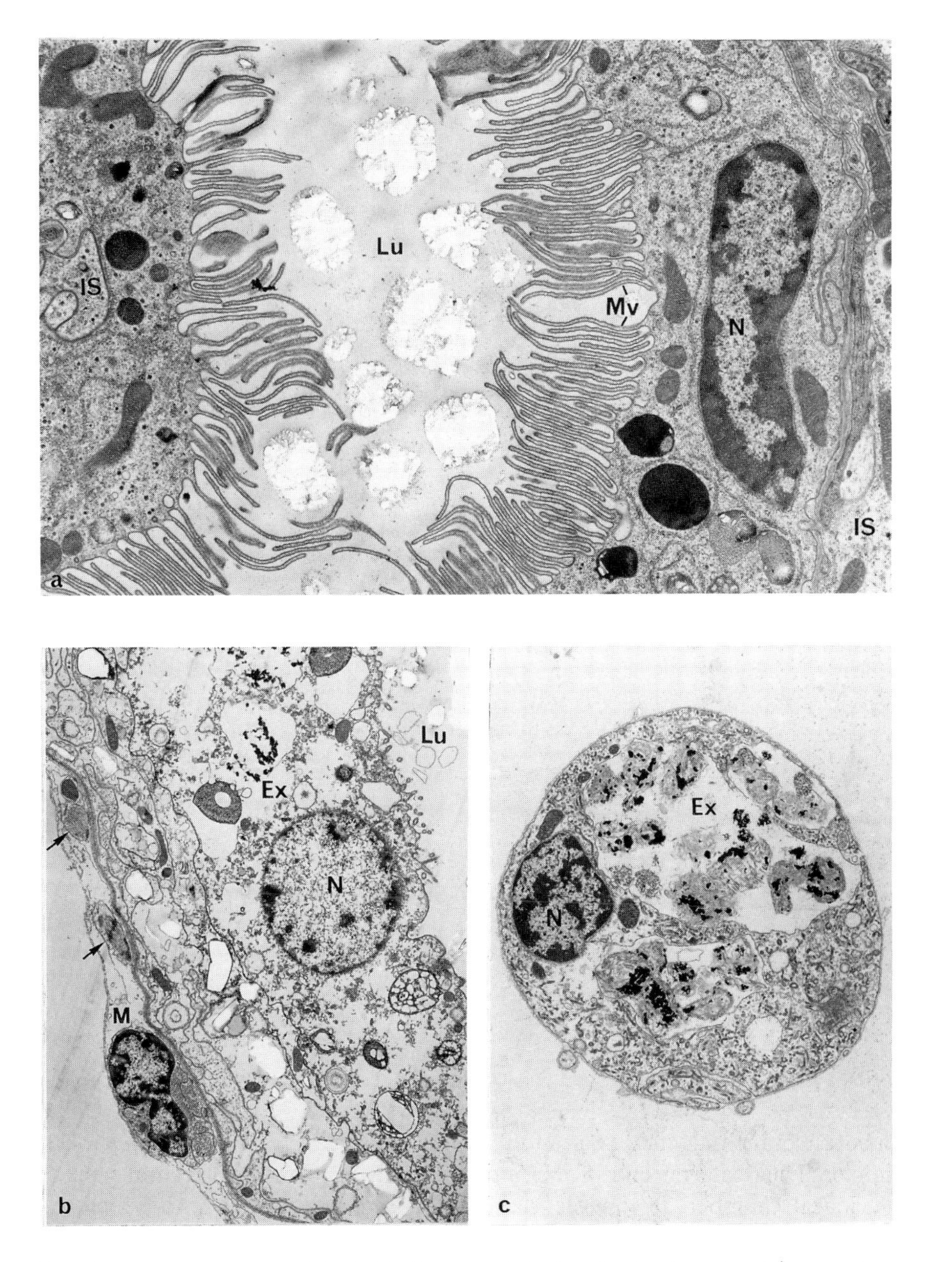

FIG. 45. Excretory organs. (*a*) Malpighian tubule in longitudinal section (*Tegenaria*). The flat tubule cells possess typical brush borders (Mv). Empty spaces in the lumen (Lu) represent excretions that have dissolved during tissue preparation. IS = interstitial tissue. 8,000 ×. (*b*) Coxal gland (*Zygiella*). The interior of the gland cell is filled with excretions (Ex) that are transported toward the lumen (Lu). Muscle cells (M) and their processes (arrows) surround the base of the excretory cells and apparently aid in dispersing excretions. 4,400 ×. (*c*) Nephrocyte (*Zygiella*). The excretory products (Ex) are accumulated in intracellular compartments. N = nucleus. 4,700 ×.

organs, however, are the Malpighian tubules. These originate from either side of the stercoral pocket as two thin evaginations and are thus entodermal derivatives, quite in contrast to insect Malpighian tubules, which are of ectodermal origin. Histologically, the Malpighian tubules do not differ much from the stercoral pocket or from the hindgut (Seitz, 1975). The excretory cells form a flat epithelium with a typical apical brush border (Figs 43a, 45a). Below the excretory cell bases lie delicate muscle cells that probably assist in the transport of the excretion products. The excretory substances are first taken up at the base of the cell. There the cell membrane is highly infolded, a typical feature of cells with augmented transport activity. Inside the cytoplasm the metabolites are concentrated and stored before they are expelled into the lumen of the Malpighian tubule. The main excretory products are guanine, adenine, hypoxanthine, and uric acid. All these substances are nearly insoluble in water, a property reflected in their tendency to crystallize.

Less conspicuous excretory organs, the segmentally arranged *coxal glands*, lie in the prosoma (Fig. 45b). They open to the outside at the coxae of the walking legs. In "primitive" spiders (Mesothelae, Orthognatha) two pairs of coxal glands open onto the posterior side of the first and third coxae (Buxton, 1913). They release a fluid only during feeding and seem to play an important role in ion and water balance (Butt and Taylor, 1991). The Labidognatha have retained only the anterior pair, and even these show gradual stages of regression in the various families. The original type of coxal gland consists of four parts: a saccule, a collecting duct, a labyrinth, and an excretory duct. This type becomes substantially reduced in orb-web spiders, in which a collecting duct is lacking and the labyrinth apparently no longer has an excretory function. Additional excretory "organs" are represented by the many *nephrocytes*. These huge cells tend to accumulate at specific sites in the prosoma, for instance, beneath the subesophageal ganglion. With a diameter of 30–80 μm they are the largest cells of the spider's body. Presumably the nephrocytes take up metabolites from the hemolymph and store them intracellularly (Fig. 45c).

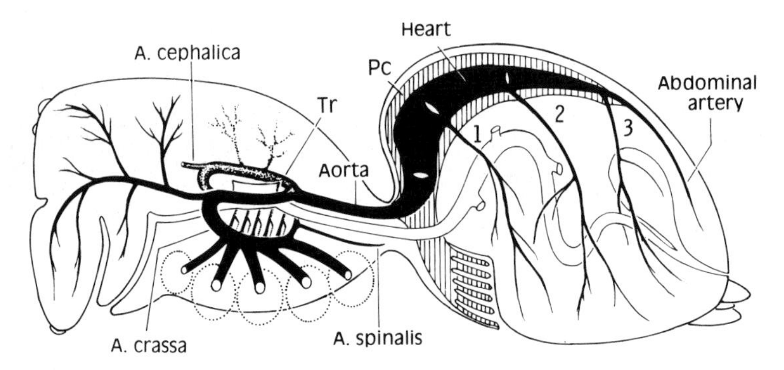

Fig. 46. Diagram of the arterial circulatory system. A = arteria, A. ceph. = arteria cephalica, Pc = pericardial sinus; 1, 2, 3 = lateral opisthosomal arteries; Tr = trunci peristomacales. (After various authors.)

Circulatory System

Spiders, like most invertebrates, have an open circulatory system. This does not mean that they lack blood vessels. On the contrary, they have quite distinct arteries that originate from the heart and that branch through the entire body (Figs 46, 47). The artery walls contain myofilaments (Pack, 1983) and can change their diameter in response to certain ions (*vasoconstriction/dilatation*; Paul et al., 1994). The leg arteries even reach into the most distal segments (tarsi),

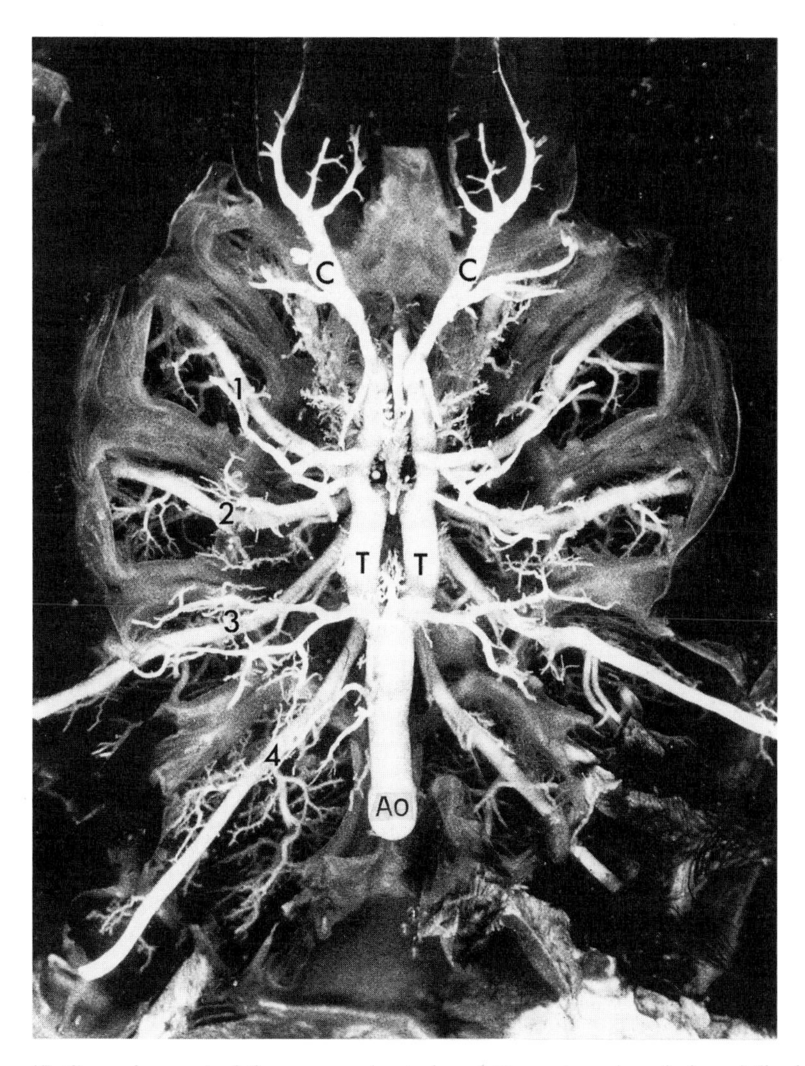

FIG. 47. Corrosion cast of the prosomal arteries of *Eurypelma*, dorsal view. A liquid plastic solution (Mercox) had been injected into the heart and had spread into the aorta (Ao), the two trunci peristomacales (T), the head arteries (C) and the leg arteries (1–4); after polymerization and removal of the soft tissues all blood vessels stand out clearly. (Photo: Paul.)

where they end openly like the needle of a syringe. There the hemolymph and blood cells can be seen seeping freely between the tissues. The flow of the hemolymph back to the heart follows the gradient of decreasing pressure. Its course is by no means random but is guided along specific pathways (Colmorgen and Paul, 1995). The blood eventually collects in lacunae on the ventral side of the body before passing the respiratory organs (book lungs) and returning via the lung veins into the heart.

In summary, spiders have a "closed" arterial system, no capillaries, and an open venous system, in which the gas exchange takes place.

Heart and Blood Vessels

The heart lies dorsally in the anterior part of the opisthosoma (Figs 28, 48). It is formed by a muscular tube suspended in a wide chamber (the *pericardial sinus*) by dorsal, ventral, and lateral ligaments. The heart tube itself has two or three pairs of buttonhole-shaped slits (*ostia*), which open to the pericardial sinus. This sinus is covered by a thin layer of smooth musculature. When the heart tube contracts (*systole*), the ostia, acting as passive valves, close automatically (Fig 49a). During this contraction the heart elongates and the hemolymph is pushed forward into the anterior aorta in the prosoma and backward into the posterior aorta in the opisthosoma. Special valvelike structures at either end of the heart tube (in addition to the closed ostia) prevent a backflow of the hemolymph (Fig. 49b,c).

The course of the various arteries through the body can be followed after carefully injecting a spider's heart with carmine, India ink, or liquid plastic resins (Petrunkevitch, 1910; Crome, 1953; Bihlmayer et al., 1989). The anterior aorta passes through the pedicel and then, close to the sucking stomach, splits into two lateral branches (*trunci peristomacales*; Figs 46, 47). Each of

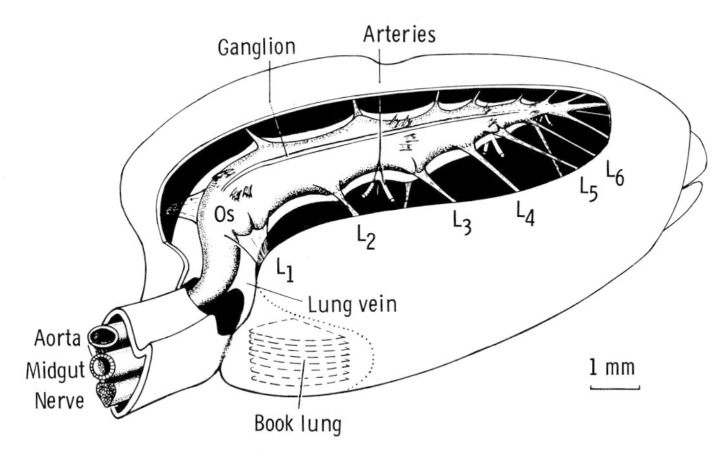

FIG. 48. Position of the heart in the opisthosoma of the banana spider *Heteropoda*. L_{1-6} = lateral ligaments, Os = ostium. (After R. S. Wilson, 1967.)

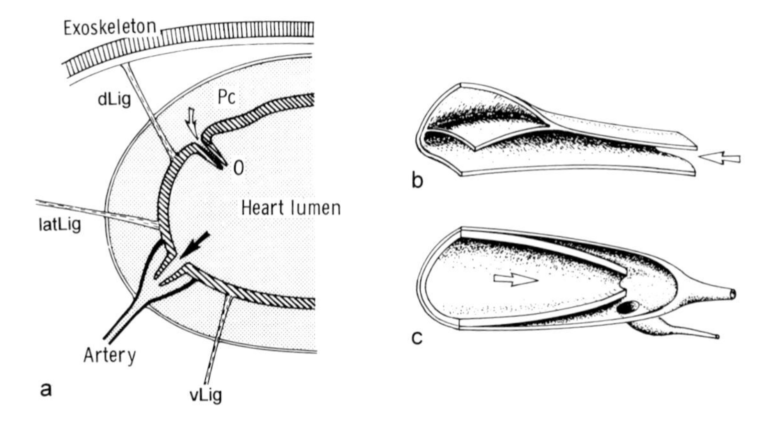

FIG. 49. (*a*) Diagrammatic cross-section of the heart tube. Arrows indicate direction of hemolymph flow. dLig, latLig, vLig = dorsal, lateral, and ventral ligaments; O = ostium; Pc = pericardial sinus. (*b* and *c*) Heart valves: the valve leading into the aorta is a flapvalve (*b*), whereas the one at the caudal end of the heart (*c*) is funnel-shaped. (After Crome, 1953.)

these branches divides again, into the *arteria cephalica*, which supplies the supraesophageal ganglion, the chelicerae, and the mouth parts, and the *arteria crassa*, which sends off one artery per extremity. The two arteriae crassae are cross-connected horizontally by means of six *rami transversales*; from these vessels smaller arteries branch off ventrally and penetrate the subesophageal ganglion. From the last ramus transversalis emerges horizontally the *arteria spinalis*, which proceeds along the upper side of the subesophageal ganglion. Most of the arterial hemolymph is directed to the central nervous system and to the skeletal muscles (Millot, 1949). Diffusion distances are about 60 μm in the CNS (Kosok and Seyfarth, 1994) and 60–90 μm in muscle tissue (Bihlmayer et al., 1990). It seems that muscle cells with a high oxidative capacity are much better supplied with hemolymph (Paul et al., 1994). Since they are completely bathed in lymph, they have a relatively large exchange area per volume of tissue and this makes up for the rather large diffusion distances.

In most spiders the organs of the opisthosoma receive a direct tracheal supply and thus are less dependent on oxygen carried by the hemolymph. Still, the opisthosoma has two or three arteries of its own that originate laterally from the heart tube (Fig. 46); their degree of ramification is quite high and the terminal diameter of the arterioles (50 μm) is similar to those in the prosoma (Zahler et al., 1990).

The expansion of the heart (*diastole*) is caused by the ligaments that connect the heart tube with the exoskeleton. When the heart tube is in the contracted stage (systole), these ligaments are stretched and under tension. When the contraction subsides, the ligaments shorten passively owing to their inherent elasticity. This in turn causes the ostia to open, and hemolymph enters the heart lumen from the pericardial sinus (R. S. Wilson, 1967; Figs 49, 54b). In other

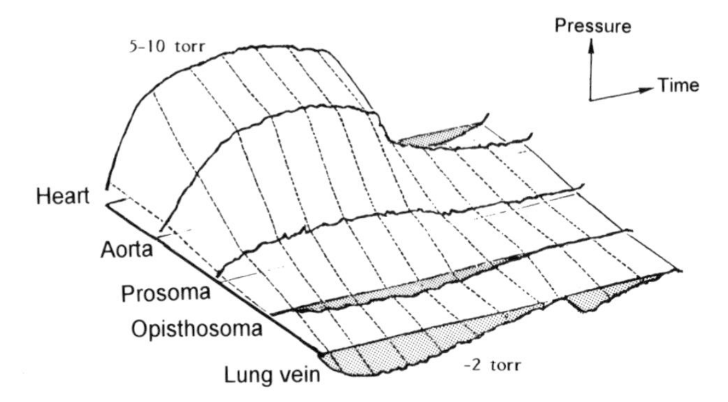

FIG. 50. Time course of the hemolymph pressure in various parts of the circulatory system during one heart beat of *Eurypelma*. Note the negative pressures in opisthosoma and lung vein, which lead to a venous backflow due to a suction effect. (After Bihlmayer, 1991 and Paul, 1994.)

words, heart tube and pericardium work together as a pressure and suction pump (Paul, 1990).

Spider circulation operates as a low-pressure system. The heart generates between +5 and +15 torr during systole. These positive pressures diminish little in the aorta and in the prosomal arteries, but drop to slightly negative values in the lung veins and in the pericardium (Fig. 50). During diastole even the heart lumen shows a negative pressure (*suction*), which leads to a refilling of the heart tube through the ostia. This pumping mechanism of pressure and suction results in a continuous flow of hemolymph through the book lungs, which ensures an efficient gas exchange (Paul et al., 1994).

Structure and Function of the Heart

The heart tube consists of a thin outer layer (*adventitia*) composed of connective tissue and of the actual muscle layer (*myocardium*) that forms the circular heart lumen (Seitz, 1972a). An inner layer (*intima*) seems to be lacking in spiders; instead, the muscle cells are in direct contact with the hemolymph (Sherman, 1973a). As in vertebrates, the heart muscle cells are cross-striated, branched, and multinucleate. In contrast to regular skeletal muscles (Fig. 23), they contain many mitochondria (Fig. 52; Midttun, 1977), which provide for the continuous energy demand of the heart.

In spiders, as in most other arthropods, the heartbeat is controlled neurogenically. A threadlike ganglion on the dorsal side of the heart tube (Fig. 48) gives off axons that innervate the heart muscle cells (R. S. Wilson, 1967; Legendre, 1968; Sherman et al., 1969; Ude and Richter, 1974). This ganglion probably contains pacemaker neurons that control the heartbeat via intercalated motoneurons. The heartbeat itself is the result of the synchronous contraction of the entire heart tube.

Heartbeat frequency has been investigated in many spiders (Bristowe, 1932;

Sherman and Pax, 1970a; Carrel and Heathcote, 1976). In general, the heart rate decreases with increasing body size. Large "tarantulas" have a heartbeat frequency of 30–40 beats per minute, whereas small spiders show a resting frequency of over 100. During running, the heartbeat accelerates rapidly. A wolf spider, for which a resting frequency of 48 beats per minute was measured, increased its heart rate to 176 beats per minute after only 30 seconds of activity (Sherman and Pax, 1968). The volume that a spider heart pumps out with each beat (*stroke volume*) varies from 0.7 ml/min at rest to 4.3 ml/min after a high activity (*Eurypelma*; Paul, 1986).

In many spiders the heartbeat frequency can be determined simply by counting the pulsations of the abdomen under a binocular microscope. By applying a fine wire electrode to the body, directly over the heart, one can also record an electrocardiogram (ECG). This method not only automatically yields the heart rate but also proves its neurogenic origin by registering oscillatory bursts.

ECGs have been obtained even from extirpated spider hearts. The resting frequency is about the same as in the living animal and is apparently determined by the heart ganglion (Sherman and Pax, 1968, 1970a). Adding L-glutamate causes the heart to contract. L-glutamate is perhaps the natural neurotransmitter in the spider heart, for other known neurotransmitters, such as acelylcholine or gamma-aminobutyric acid (GABA), have no effect on heart action (Sherman and Pax, 1970b,c).

The CNS can also exert an excitatory or an inhibitory effect on the heartbeat (R. S. Wilson, 1967). This is obvious because even minor sensory input influences the heart rate: spiders that are physically restrained for observation show a markedly faster heartbeat than normal. Indeed, cardioregulatory neurons have been demonstrated in the CNS by morphological and electrophysiological methods (Gonzalez-Fernandez and Sherman, 1984).

Blood Cells and Hemolymph

The spider's hemolymph exhibits a large variety of blood cells (*hemocytes*). Although very little is known about the function of specific hemocytes, it seems likely that they play a role in blood clotting, wound healing, and fighting off infections. Structurally, at least four different types can be distinguished (Seitz, 1972a,b; Sherman, 1973b, 1981). Most common are the granular hemocytes, which have many dense granules concentrated in the cytoplasm (Fig. 51); it has been suggested that the granular hemocytes contribute to the sclerotization of the exocuticle (Browning, 1942). Some hemocytes, called *leberidiocytes*, enclose a single secretory vacuole. Other blood cell types are believed to act as phagocytes or as storage cells. During molting the relative percentage of the different hemocyte types changes drastically.

The curious origin of the hemocytes should be mentioned here: they derive directly from myocardium cells of the heart wall (Franz, 1904; Millot, 1926). This is easily seen in young spiders, where immature "prehemocytes," or hemoblasts, sit on the luminal side of the heart tube, where they undergo mitotic

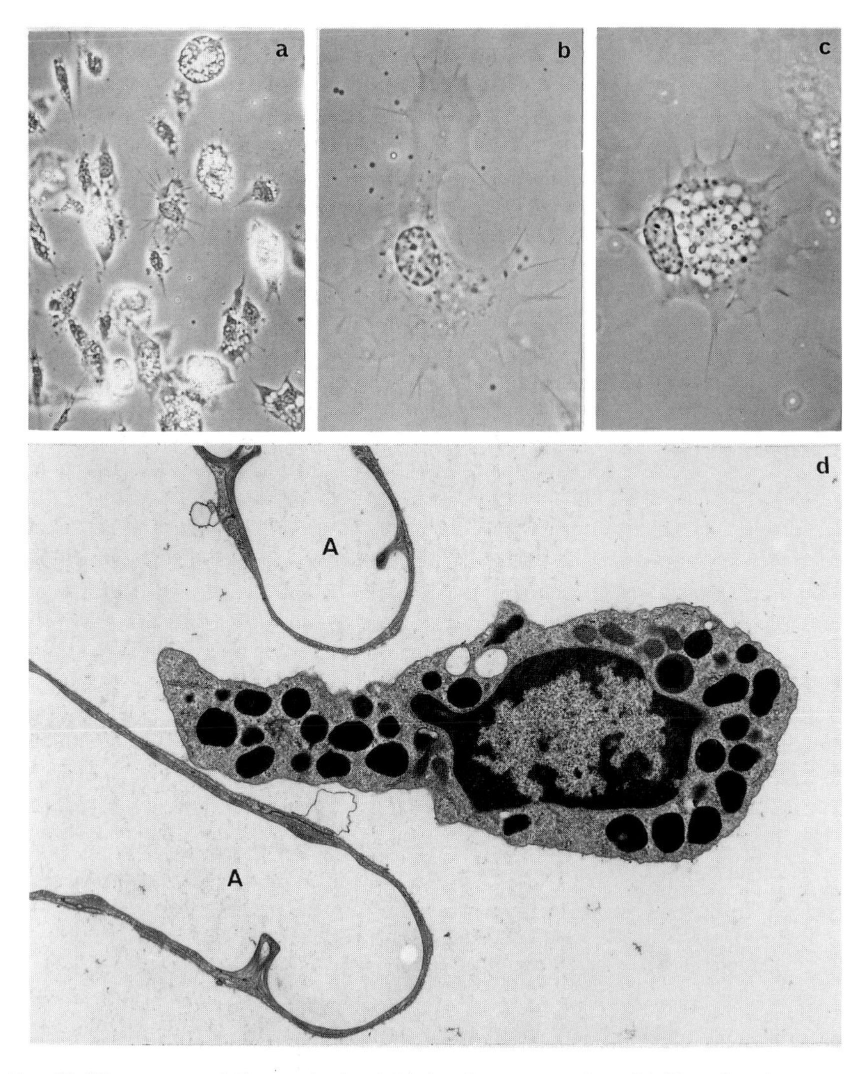

FIG. 51. Hemocytes of *Tegenaria*. (*a–c*) Living hemocytes show highly refractive granules when viewed in a phase contrast microscope; pseudopodia are the first sign of blood clotting. (*a*, 260 ×; *b* and *c*, 650 ×.) (*d*) A granular hemocyte just entering the hemolymph space between two air spaces (A) of a book lung. 7,000 ×.

divisions (Fig. 52). These hemoblasts develop into the different types of hemocytes by acquiring specific cell organelles (Seitz, 1972a).

Fresh spider hemolymph appears bluish because of the copper-containing respiratory pigment *hemocyanin*. It is a large protein similar to hemoglobin, but instead of iron there are two atoms of copper in the center of the molecule, which bind to oxygen. Its chemical structure has been analyzed in detail in the "tarantula" *Eurypelma* (Linzen et al., 1985; Paul et al., 1992). One hemocyanin molecule consists of 24 subunits, and each subunit is made up by more than

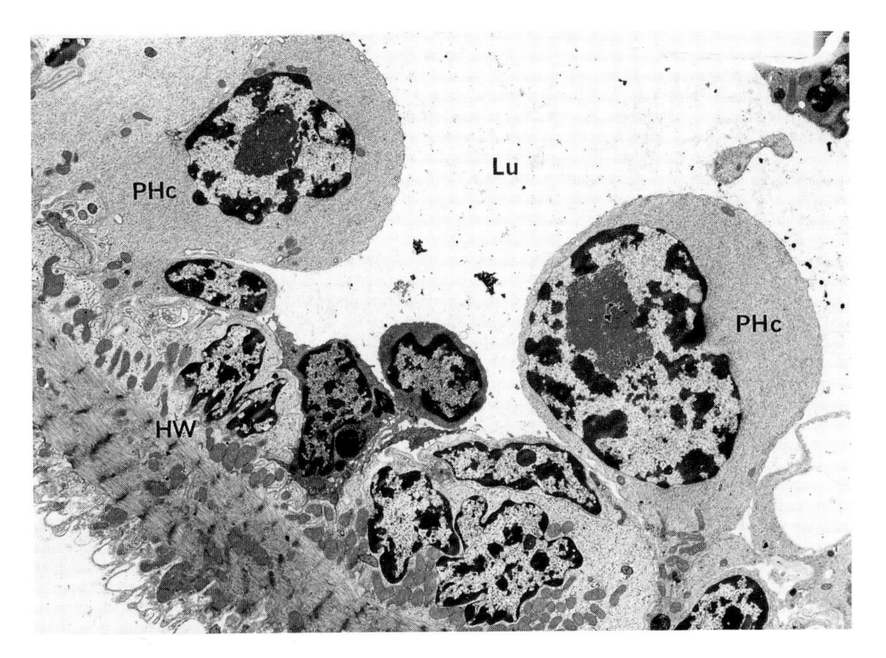

FIG. 52. Two prehemocytes (PHc) in the process of being pinched off from the heart wall (HW). Lu = heart lumen. 3,100 ×.

600 amino acids. Since each subunit contributes a mass of about 72,000, the total mass exceeds 1.7 million—a huge molecule. For comparison, hemoglobin has a mass of 66,400, with subunits of only 16,000. However, hemoglobin can bind 210 cm^3 O_2/l, whereas hemocyanin in spider blood can hold only 12 cm^3/l. What is the reason for this large difference? Firstly, hemoglobin is highly concentrated in our blood (150 g/l), because it is packed into special blood cells, the *erythrocytes.* The concentration of hemocyanin in spider blood is much lower (40 g/1) as it is freely dissolved in the hemolymph. Secondly, the subunits of hemoglobin that can bind one molecule of oxygen are about five times smaller (16,000 vs. 72,000) than in hemocyanin—and are therefore more efficient in transporting oxygen. However, spiders do not really need a highly efficient respiratory pigment, since they are inactive most of the time and oxygen consumption is very low at rest.

Hemocyanin is synthesized in special blood cells, the *cyanocytes* (Fig. 53). Initially the hemocyanin is stored in crystalline form, but mature cyanocytes may deliver it to the hemolymph (see Fahrenbach, 1970). Since hemocyanin has a strong affinity for oxygen, this element can be given off to the surrounding tissue only if the local oxygen pressure is low there ($pO_2 < 10$ torr). Values of 1–10 torr were recorded in muscle tissue of *Eurypelma helluo,* whereas the "arterial" hemolymph inside the heart lumen measured 31 torr (Angersbach, 1975). However, hemocyanin may be more of a storage site for oxygen than a carrier of oxygen.

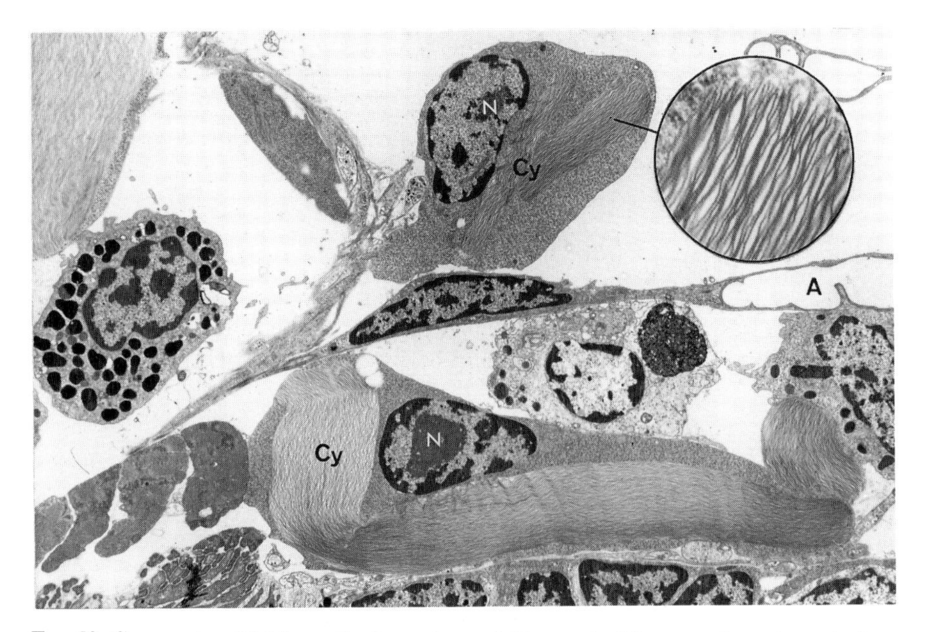

FIG. 53. Cyanocytes (Cy) from the lung sinus of *Tegenaria*. The cytoplasm is packed with bundles of hemocyanin crystals (Cy). A = air space, N = nucleus. 3,200 ×. Inset: Hemocyanin bundles at higher magnification. 38,000 ×.

Only a few studies have concentrated on the composition of spider hemolymph (Loewe et al., 1970). For the "tarantula" Eurypelma the inorganic ions have been determined many years ago (Rathmayer, 1965c) and lately the organic components were added (Schartau and Leidescher, 1983). Among the cations, Na^+ is predominant (200 mM/l) as compared to Ca^{2+}, Mg^{2+} (4 mM/l each), K^+ (2 mM/l) and Cu^{2+} (1.3 mM/l); among the anions Cl^- (240 mM/l) is prevalent with respect to $PO_4{}^{3-}$ (3 mM/l) and $SO_4{}^{2-}$ (0.7 mM/l). The pH value is about 7.5 (Angersbach, 1978), the osmolarity (480 mosmol/l) lies in the same range (400–600 mosmol/l) as in other spiders (Pinkston and Frick, 1973; Cohen, 1980). Most of the organic substances of the hemolymph are proteins, and of these hemocyanin makes up 80%. Others are free amino acids (mainly proline), carbohydrates (mainly glucose), and fatty acids (palmitic, linoleic, and stearic acid).

Outside the body the hemolymph coagulates quickly. Coagulation is initiated by many fine pseudopodia that grow out from the hemocyte (Fig. 51; Deevey, 1941; Grégoire, 1952). The hemolymph accounts for about 20% of the body weight of a spider (Stewart and Martin, 1970). It is surely imperative for a spider to maintain a certain hemolymph volume, since the hydraulics of the leg joints depend on the pressure, and thus on the quantity, of the hemolymph. Under normal conditions the exoskeleton protects the spider from desiccation. If extremely dry spells occur, or if the spider suffers direct blood loss through injury or autotomy, the animal attempts to replace lost fluid by drinking.

Respiratory Organs

Most spiders possess two entirely different kinds of respiratory systems: one pair of strictly localized book lungs, and one or two pairs of tubular tracheae, which can branch throughout the body. The "primitive" spiders (Mesothelae, Orthognatha, Hypochilidae) have book lungs only, but always *two* pairs. These are situated on the second and third abdominal segments. In the majority of labidognath spiders, only the first pair of book lungs has been retained, while the second pair has been modified into tubular tracheae. The book lungs are structurally very uniform in all spiders (Fig. 54), but the tubular tracheae vary considerably in relative size and pattern of distribution (Lamy, 1902; Kaestner, 1929).

Book Lungs

The book lungs lie ventrally in the anterior opisthosoma (Fig. 28). Their location is often visible from the outside as a hairless patch of cuticle that borders posteriorally on a narrow slit, the *lung slit* (Fig. 1), which leads to the interior of the abdomen. There a small atrium enlarges and then extends into many horizontal air pockets which are in contact with blood-filled lamellae (Figs 55,56). In other words, stacks of air-filled spaces alternate with leaflets of hemolymph spaces—hence the descriptive term book lung. The leaflets as well as the atrium are covered by a very thin cuticle (0.03 μm; Reisinger et al., 1991).

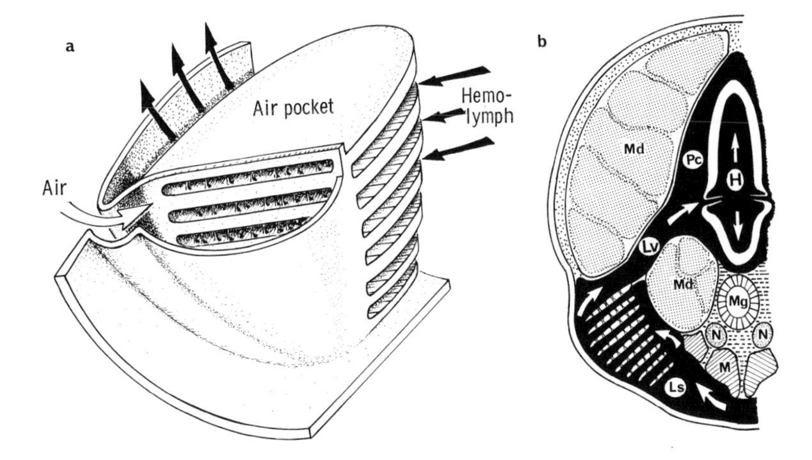

FIG. 54. Structure of a book lung. (After Kaestner, 1929.) (*a*) Three-dimensional representation of the lower part of a book lung. The white arrow indicates the air entering the atrium of the book lung, the black arrows show the direction of hemolymph flow. (*b*) Cross-section of the opisthosoma at the level of the book lungs. The arrows show the hemolymph coming from the lung sinus (Ls), passing through the book lung, and then entering the heart (H). Lv = lung vein, M = abdominal muscles, Md = midgut diverticula, Mg = midgut. N = abdominal nerves, Pc = pericardial sinus.

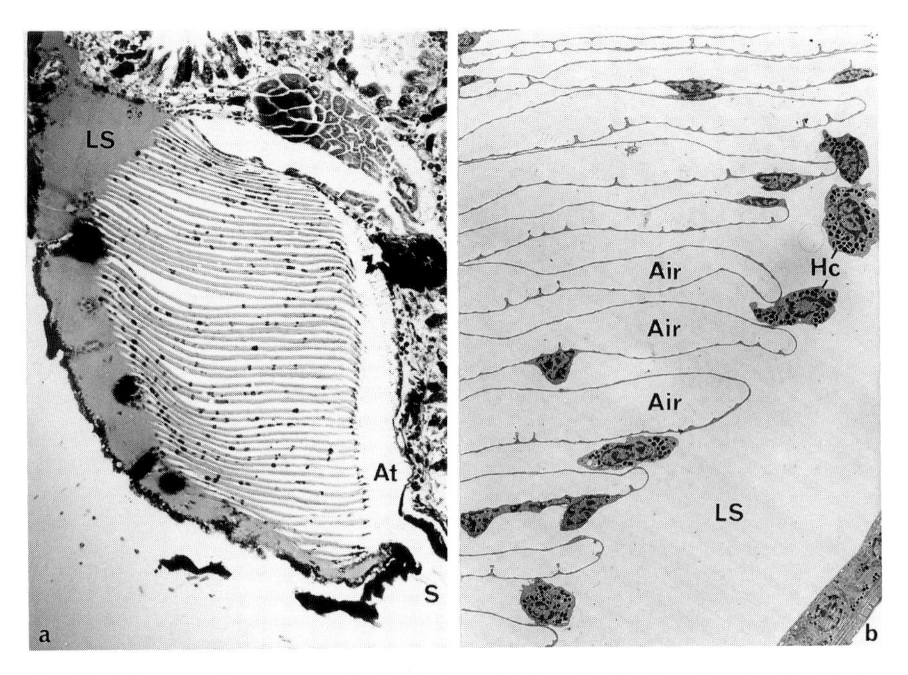

FIG. 55. Microscopic structure of a book lung. (*a*) Longitudinal section (wolf spider). At = atrium, LS = lung sinus with stained hemolymph, S = lung slit. 85 ×.
(*b*) Cross-section (*Tegenaria*). The hemolymph is unstained in this preparation, but hemocytes (Hc) are clearly recognizable in the lung sinus (LS). Note the extremely thin walls of the air pockets (Air). 1,100 ×.

Short vertical extensions of this cuticle form minute columns that act like struts to prevent collapse of the narrow air pockets (Fig. 56b,c). Each leaflet is formed by a doubling of the thin hypodermis (0.02 μm) leaving an open space in between where the hemolymph can circulate freely. In regular histological preparations the acidophilic hemolymph stains distinctly, yet the thin hypodermis extensions are hardly visible (Fig. 56a). Only the nuclei of the hypodermis cells are apparent; they become more conspicuous if they lie exactly opposite each other within one leaflet. In such cases, the two hypodermis cells seem to form a pillar between two air spaces (Fig. 56b).

For the house spider *Tegenaria*, the entire surface area of the two book lungs was determined as 70 mm^2, the corresponding lung volume as 0.6 mm^3 (Strazny and Perry, 1984). In the large "tarantula" *Eurypelma*, the surface of all (four) book lungs is considerably higher (70 cm^2), its volume lies between 10 and 20 mm^3 (Reisinger et al., 1990).

It is interesting that the two pairs of book lungs in "tarantulas" seem to serve two pathways of blood circulation: hemolymph from the prosoma passes only through the anterior pair of book lungs, while hemolymph from the opisthosoma flows only through the posterior pair (Paul et al., 1989).

Practically all the hemolymph that returns from the lacunae to the heart

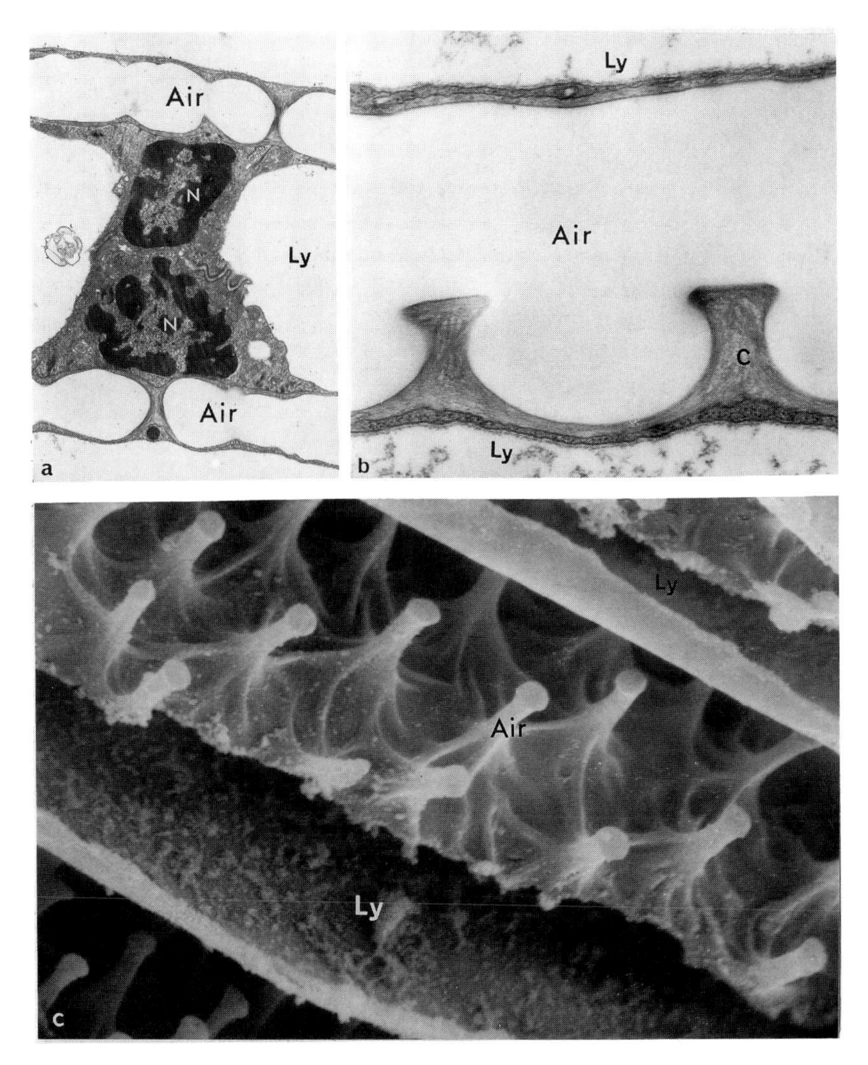

FIG. 56. Ultrastructure of a book lung. (*a*) The hemolymph space (Ly) often contains two cells stacked on each other and thus forming a minute supporting column between neighboring air pockets (Air) (*Tegenaria*). N = nucleus. 5,000 ×. (*b*) To reach the hemolymph (Ly), oxygen must diffuse through the cuticular lamella (C) lining the air pocket and through the underlying cellular layer (*Zygiella*). 22,000 ×. (*c*) Cryofracture of the book lung of *Phidippus* showing the cuticular struts projecting into the air pockets (Air). 9,000 ×. (Photo: Hill.)

has to pass through the book lungs. The hemolymph coming from the prosoma flows along two lateral lacunae in the pedicel and enters the "lung sinus" in the anterior part of the abdomen. Lacunae are also present in the posterior opisthosoma and the flow of hemolymph there is directed forward toward the lung sinus. Streaming along the midline of the abdomen, the flow of the hemolymph divides to enter the leaflets of each book lung laterally (Fig 54b). Oxygen from the air spaces must penetrate the thin cuticular lining and the hypodermis cells of the book lung in order to reach the hemolymph. Since both the cuticle and hypodermis are extremely thin (Fig. 56b), they pose no significant barrier to the process of diffusion.

After the hemolymph has become oxygenated, it leaves the book lungs laterally and continues upward into another sinus, the lung vein, which then connects with the pericardial sinus. At diastole, the oxygenated hemolymph is sucked into the heart tube via the opened ostia. During the following systole, the exiting hemolymph can meet quite different pressures in the prosoma and the opisthosoma. During quick runs, for instance, there is a much higher hemolymph pressure in the prosoma than in the opisthosoma, owing to the contracted entosternal and lateral muscles (see p. 22). For brief periods the hemolymph cannot be pushed through the pedicel into the prosoma, which will lead to a deficit in oxygen (Paul et al., 1987). This is probably one of the main factors for the exhaustion that spiders show after several minutes of running. Following strenuous activity, the respiration and heart rate is markedly increased. Hemocyanin becomes fully oxygenated and the oxygen transport capacity of the hemolymph is fully used. During the long recovery (several hours) D-lactate needs to be oxidized, and the phosphagen and glycogen stores have to be refilled in the muscle tissue. The energy necessary for these processes comes from an oxidation of lipids (Paul, 1990).

The lung slits can be enlarged by the action of specific muscles, thus actually increasing the gas exchange of oxygen and carbon dioxide (Fincke and Paul, 1989). However, this is not a true ventilation of the book lungs, which has been deduced from the observation of slightly pulsating book lung lamellae (Willem, 1918; Hill, 1977b). After careful physiological examinations it seems safe to say that, despite small volume changes, the main exchange of gases in spider book lungs is the result of diffusion (Paul, 1992a,b).

The importance of the book lungs becomes clear after one closes the lung slits with vaseline; after only 2 minutes the animals become severely paralyzed and after several hours most of them are dead (Kaestner, 1929). In contrast, water spiders (*Argyroneta*) can stay submerged for several hours, if their abdomen is enclosed by an air bubble (Heinzberger, 1974).

Tubular Tracheae

The tubular tracheae lie behind the book lungs in the third abdominal segment. Externally they are barely visible as one or two small openings, the stigmata, or spiracles. Most spiders have only a single stigma, which is located in front of the spinnerets (Figs 1, 28). The tubular tracheae have developed quite differently among various spider families (Opell, 1987). Some families (for example, Filistatidae) possess only very short tubes; others, such as Salticidae and Thomisidae, have highly branched tubes that pervade the prosoma and even the extremities (Fig. 57). Spiders with prosomal tracheae (*Argyroneta, Dysdera, Segestria*) usually have a small heart and a low heart frequency,

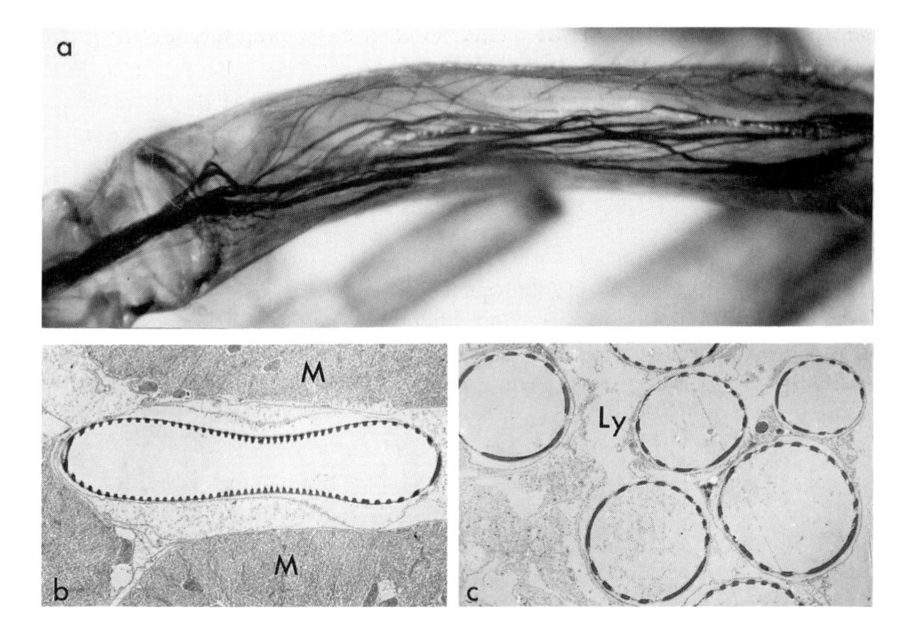

FIG. 57. Tubular tracheae in the leg of *Dysdera*. (*a*) The branching tracheae are clearly visible because they were filled with black cobalt sulfide. 20 ×. (Photo: Bromhall.) (*b*) Longitudinal section of a tracheal tube showing the cuticular reinforcement on the inside of the tube. M = leg muscles. 2,800 ×. (*c*) Cross-section of several tracheal branches. Ly = hemolymph. 1,200 ×.

presumably because the gas exchange is more efficient, owing to the additional tubular tracheae (Bromhall, 1987).

Usually, a median stigma leads into a small atrium from which two lateral and two medial tubes arise. The two lateral tubes correspond to the second pair of book lungs of some "primitive" spiders, but the two median tubes are derived secondarily. They develop from muscular insertion points (entapophyses) in the third abdominal segment; presumably they became hollowed out and can now function as respiratory organs (Purcell, 1909, 1910). The lateral and median tubes of the tracheal system are hence of completely different origin. Furthermore, tracheal tubes can either split into many branches or be unbranched. If unbranched, they remain restricted to the opisthosoma; if highly branched (as they are, for example, in Dictynidae), they also enter the prosoma and supply the various organs there with oxygen. The tracheae in spiders, unlike those in insects, always terminate in an open end without contacting a cell (Fig. 58), so oxygen is not delivered directly. Instead, it is always the hemolymph that is responsible for the final transport of oxygen to its destination. The uptake of oxygen increases by a factor of 2.8 shortly after a fast run. This is still much less than in insects (a factor of 35), presumably

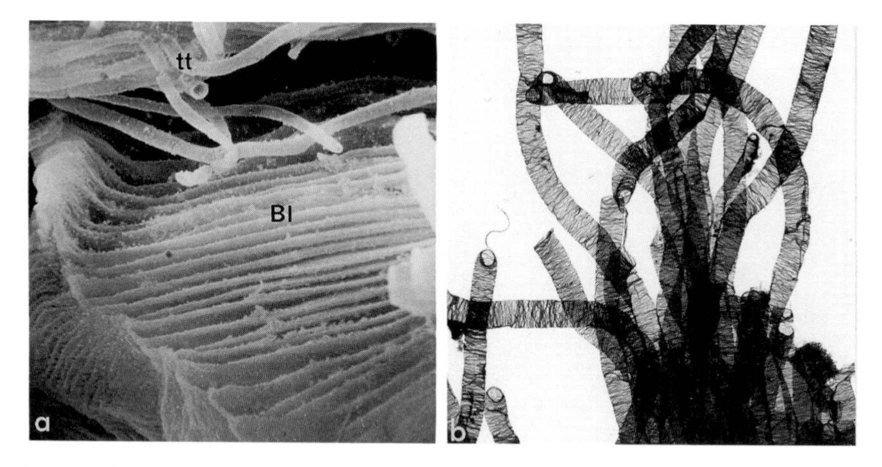

FIG. 58. (*a*) Tubular tracheae (tt) and adjacent book lung (Bl) of a water spider (*Argyroneta*). All soft tissue has been removed by treatment with potassium hydroxide, but the cuticular parts remain intact. 240 ×. (*b*) Similar preparation of tube tracheae from the opisthosoma of a jumping spider (*Salticus*). 350 ×.

because their tracheal system works more efficiently than that in spiders (Anderson and Prestwich, 1985).

Comparison of the Two Respiratory Systems

The question of which of the two respiratory organs is more efficient cannot be answered clearly, since comparative physiological studies are lacking. It is usually assumed, however, that a highly branched system of tubular tracheae would provide the better solution (Levi, 1967). In species with branched tubular tracheae the circulatory system is normally reduced: the heart is shorter, and the number of ostia and arteries is smaller. Such is the case for the water spider *Argyroneta aquatica* (Crome, 1953). The small book lungs of a water spider cannot store much oxygen, compared to the extensively developed tubular tracheae (Braun, 1931). If both the lung slits and the stigma of a water spider are sealed, it takes days before any effects (disturbed motor behavior) become noticeable. In comparison, a relative of *Argyroneta*, the house spider *Tegenaria*, is paralyzed within a few minutes after such an operation. The question of relative efficiency is further complicated by the fact that the water spider can apparently also exchange gases through its integument. Carbon dioxide can pass directly through the abdominal cuticle and into the captive air bubble, thus preventing carbon dioxide poisoning (Crome, 1953). Some lesser-known spider families (Caponiidae, Symphytognathidae) have no book lungs at all. Instead these possess, in the second abdominal segment, special tubular tracheae called *sieve tracheae*. Sieve tracheae are delicately branched and arise immediately from the atrium.

The following simplified summary constitutes an overview of the types of respiratory organs and their respective development in different kinds of spiders (Kaestner, 1929):

1. One pair of book lungs only (for example, Pholcidae).
2. Two pairs of book lungs only (Mesothelae, Orthognatha).
3. One pair of book lungs and one pair of tubular tracheae (Araneidae, Lycosidae).
4. One pair of sieve tracheae and one pair of tubular tracheae (Caponiidae).
5. One pair of sieve tracheae only (Symphytognathidae).

Phylogenetically, book lungs must be regarded as more ancient ("primitive") than tubular tracheae. This can be concluded from embryological studies (because book lungs are homologous to appendages), and also from the observation that all "primitive" spiders have only book lungs and no tracheae. Elaborate tubular tracheae occur only in small spiders, which are more prone to desiccation; perhaps tubular tracheae have evolved as an adaption to provide the necessary protection (Levi, 1967).

4

Neurobiology

The behavior of spiders, like that of all animals, is controlled by the central nervous system. Of the various types of sensory organs that collect information about the environment—for example mechanoreceptors, chemoreceptors, and visual receptors—the mechanoreceptors are the most important among spiders.

Mechanical Senses

Mechanoreceptors respond to external stimuli such as touch, substrate vibrations, and air currents, and also keep the spider informed about leg and joint positions. Usually we find different receptor types for the various stimuli. On the other hand, structurally similar receptors may subserve quite different functions. The most common mechanoreceptor is the hair sensillum. It may appear as a simple tactile hair or as a more complex filiform hair (a trichobothrium, Fig. 59). Whereas tactile hairs are distributed over the entire body surface, the trichobothria occur only on the extremities. Other less conspicuous but nevertheless characteristic mechanoreceptors of spiders are the slit sensilla (Fig. 63). These small sensory organs are embedded in the exoskeleton at "strategic" points. Their function remained obscure for many years, but during the last decades their mechanosensitive capacity has been proved beyond doubt (Barth, 1976). In addition to these external mechanoreceptors there are also internal proprioceptors inside the legs; these measure the relative positions of the joints.

Tactile Hairs

The hairiness of spiders is proverbial, and it is certainly this characteristic that contributes most to the general dislike many people have of spiders. For those interested in the biology of these animals, however, this hairiness sparks some scientific questions. Why are spiders hairy? Do the thousands of hairs merely decorate the exoskeleton, or do they fulfill some vital functions? The fact that most of these hairs are innervated proves their nature as sensory organs. This can also be demonstrated in a simple behavioral experiment: touching only a single hair triggers escape or aggressive reactions from the spider. If certain tactile hairs on the coxa are bent, the spider (*Cupiennius*) reacts by abruptly

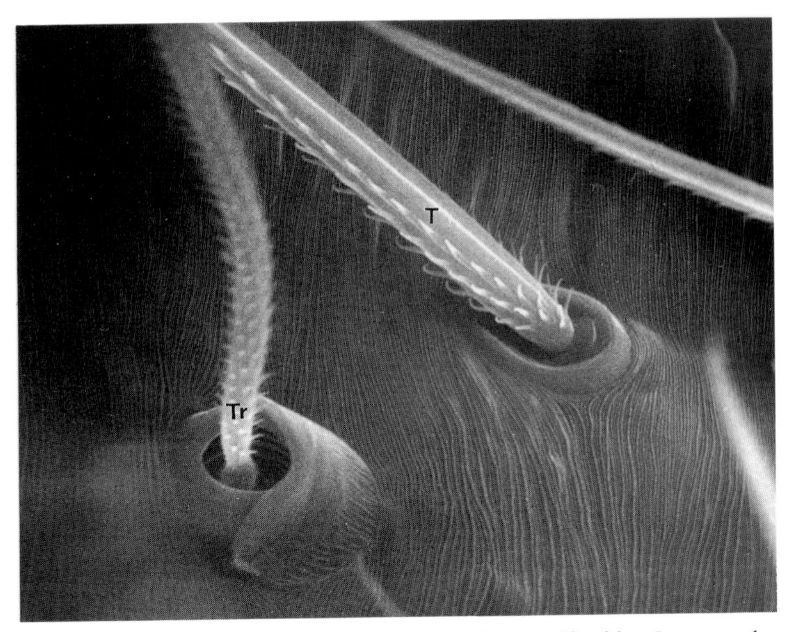

FIG. 59. Mechanosensitive hair sensilla from a leg of the wolf spider *Lycosa gulosa*. The slender trichobothrium (Tr) arises vertically from a prominent socket; the stronger tactile hair (T) emerges obliquely from a less-developed socket. 1,200 ×. (Photo: Chu-Wang and Foelix.)

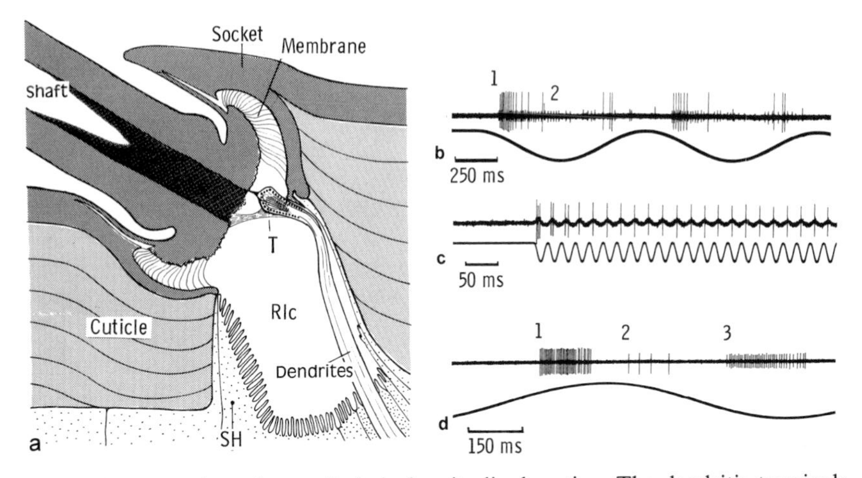

FIG. 60. (*a*) Innervation of a tactile hair, longitudinal section. The dendritic terminals (T) attach to the base of the movable, suspended hair shaft. Rlc = receptor lymph cavity, SH = sheath cell. (After Foelix and Chu-Wang, 1973a and Harris, 1977.) (*b–d*) Electrophysiological recordings from tactile hairs. (*b*) Displacement of the hair shaft (lower tracing) elicits nervous impulses from two sensory cells (1, 2). (*c*) Stimulation with a 50 Hz vibration. (*d*) Responses from three sensory cells of a trichobothrium after lateral displacement of the hair shaft. (*b–d* after Harris and Mill, 1977a.)

lifting the entire body, thus gaining more distance from the ground (Eckweiler and Seyfarth, 1988).

It is hard to imagine that the thousands of hairs on a spider's body are all innervated, yet examination of histological sections shows that indeed all large, movable, articulated hairs have a triple innervation. Only the short body hairs and most adhesive scopula hairs lack innervation. The regular tactile hair consists of a long, exocuticular hair shaft that is suspended in a slipper-shaped socket in which it can move (Figs 59, 60). Three dendritic nerve endings are attached to the base of the hair shaft (Foelix and Chu-Wang, 1973a) and monitor movements of the shaft. Electrophysiological recordings show that the tactile hairs react like typical phasic receptors; that is, only changes from the resting position are answered with nerve impulses (Den Otter, 1974). Furthermore, the initial bending of the hair elicits more nerve impulses than its return to the resting position (Fig. 60b).

The large leg spines are also triply innervated (Harris and Mill, 1977a). These spines become erect when hemolymph pressure increases. Nerve impulses of the leg spines, unlike those of simple tactile hairs, can be recorded only during the erection phase, but not during the return to the flat resting position; the spines might therefore be hemolymph pressure receptors rather than touch receptors.

Although a triple innervation of tactile hairs is the rule in spiders, there are also some exceptions. For instance, the sensilla of the coxal hair plates (see p. 77) are only singly innervated (Seyfarth et al., 1990). The adhesive scopula hairs usually are not innervated at all, but some were found to be attached to a single neuron (Foelix et al., 1984).

Aside from specific sensory functions, many hairs also serve purely mechanical tasks (Fig. 61). The row of specialized bristles forming the *calamistrum* on the fourth leg of cribellate spiders is used to comb out fine catching threads. Actually, each bristle is serrated on one side and thus acts as a comb (Foelix and Jung, 1978). Theridiid spiders also have a row of serrated bristles on their fourth legs. These tarsal bristles comb out sticky silk threads from the spinnerets as the spider wraps its prey. Some spiders (for example, *Zelotes*, Gnaphosidae) have specialized brushlike hairs that are used as cleaning devices during grooming (Berland, 1932). Finally, the scopula hairs must be mentioned here, since they provide for increased adhesion to the substrate (see chapter 2).

The short hairs covering the opisthosoma of many spiders may have special functions. In the water spider *Argyroneta aquatica*, for instance, these abdominal hairs are necessary for establishing the air bubble that surrounds the opisthosoma (Braun, 1931). Some female wolf spiders have special "knobbed" abdominal hairs, used by the young spiderlings to hold on to their mother (Rovner et al., 1973). A more impressive example of how a spider can use its abdominal hairs is found in the New World "tarantulas." When these animals feel threatened, they can quickly brush off clouds of abdominal (urticating) hairs (Fig. 61a) with their hind legs. Each hair is covered by hundreds of little hooks, which cause severe itching when in contact with the skin, especially in the nose and eye region. Experiments show that these

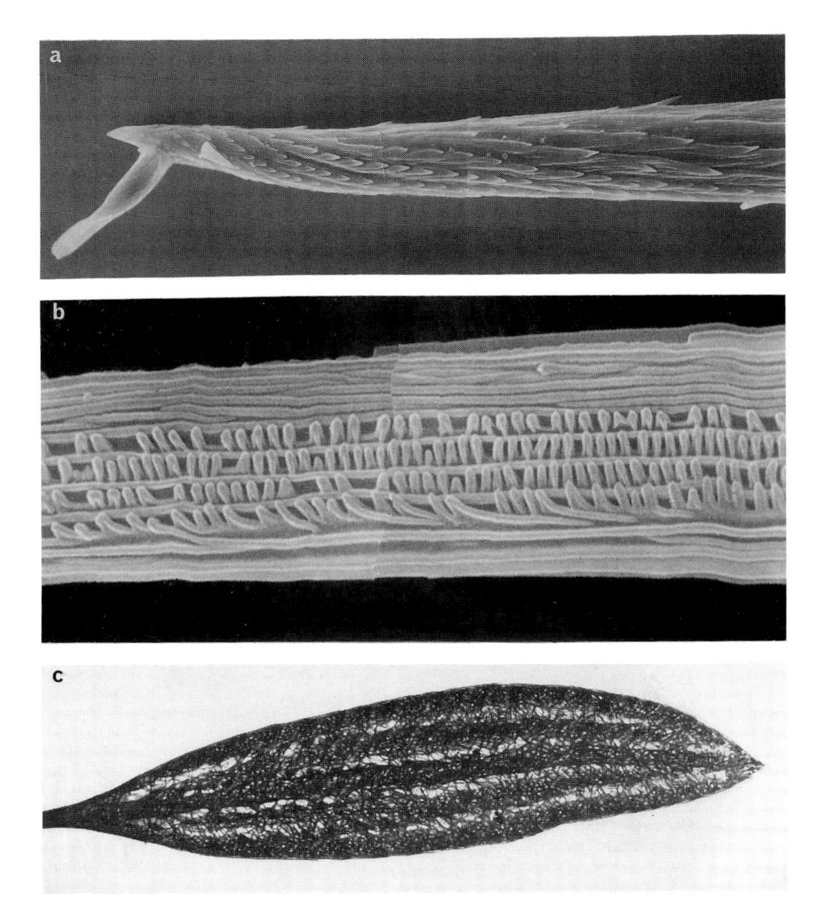

FIG. 61. Fine structure of various spider hairs. (*a*) Urticating hair from a "tarantula." The hair base (left) breaks off easily; the hair shaft exhibits spirally arranged barbs. 890 ×. (*b*) Surface structure of a calamistrum bristle of *Hypochilus*. The hair shaft bears multiple rows of cuticular teeth on one side. 5,300 × . (*c*) Scale hair from the jumping spider *Salticus*. The leaflike hair shaft consists of a delicate cuticular meshwork. 940 ×.

urticating hairs can work themselves 2 mm deep into human skin (Cooke et al., 1972, 1973). Sometimes this happens quite inadvertently when one works with dead "tarantulas" (preserved in alcohol or as dry exuvia)—as many museum curators can testify!

Trichobothria

The filiform hairs, or trichobothria, are extremely fine hairs within special sockets. They are much less numerous than the common tactile hairs, and are arranged in straight lines or in small clusters on certain leg segments. In *Tegenaria*, for instance, they form one dorsal row on each tarsus and metatarsus, with the length of their respective hair shafts gradually increasing toward the

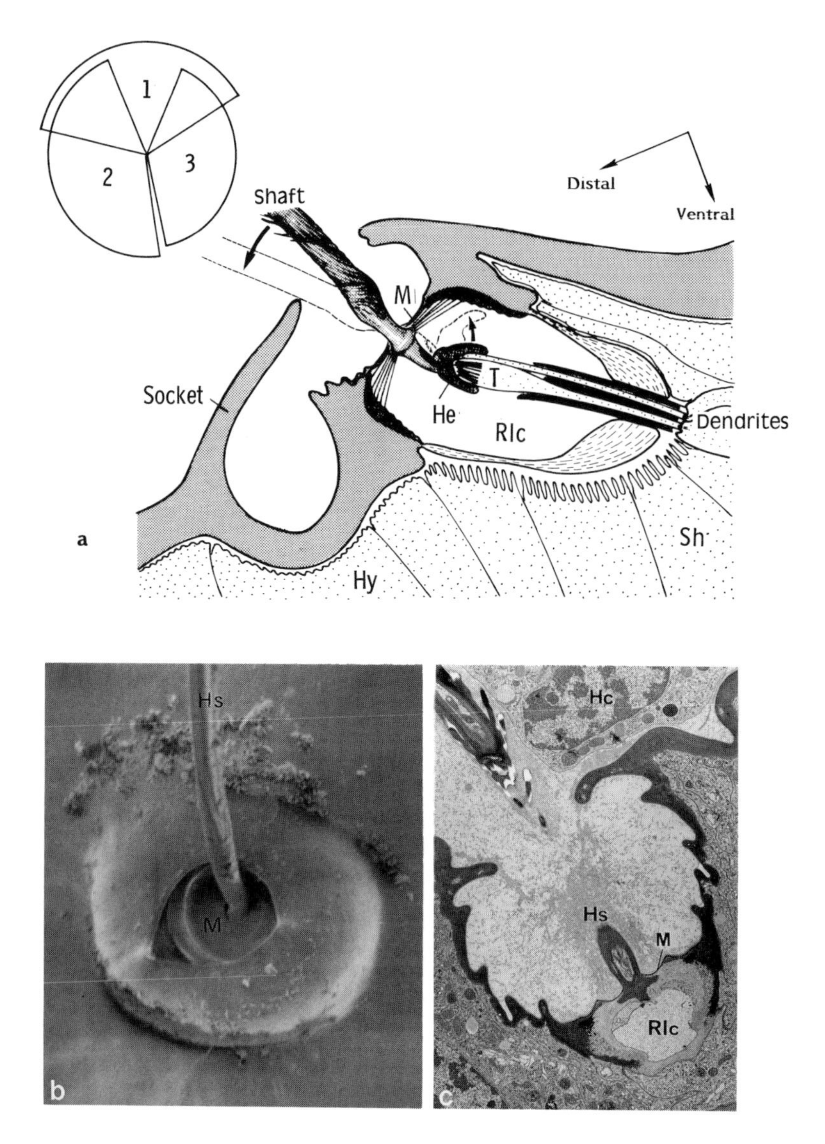

FIG. 62. (*a*) Innervation of a trichobothrium of the house spider *Tegenaria*, longitudinal section. When the hair shaft is deflected (dashed line), the helmet structure (He) pushes against the dendritic terminals (T), triggering the generation of nerve impulses. Rlc = receptor lymph cavity, Sh = sheath cells, Hy = hypodermis, M = articulating membrane. (After Christian, 1973.) *Inset*: the directional sensitivity of a trichobothrium as seen from above; displacements of the hair shaft within each of the three sectors indicated are preferentially answered by one of three sensory cells (1, 2, 3). (After Görner, 1965.) (*b*) Trichobothrium of the cellar spider *Nesticus*. Note the insertion of the hair shaft (Hs) in the articulating membrane (M). 1,200 ×. (Photo: Foelix and Rohner.) (*c*) Slightly oblique section of a trichobothrium in a young *Latrodectus*. This micrograph was taken just before the animal molted; thus the "outside" (exuvial space) is filled with hemolymph and hemocytes (Hc). 1,500 ×.

tip of the leg. On the tibia, in contrast, we find four rows of trichobothria with hair shafts of different lengths (Görner, 1965; Christian, 1971, 1973). The arrangement of trichobothria is constant within a species and is hence often used by systematicists to characterize certain spider species and to clarify taxonomic relationships (termed *trichobothriotaxy*; Emerit and Bonaric, 1975; Lehtinen, 1980). The large wandering spider *Cupiennius* has almost 1,000 trichobothria: about 100 on each leg and about 50 on each palp (Barth et al., 1993). Even in small hunting spiders like *Philodromus* or *Pardosa* there are 20–40 trichobothria per leg. Such a large number of trichobothria and their concentration on the distal leg segments certainly suggest a behavioral significance. Interestingly, most web spiders exhibit much fewer trichobothria (about 10 per leg, mainly on the tibiae, none on the tarsi) than ground spiders (Peters and Pfreundt, 1986). On the other hand, web spiders have more slit sensilla (see below), which is probably related to the way they perceive prey, namely through vibrations of the web.

The most striking feature of the trichobothria is their extreme sensitivity. The long slender hair shaft is suspended in a very thin (0.5 μm) cuticular membrane (Fig. 62) so that the slightest air current (1 mm/s!) will make it quiver. The surface of the hair shaft has a feathery appearance owing to its many fine cuticular extensions. They cause an increase of drag forces and thus a higher mechanical sensitivity toward air movements. Many trichobothria show a preferred direction of deflection, i.e., along the leg axis or perpendicular to it. The deflection angle is limited by the socket rim and was measured as 25°–35°; it varies with the stimulus frequency (Barth et al., 1993). Four dendritic nerve endings attach to the specialized hair base (the "helmet"), and three of the four have been found to have a specific directional sensitivity (Fig. 62a, inset). Each sensory cell reacts to displacements within a defined sector, and there is little overlap between the three neurons (Görner, 1965; Harris and Mill, 1977a). In the field, normal stimuli are air currents and low-frequency air vibrations (sound). The air vibrations that an insect produces with its wings suffice to trigger a directed capture response from the spider. Even a totally blinded spider can accurately locate a buzzing fly as much as 30 cm away (Görner and Andrews, 1969; Barth et al., 1993). For this reason trichobothria have also been called "touch-at-a-distance" receptors. Later it turned out that a localization of a vibrating prey was still possible, even after all trichobothria had been plucked off (Hergenröder and Barth, 1983a; Reissland and Görner, 1985). This means that some other receptor must be involved. Most likely single slit sensilla on the tarsi elicit an oriented prey-catching behavior, whereas the trichobothria signal only a general alertness. In ground spiders, the trichobothria probably trigger either flight or attack; in web spiders they are believed to play a role in the localization of prey or enemies.

Slit Sense Organs

The exoskeleton not only serves as a protective shield and as the insertion site for the muscles, but also transmits mechanical stress that may be caused by substrate vibrations, by gravity, or by the spider's own movements (Barth, 1976, 1985a). The specific stress (strain) receptors are the *slit sense organs*, or sensilla, which are embedded in the exoskeleton. They are distributed over the entire body surface but are most numerous on the legs. An adult *Cupiennius*, for instance, has more than 3,000 slit sense organs. Of these, 86% lie in the hard exocuticle of the extremities, whereas only 3% are found in the soft abdominal cuticle (Barth and Libera, 1970).

Slit sense organs may occur either singly or in groups (Fig. 64). Most conspicuous are those groups where the slits run strictly parallel to each other.

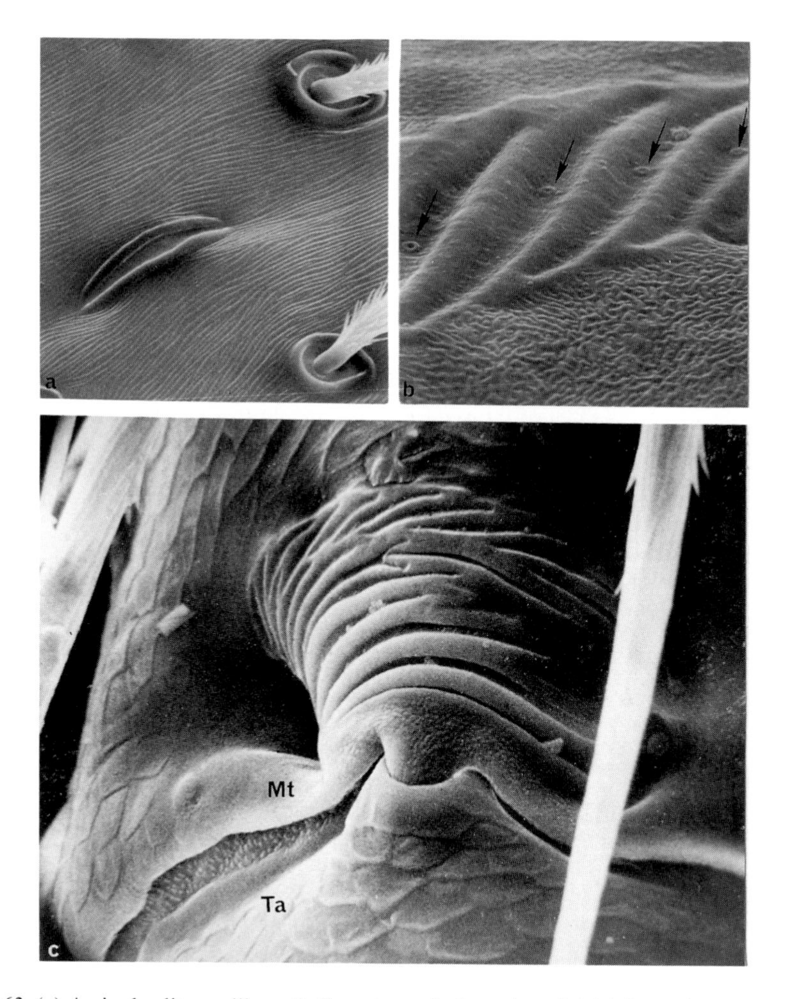

FIG. 63. (*a*) A single slit sensillum (*left*) and two hair sockets (*right*) from the sternum of the social spider *Mallos gregalis*. 1,350 ×. (*b*) Lyriform organ from a walking leg of *Zygiella x-notata*. The attachment sites of the dendrites are indicated by arrows. 1,600 ×. (Photo: Jung and Foelix.) (*c*) Metatarsal lyriform organ in *Araneus diadematus*. The slits are arranged perpendicular to the leg axis at the border of the metatarsus (Mt) and the tarsus (Ta). 1,070 ×. (From Foelix, 1970a.)

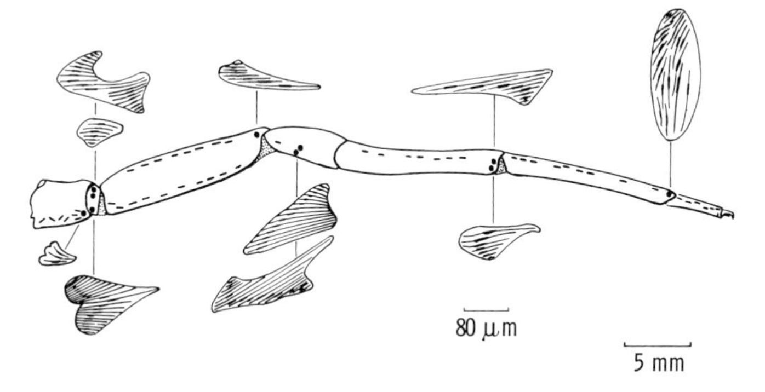

Fig. 64. Distribution of lyriform sensilla on the back side of the first leg of Cupiennius salei. Single slit sensilla are indicated as dashes, lyriform organs as black dots; the latter are enlarged to show greater detail. (After Barth and Libera, 1970.)

They are then called *lyriform organs*, because their shape is reminiscent of a lyre. Lyriform organs are found mostly on the extremities, particularly near the joints (Figs 63, 64). Single slit sensilla can be found anywhere on the body surface, yet some specific localization and orientation of these organs is still noticeable. On the legs, for instance, the single slits are generally oriented parallel to the leg axis (Fig. 64).

There is no difference in the basic design of single slit sensilla and that of lyriform organs (Barth, 1971). Each slit is only 1–2 μm wide, yet may be from 8 to 200 μm long. The slit is bordered by a cuticular lip on either side, and the gap between these ridges is spanned by a thin cuticular membrane (Fig. 65).

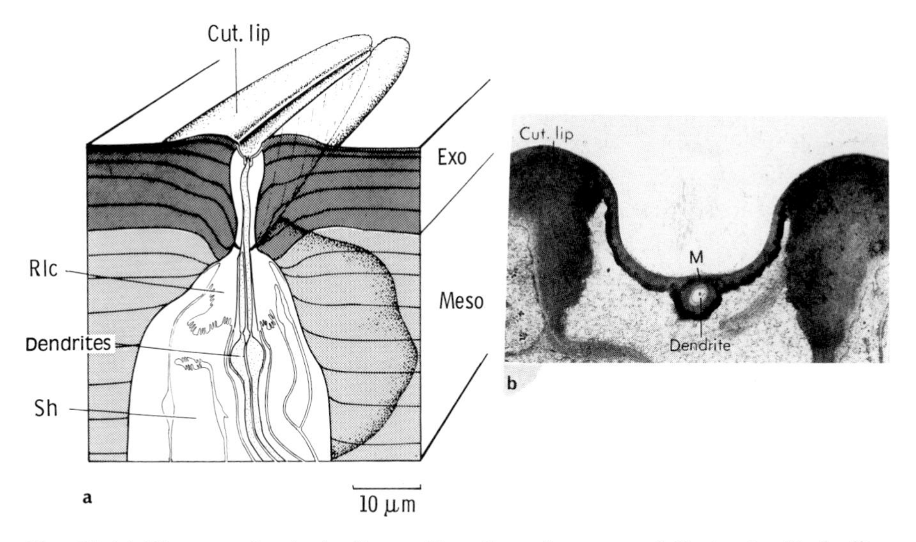

Fig. 65. (*a*) Diagram of a single slit sensillum from the tarsus of *Cupiennius*. Each slit is supplied by two dendrites, but only one of these extends to the covering membrane. Exo, Meso = exo-, mesocuticle, Rlc = receptor lymph cavity, Sh = sheath cells. (After Barth, 1971.) (*b*) Covering membrane (M) of slit sensillum with dendritic attachment. 14,000 ×.

Beneath this covering membrane a trough extends downward and widens into a bell-shaped structure at the border of the exo- and mesocuticle. Each slit sensillum has two dendrites, but only one of these traverses the trough and attaches to the covering membrane. The dendritic tip contains a "tubular body," which is a microtubule assembly characteristic of arthropod mechanoreceptors.

The cuticular structures of the slit sensilla are quite important for transmitting the adequate stimuli to the nerve endings. Even very small loads or slight strains on the cuticle lead to a deformation of the slit. Most effective are those forces that are oriented perpendicular to the slit (Barth, 1972a,b). In technical terms one could say that the slits function as mechanical filters. When the slit narrows, the covering membrane is indented and the tip of the dendrite becomes deformed. Only compression, not dilatation, of the slit elicits nerve impulses (Barth, 1973b). In the lyriform organ all the slits are of different lengths, and peripheral slits can be deformed more easily than central ones (Barth and Pickelmann, 1975); this implies that they are differentially sensitive. The site of maximal compression usually coincides with the attachment site of the dendrite, which means that the system works with maximum sensitivity.

The function of the slit sensilla cannot be generalized, since different slit sensilla are used for different tasks. Usually, the slit sensilla are categorized as proprioceptive mechanoreceptors, but this is not entirely accurate since the stimuli need not be related to the spider's own movements. For a single tarsal slit sensillum, responses to airborne sound have been shown electrophysiologically (Barth, 1967); the highest sensitivity was noted between 300 and 700 Hz. Some slit sensilla on the pedicel may act as gravity receptors, since they probably register the relative movements between prosoma and opisthosoma (Barth and Libera, 1970)

The metatarsal lyriform organ (Fig. 63c) is a highly sensitive vibration receptor (Walcott and van der Kloot, 1959; Liesenfeld, 1961). In the web spiders *Zygiella* (Araneidae) and *Achaearanea* (Theridiidae), a displacement of the leg tip by 0.1–0.25 μm (at 2–5 kHz) sufficed to elicit a response. However, if a vibrating needle is hooked to the web, the spider will only attack within a narrow frequency range (400–700 Hz). It has been known for more than a hundred years that spiders react to a vibrating tuning fork in the same manner as to a buzzing insect (Boys, 1881). Even small vibrations on the water surface can be perceived by slit sense organs: insects which have fallen onto the water and begin to struggle, are quickly and precisely located by the pirate spider *Dolomedes* (Bleckmann and Barth, 1984). Vibrations, incidentally, are important not only for capturing prey but also during courtship, when the male signals his presence to the female (see chapter 7).

Tibial and femoral lyriform organs are used for kinesthetic orientation. Wandering spiders (*Cupiennius, Pardosa*) can normally orient themselves according to their previous motility pattern; that is, they can retrace the direction and distance of a previous run without the help of external stimuli (Fig. 66). This is no longer true if the tibial or femoral lyriform organs have been destroyed (Seyfarth and Barth, 1972; Görner and Zeppenfeld, 1980; Seyfarth et al., 1982).

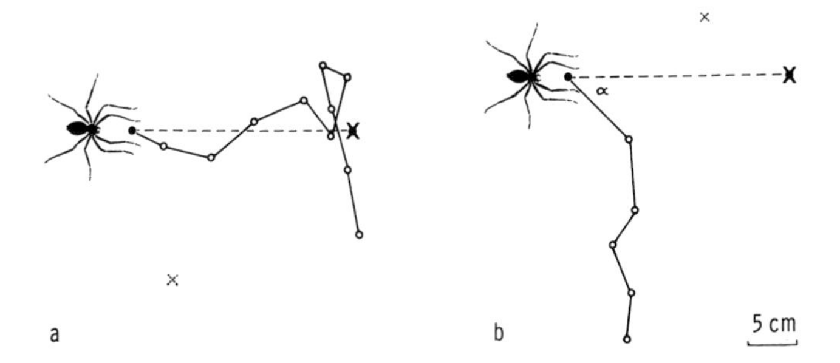

FIG. 66. Involvement of lyriform organs in kinesthetic orientation. The spider (*Cupiennius*) caught a fly at X and was then chased away from it; the dead fly was then moved to x. In (*a*), a normal spider returns in a direct course to the capture site. In (*b*), all tibial lyriform organs had been destroyed; the direction of the return run (indicated by the angle α) deviates markedly from the ideal course (dashed line). (After Seyfarth and Barth, 1972.)

Slit sense organs are typical of all arachnids, but in none of the arachnid orders (harvestmen, whip spiders, whip scorpions, and scorpions) are they as diversified as in the "true" spiders (Araneae). Whereas spiders have about 300 slit sensilla per leg, whip spiders and harvest men have only 45 or 58, respectively (Barth and Stagl, 1976). It is interesting that similar sense organs, the *campaniform sensilla*, are found in insects; these likewise measure strain-induced deformation of the cuticle.

Proprioceptors

Aside from the lyriform organs, spiders possess a number of other sensory organs that provide information about their own bodies, e.g., the position or the movements of a joint. Such *joint receptors* may be long tactile hairs, which stand close to an articulation and become bent when the leg is flexed (Seyfarth, 1985). A special case of such sensilla is represented by the so-called *hair plates*, which have recently been discovered in several spiders (Seyfarth, 1985; Seyfarth et al., 1990): many short tactile hairs occupy a small area on the coxae. During locomotion they are pressed down by the overlying pleural membrane (stippled area in Fig. 1). Such proprioceptive hair plates have been known in insects for a long time (Pringle, 1938); they control body posture and also act as gravity receptors (Markl, 1962).

An important group of proprioceptors in spiders are the *internal joint receptors*. Several groups of sensory cells (*ganglia*) lying inside the palps and legs of spiders (Fig. 67) provide the animal with information about each joint (Rathmayer, 1967; Rathmayer and Koopmann, 1970). These proprioceptors register not only the position, but also the beginning, direction, and velocity of changes in position of a joint. The more freedom of movement a joint has, the more proprioceptor ganglia are present. The coxa–trochanter joint, which can

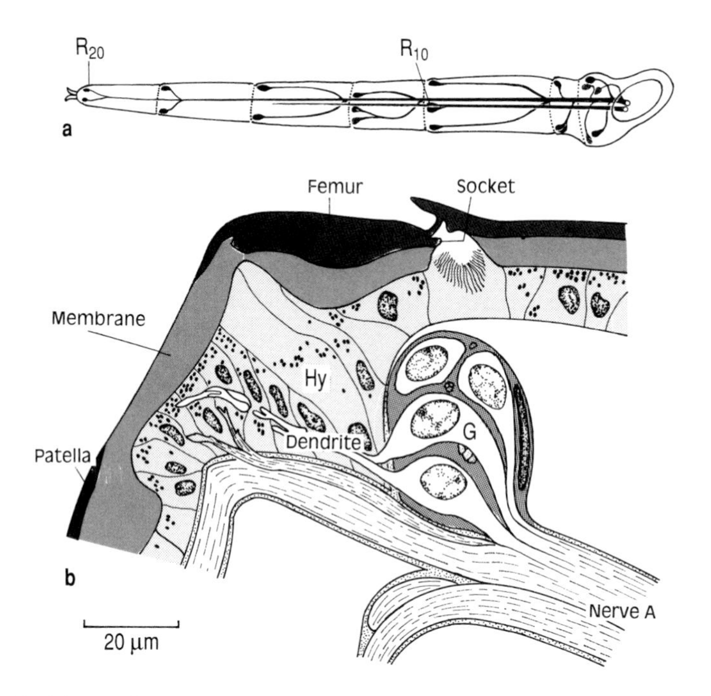

FIG. 67. Position and structure of proprioreceptors in a spider leg. (*a*) Each leg contains 20 receptor groups, which invariably lie near the joints. (After Rathmayer and Koopmann, 1970; Seyfarth et al., 1985.) (*b*) Longitudinal section of R10, a receptor group located at the femur–patella joint. Ten sensory cells form a compact ganglion (G), which gives off branched dendrites. The dendritic terminals fan out into the hypodermis cells (Hy) underlying the joint membrane. (After Foelix and Choms, 1979.)

move in all directions, has five proprioceptive ganglia. All the other joints possess only two or three ganglia, which are located near the pivoting points of the joint. Within a given joint, one ganglion may be sensitive only to bending, while the other may respond only to stretching of the joint (Mill and Harris, 1977). The sensory cells of the ganglion are multipolar and lie directly below the joint membrane (Fig. 67b; Parry, 1960). Their dendrites fan out into the hypodermis beneath the joint membrane and terminate there as delicate free nerve endings. The shearing forces generated during leg movement are presumably transmitted to the dendritic terminals via rigid hypodermis cells (Foelix and Choms, 1979). The situation is further complicated by small nerve fibers that make synaptic contacts (Fig. 80) with the sensory cells and their dendrites. These efferent nerve fibers possibly originate in the CNS and might control (inhibit) the activity of the proprioceptor.

Chemical Senses

It has been known for a long time that spiders react to chemical stimuli. Just as humans can differentiate between the sense of smell (olfaction) and taste, so can spiders. *Taste* implies contact with a substance, usually at high concentration, whereas *olfaction* denotes that volatile substances affect receptors over relatively large distances, and often at very low concentrations. With many highly volatile substances (such as essential oils), one can easily elicit a behavioral reaction from the spider simply by putting the compound close to the animal. Strongly odoriferous substances usually cause the spider to run away, or at least to exhibit a local reaction such as withdrawing a leg. Although such experiments prove the existence of an olfactory sense, they tell us little about the significance of olfaction under natural conditions.

In the natural environment the spider most likely uses olfaction to find a mate during courtship, and perhaps to recognize prey and enemies (see chapter 9). It has been shown that male spiders are apparently attracted by the "scent"

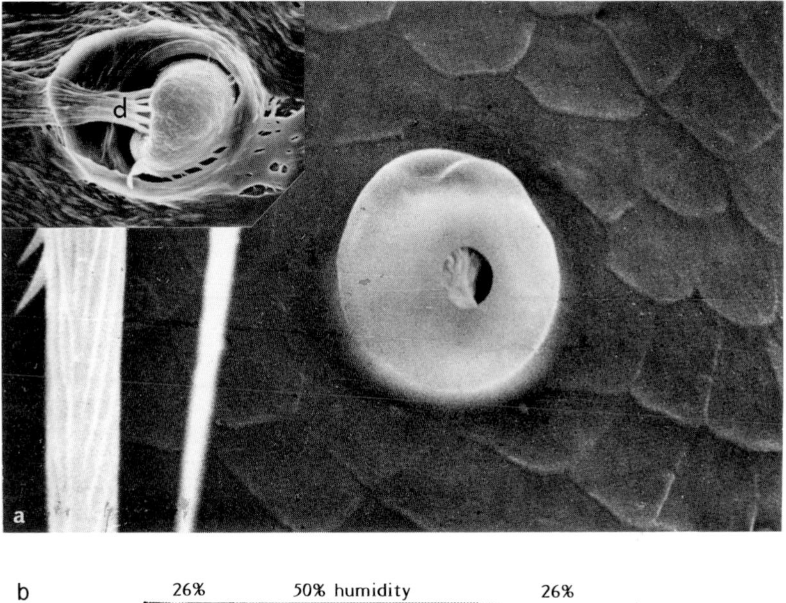

b

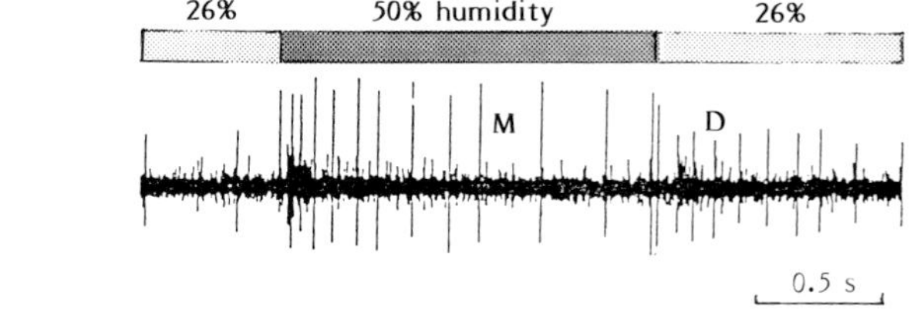

FIG. 68. (*a*) Tarsal organ of *Araneus diadematus*. The dendritic endings of seven sensilla lie within the cuticular pit. 1,250 ×. *Inset*: Internal view of the tarsal organ in *Cupiennius* (exuvium), showing the dendritic sheaths (d) leaving the cuticular pit. 950 ×. (Photo: Anton and Tichy). (*b*) Electrophysiological recording from the tarsal organ (*Cupiennius*). Brief exposure to moist air (50% humidity) elicits responses from a moist cell (M), whereas dry air triggers a dry cell (D) of smaller amplitude. (After Ehn and Tichy, 1994.)

given off by the female (Blanke, 1972, 1975b; Willey and Jackson, 1993). A male will not be attracted to a female until just before her last molt; immediately after she molts, however, often several males will flock around her web. Presumably the mature female produces sex-specific substances (*sex pheromones*) that attract the male. Such pheromonal effects are, of course, well known from studies of insects (such as butterflies), and the pheromones involved have been chemically analyzed in great detail. In spiders, only the sex pheromone of some linyphiid spiders has been identified chemically (R-3-hydroxybutyric acid; Schulz and Toft, 1993).

Although there can be little doubt about the spider's ability to smell, the location of the actual olfactory organs has been under debate for a long time. The *tarsal organs* (Fig. 68) have remained likely candidates since they were first described in detail (Blumenthal, 1935). Usually the tarsal organs are small pits on the dorsal side of each tarsus, although in some spider families they may appear rod-shaped, like a hair (Forster, 1980). In either case they are multiply innervated (by about 20 neurons) and are connected to the external environment by seven small pores (Foelix and Chu-Wang, 1973b). The first electrophysiological investigations have shown that different volatile substances can excite the tarsal organs (Dumpert, 1978). However, according to more recent studies (Ehn and Tichy, 1994), the tarsal organs are primarily *hygroreceptors*. Each of the seven sensilla consists of three sensory cells, two responding to changes in humidity (Fig. 68b) and the third one reacting to changes in temperature. The sensitivity of these sensory cells is noteworthy: while the "warm" cell can resolve temperature differences of 0.4°C, the "moist" and "dry" cells will discriminate differences of 10% in humidity. Although the tarsal organ does respond to certain pungent odors such as acids or ammonia vapor, it appears to be unresponsive to alcohols, aldehydes, esters, etc., and to natural scents coming from other spiders or insects. Thus we are still left with the question of where the spider's *odor receptors* are actually located. A putative olfactory organ was described between the rostrum and the palps (Legendre, 1956), but its function has never been clarified.

The most important chemoreceptors of spiders are the *contact chemoreceptors*, or taste hairs, which are found mainly on the distal segments of the legs and palps (Foelix, 1970b; Foelix and Chu-Wang, 1973b; Harris and Mill, 1973). At first glance these hair sensilla resemble the normal tactile hairs, yet they differ from them in three respects: (1) they arise at a steep angle (about 70°) from the leg surface; (2) the hair shaft is S-shaped; and (3) the hair tip is open to the outside. Each taste hair is usually innervated by 21 sensory cells: two nerve endings (mechanoreceptors) with tubular bodies terminate at the hair base, while the other 19 (chemosensitive) dendrites traverse the hair shaft to the opening near the hair tip (Fig. 69a).

An adult spider (*Araneus*) can possess over 1,000 chemosensitive hairs, most of which are concentrated on the tarsi of the first legs. This distribution helps explain many observations of spider behavior. For instance, spiders readily accept freshly killed flies after briefly touching them with their legs. Old dead flies, in contrast, are quickly discarded (Heil, 1936; Eberhard, 1969). Apparently

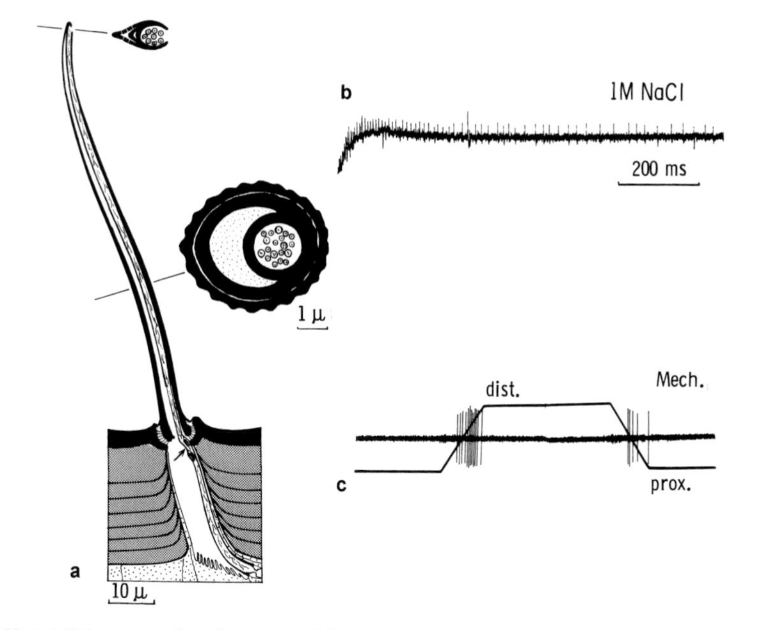

FIG. 69. (*a*) Diagram of a chemosensitive hair. Two mechanosensitive dendrites (arrow) terminate at the hair base, while the chemosensory dendrites traverse the hair shaft until they reach the terminal pore opening. (After Foelix and Chu-Wang, 1973b.) (*b* and *c*) Electrophysiological recording after stimulation of a single hair sensillum. (After Harris and Mill, 1977a,b.) (*b*) Stimulation with a 1 M salt solution causes several neurons to respond with spikes. (*c*) Displacement of the hair shaft in a distal (dist.) or proximal (prox.) direction triggers the response of a mechanoreceptor at the hair base.

spiders can determine the chemical properties of a substrate merely by probing it with their tarsi. Bristowe (1941, 1958) noted this ability in many different spiders and coined the term "taste-by-touch" sense. The observation that most spiders drop certain bugs or ticks after briefly touching them can be easily explained by the presence of chemicals detected by the taste hairs. The chemical sensitivity of these sensilla has also been proved directly by using electrophysiological methods (Drewes and Bernard, 1976; Harris and Mill, 1977b).

In addition to testing the quality of food, contact chemoreceptors are also used during courtship. Many male ground-living spiders follow the trail of a conspecific female and often begin their courtship rituals (such as palp drumming) when they encounter threads laid down by her (Bristowe and Locket, 1926; Kaston, 1936; Rovner, 1968a; Dijkstra, 1976). While walking along the female's dragline, the male will keep the inside of both palps in close contact with her thread (Fig. 163; Tietjen, 1977). Interestingly, male palps bear about three times as many chemosensitive hairs as females. Such a sexual dimorphism is well known from insect antennae, which also exhibit many more sensilla in the male. Most likely, some specific substances (sex pheromones) are associated with the female's silk, since it can trigger the male's courtship behavior. If the female's spinnerets are sealed with wax so that no silk can be deposited on the ground, then the male fails to court (Dondale and Hedgekar, 1973). So far, only little is known about pheromones in spiders. For instance, it is not clear where exactly they are produced and whether they are perceived by contact chemoreception alone or, eventually, also by olfaction. It seems that olfaction is involved mainly in web spiders (Blanke, 1972), but also in some ground spiders (Tietjen, 1979a,b). The borderline between olfaction and taste is not very distinct, and it may well be that the taste hairs also sense certain smells.

Contact chemoreception has been extensively studied in insects. The hair receptors of insects are very similar to those described for spiders, and they are similarly located on the legs and mouth parts. In flies, stimulation of a single tarsal hair with sugar will trigger the extension of the proboscis. A comparable behavior reaction was elicited from wolf spiders after a single chemosensitive hair on the palp was touched with a 0.2 M salt solution: the spider responded by flicking the entire palp in withdrawal (Drewes and Bernard, 1976).

Some chemosensitive hairs do occur on the mouth parts (on the maxillae and labium), but the presence of "taste cells" in the pharynx (Millot, 1936, 1946) remains questionable. More recent microscopic investigations of the gullet have shown that the supposed "taste cells" are in fact gland cells (De la Serna de Estaban et al., 1985). On the other hand, it is known that spiders wipe their mouth parts extensively after having bitten into distasteful prey (Bristowe, 1941).

Vision

In most spiders the sense of vision plays only a minor role in behavior. Many spiders are active at night, and are thus more dependent on tactile and chemical cues than on sight. For web spiders visual stimuli seem to be especially unnecessary, since these spiders can build their webs at night and can catch their prey in total darkness, yet to conclude from these observations that vision is irrelevant for such spiders would be premature. For instance, the orb weaver *Araneus sexpunctatus* (synonym *Nuctenea umbratica*) can detect very subtle changes in light intensity (Homann, 1947); leaving its retreat in the evening is apparently closely related to the diminishing light at dusk. Other orb weavers drop very quickly out of the hubs of their webs when an observer approaches them, and this behavior is most likely a response to visual stimuli alone.

Vision is, however, important to some spiders other than orb weavers. Sheet-web spiders (Agelenidae) and wolf spiders (Lycosidae) can perceive polarized light, and use it to orient themselves. For most "modern" hunting spiders (Lycosidae, Thomisidae, Salticidae), the sense of vision is vital, not only for capturing prey but also for recognizing the opposite sex during courtship. Jumping spiders kept in the dark (or under dark red light) cannot catch prey or notice a conspecific (Jackson, 1977). Even if a fly bumps into the spider (*Phidippus*), she will not attempt to seize it. However, this is not true for all species: *Trite planiceps* is capable of capturing prey under red light and even in the dark (Forster, 1982, 1985)—apparently it uses vibratory cues for a successful orientation. In a later section we will look in more detail at the highly developed eyes of salticids.

Eye Structure and Function

Most spiders possess eight eyes arranged in two or three rows on the frontal carapace. According to their position, one can differentiate between anterior median eyes (AME), anterior lateral eyes (ALE), posterior median eyes (PME),

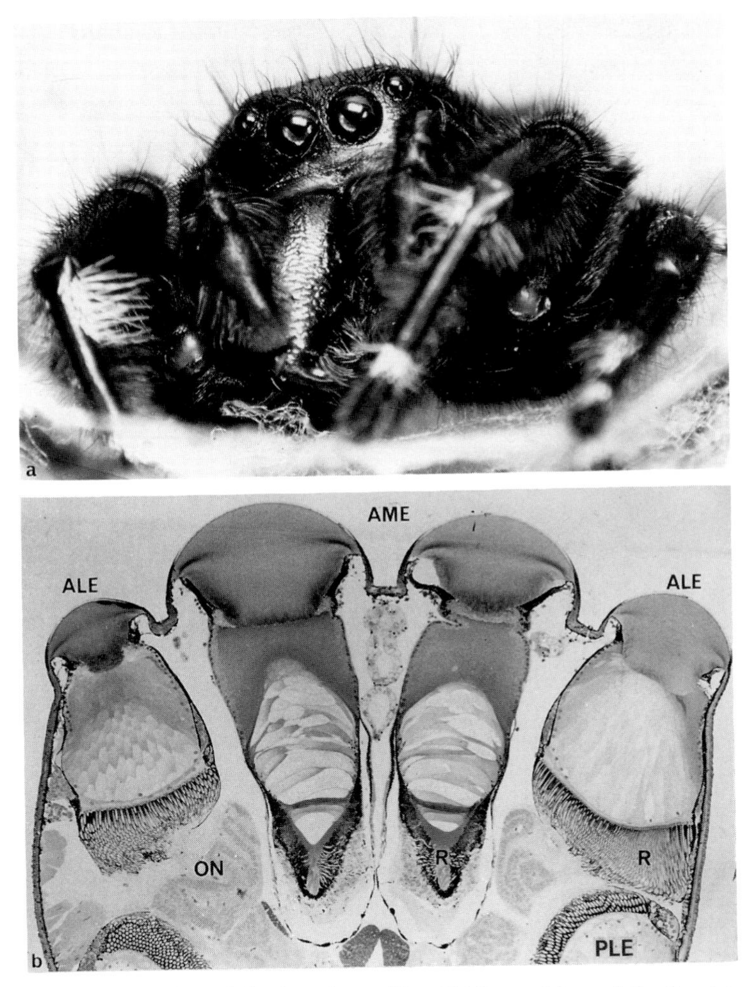

FIG. 70. (*a*) Head-on view of the jumping spider *Phidippus*. Most of the frontal carapace is occupied by the anterior row of eyes. (Photo: Troyer and Foelix.) (*b*) Horizontal section through the anterior row of eyes. AME = anterior median eyes, ALE = anterior lateral eyes, PME = posterior median eyes, ON = optic neuropil of the CNS, R = retina. 75 ×.

and posterior lateral eyes (PLE) (see Figs 70, 73). All eyes are *ocelli*, or so-called simple eyes: beneath a single cuticular *lens* lie a cellular *vitreous body* and the *visual cells*, which together with *pigment cells* compose the *retina* (Fig. 71; Blest, 1985). Structurally the eyes are of two different types, *main eyes* and *secondary eyes* (Homann, 1928; Land, 1985).

The main eyes are always the AME. The light-sensitive parts (*rhabdomeres*) of the visual cells lie distally; that is, they point toward the light, and are thus called everted eyes. Since the main eyes lack a reflecting layer (*tapetum*), they appear black. In some spiders the retina can be displaced sideways by the action of 1–6 muscles, which leads to a considerable increase of the visual field. Actual focusing of the image does not occur; this is not really necessary, since the lenses are small and of short focal length, yielding a large depth of field for

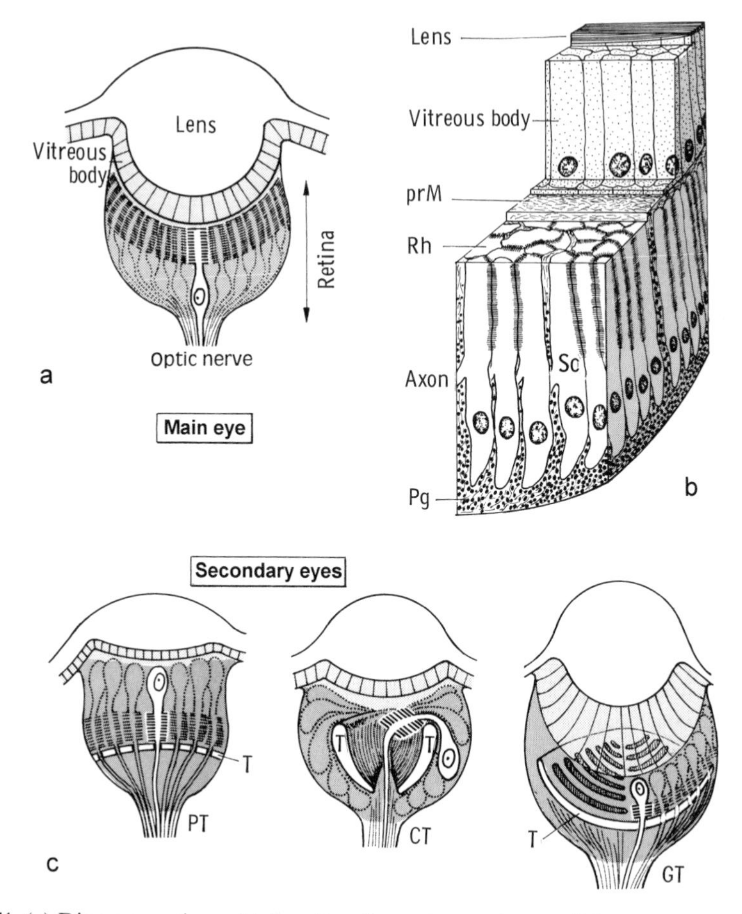

Fig. 71. (*a*) Diagrammatic sagittal section through a *main eye*. (After Homann, 1971.) (*b*) Fine structure of a main eye of *Agelena*. Pg = pigment cells, prM = preretinal membrane, Rh = rhabdome, Sc = sensory cell. (After Schröer, 1977, unpubl.) (*c*) The three types of *secondary eyes* differ in the structure of their tapetum (T). PT = primitive type, CT = canoe-shaped type, GT = grate type. (After Homann, 1971.)

objects at close range. The optic nerve usually arises from the middle of the eye cup. In some spiders (such as Agelenidae), however, the optic nerve emanates from one side of the eye cup, because the axon of each visual cell originates laterally (Fig. 71b). The main eyes of most spiders are small and have rather few visual cells (Homann, 1971). Jumping spiders (Salticidae) and crab spiders (Thomisidae) are an exception; their specialized main eyes will be discussed below. Those spiders with only six eyes (Dysderidae, Sicariidae, Oonopidae) lack main eyes.

The secondary eyes are "inverted," that is, the light-sensitive rhabdomeres point away from the incoming light (Fig. 71c); this construction is similar in principle to that of the vertebrate retina. Most secondary eyes possess a light-reflecting tapetum of crystalline deposits and therefore appear light. It has been assumed that such eyes are especially suited for seeing at night or in dim light, but this has never been substantiated experimentally. Although the tapetum varies a great deal among the different spider families, we can distinguish three basic types (Fig. 71c): (1) a primitive tapetum (PT), which fills the entire eye cup and leaves holes only for nerve fibers; (2) a canoe-shaped tapetum (CT), which consists of two lateral walls and a medial gap to allow nerve fibers to exit; and (3) a grated tapetum (GT), which resembles the grill of an oven. In some spider families (for instance, Salticidae and Oxyopidae) a tapetum is altogether lacking.

Whereas the main eyes are structurally rather uniform in most spiders, the secondary eyes differ considerably among different spider families. Since these differences are small among members of one family, the anatomy of the secondary eyes has been used for the systematics of spiders (Homann, 1950, 1952).

PT-type secondary eyes are typical for the Mesothelae, Orthognatha, and haplogyne spiders (the "primitive" spiders). The CT type is common among theridiids, agelenids, clubionids, amaurobiids, and many linyphiids and araneids. The GT type is found in many hunting spiders: Lycosidae, Oxyopidae, Pisauridae, and in the cribellate Psechridae and Zoropsidae. Even within the same species different types of secondary eyes can occur together. *Araneus diadematus,* for instance, has ALE of the CT type. Its PLE, however, are only laterally of the CT type; medially, they lack a tapetum (Homann, 1971).

The most efficient secondary eyes are those with a grated tapetum. The spherical lens ensures an image of high quality (Fig. 72), and the elongated vitreous body provides the necessary distance for a sharp focus. The resolution of the image depends on the number and density of rhabdomeres in the retina. The large PME and PLE of the wolf spider *Pardosa* contain about 4,000 rhabdomeres, whereas the small anterior eyes (AME, ALE) have only 300 and 120, respectively (Foelix, unpublished). In the orb-web spider *Araneus* there are much fewer photoreceptors (ALE = 80) and the small cave spider *Speocera* (Ochyroceratidae) has only 10–20 visual cells per eye. Obviously, with such a small number of photoreceptor cells in the retina, a spider would not be expected to detect much more than movement (Homann, 1971). The secondary eyes may have some other advantages. For instance, in experiments with the

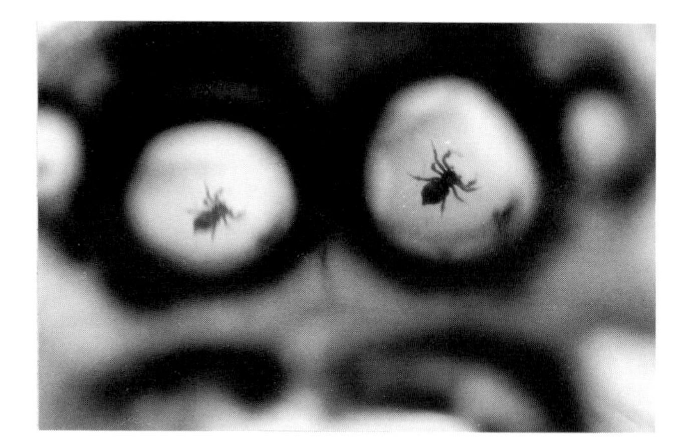

Fig. 72. The large lenses of jumping spider eyes provide amazingly detailed images. In this picture the images of a conspecific spider (*Portia*) were photographed behind the lenses of the main eyes.

funnel-web spider *Agelena* it was noted that they are 100 times more sensitive to light than the main eyes (Görner and Claas, 1985). Similarly, the extremely enlarged secondary eyes (PME) of the nocturnal hunter *Dinopis* are about 3,000 times more sensitive than the large main eyes of the diurnal jumping spider *Portia* (Land, 1985, see p. 173).

Perception of Polarized Light

Although by human standards most spiders have poor vision, some spiders can "see" something we cannot—polarized light. Sunlight becomes polarized when light waves are scattered by molecules or other small particles in the earth's atmosphere. The light waves then vibrate in a specific direction at each point of the sky. This results in a specific polarization pattern that changes with the position of the sun. Whereas the human eye, or generally speaking any vertebrate eye, cannot register the polarization pattern of the sky, arthropod eyes can. Studies of bees and ants have shown that they use polarized light for orientation (von Frisch, 1967; Wehner, 1976). Very similar observations have been made for some wolf spiders and sheet-web spiders (Papi, 1955; Görner, 1962; Henton and Crawford, 1966; Magni, 1966).

The wolf spider *Arctosa perita* dwells along the banks of rivers and the shores of lakes, and can easily walk on the water's surface (see Fig. 133). If such a spider is thrown onto the water, it returns directly to shore (Papi, 1955). However, if Arctosa is transferred from its customary northern shore to the opposite southern shore of a lake and is then thrown onto the water, it will run out onto the lake, but only if the sky is clear. If the sky is overcast the spider will return to the closest shore. Apparently, under a clear sky the spider orients itself by using the sun and the direction of the polarized light. It uses visual landmarks for its orientation if the sky is cloudy.

The sheet-web spider *Agelena labyrinthica* becomes disoriented during a return to its retreat if the normal pattern of light from the sky is changed by rotating a polarization filter above the spider (Görner, 1958). By covering specific eyes with opaque paint it has been shown that only the main eyes (AME) can analyze the plane of polarization. It has been found that the ventral visual cells have rhabdomeres with specifically aligned microvilli that are probably responsible for detecting polarized light (Schroer, 1974, 1976). The retina of the main eye seems to be divided into two functional parts: a ventral part for polarized light perception and a dorsal part for the perception of relatively coarse images.

The Visual Sense of Jumping Spiders

It is well known that the behavior of jumping spiders is strongly dependent on their visual sense; orientation, prey capture, courtship, and escape are all essentially controlled by visual stimuli. Indeed, among the most distinctive features of jumping spiders are their large eyes. Especially impressive is the anterior row of eyes, which consists of the large main eyes (AME) and the slightly smaller anterior lateral eyes (ALE) (Figs 70, 73). The posterior median eyes (PME) are also of considerable size, whereas the posterior lateral eyes

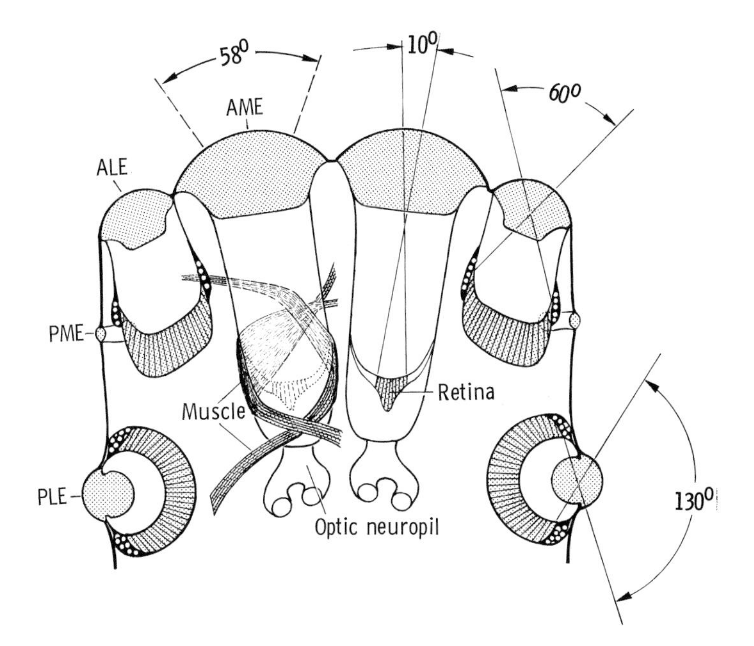

Fig. 73. Horizontal section of the prosoma of a jumping spider. Visual angles of the various eyes are indicated. The angle (58°) given for the left main eye refers to the maximum visual angle that the spider can attain by using its retinal muscles to shift the eye cup laterally. Abbreviations as in Fig. 70. (After Homann, 1928 and Land, 1969b.)

(PLE) are much reduced (some workers refer to the PME as PLE, and vice versa). Many elegant experiments (Homann, 1928; Crane, 1949) have demonstrated that the secondary eyes are primarily movement detectors, while the main eyes perceive detailed images (form perception). Before we discuss these experiments in any detail, we must become acquainted with the specialized anatomy of the jumping spider's eyes.

The efficiency of any eye is determined by the design of its optics and by the structure of the retina. Both are well developed in the eyes of jumping spiders. The main eyes are characterized by large lenses (Fig. 72), a large vitreous body with a focal length of 770 μm, and a retina with multiple layers of visual cells (Land, 1969a; Eakin and Brandenburger, 1971). The main eyes can be compared optically with a telephotographic lens, through which only a small area of an image can be seen, but at high resolution. The "movable" retinas of the main eyes partly compensate for the narrow angle of vision; several antagonistic muscles are attached to the main eye (Scheuring, 1914), and they can move the retina in all three dimensions (Fig. 73). This, of course, greatly increases the visual field.

The retina is composed of four different layers of visual cells (Fig. 74a), and it is possible that each layer is preferentially sensitive to a certain wavelength of light. De Voe (1975) and Yamashita and Tateda (1976) have shown that the retina of a jumping spider indeed has different spectral sensitivities. Maximal sensitivities were found at wavelengths of 360, 480, 500, and 580 nm, and it was concluded that each retinal layer contains a different visual pigment. In the most distal layer (layer 4) the rhabdomeres are aligned roughly perpendicular

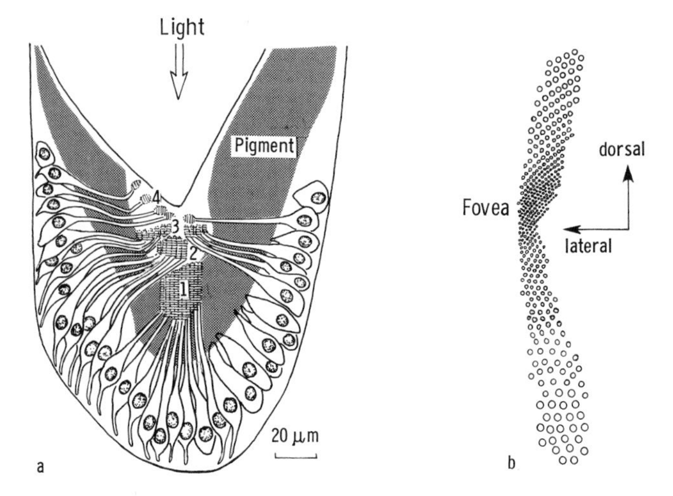

FIG. 74. Retina from the main eye of a jumping spider. (*a*) Horizontal section showing four different receptor layers (1–4), surrounded by a pigment layer. (*b*) Schematic frontal view of layer 1 of the retina; each circle represents the terminal of a visual cell. Note that the density of receptors is highest in the central retina (fovea). (After Land, 1969a.)

to the rhabdomeres of layers 1–3. Therefore, one might expect that layer 4 might detect polarized light. The specialization of layer 4 is underscored by the fact that, for optical reasons, it can no longer receive a truly focused image. This corresponds somewhat to the situation found in insect ocelli, where the focal plane of the lens also lies behind the retina.

Another interesting feature of the jumping spider's main eyes is the distribution of the 1,000 visual cells within the retina. The density of visual cells is ten times higher in the central region (the *fovea*) than in the periphery (Fig. 74b), and the angle of divergence between neighboring receptors is very small in the fovea—12′, compared to 40′ in the periphery. The spider's greatest visual acuity must therefore lie in the center of the retina (Homann, 1928; Land, 1969b). Owing to a very small angle of divergence (2.4′), the best resolution was found in the main eyes of *Portia* (Williams and McIntyre, 1980). This was convincingly demonstrated when *Portia* differentiated conspecifics from other jumping spiders, even at distances of 20–30 cm (Forster, 1982; Jackson and Blest, 1982).

The back of the retina is covered by dark pigments, as in vertebrate eyes. The visual cells of the main eyes, however, lack the individual pigment screens found in the photoreceptors of the secondary eyes.

The secondary eyes of jumping spiders are either elongated (the ALE have a focal length of 330 μm) or spherical (PME focal length of 240 μm) and possess relatively large visual fields (Figs 73, 75). The extent of the entire visual field of a jumping spider is effectively the sum of the individual visual fields of the secondary eyes (Fig. 76). It is significant that the visual fields of the ALE overlap by 40° (Homann, 1928); the resulting binocular vision enables the spider to

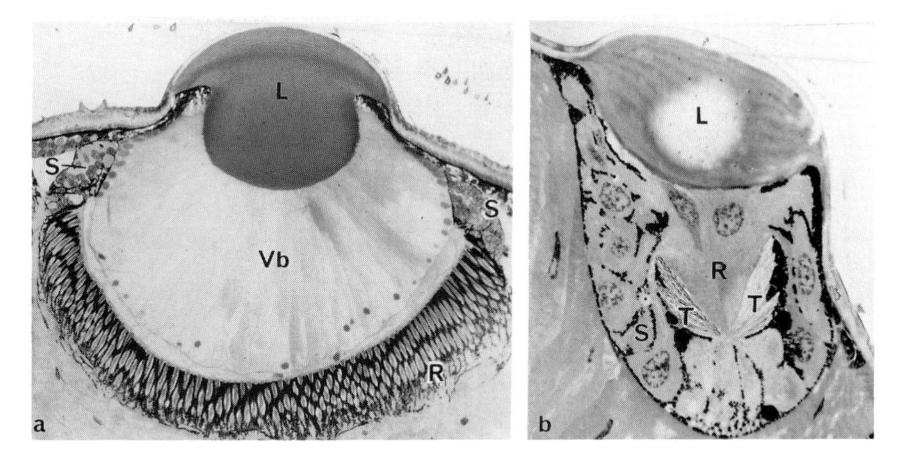

FIG. 75. (*a*) PME of a jumping spider. The bodies of the visual cells (S) are situated outside the eye cup and thus do not interfere with the incoming light. A tapetum is lacking. L = lens, R = retina, Vb = vitreous body. 150 ×. (*b*) ALE of a theridiid. The retina (R) is bordered by the canoe-shaped tapetum (T); a vitreous body is barely visible as a thin layer beneath the lens (L). The visual cells (S) are demarcated by pigment granules. 540 ×.

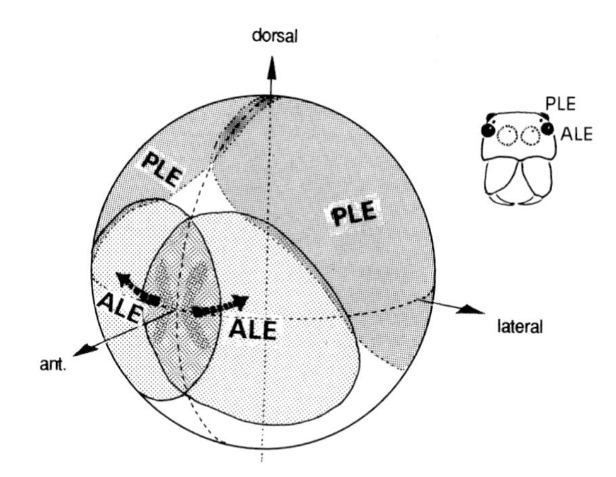

FIG. 76. The visual fields of the large secondary eyes (ALE, PLE) of a jumping spider (*Phidippus*), projected onto a sphere. Note that the visual fields of the ALE overlap, which allows binocular vision. The boomerang-shaped visual fields of the main eyes (white in the *inset*) are rather small. However, since the retinae of the main eyes can be moved by muscles, their visual fields can also be shifted (dotted arrows) and thus cover the same range as the ALE. (After Land, 1985.)

estimate distances. The very narrow visual fields of the main eyes, in contrast, do not overlap. Structural characteristics of the secondary eyes are the lack of a tapetum, the location of the visual cell bodies outside the eye cup (Fig. 75a), and the occurrence of pigmented and unpigmented supporting cells between the rhabdomeres (Figs 75a, 77; Eakin and Brandenburger, 1971). The single-layered retina has only one cell type, which is maximally sensitive to wavelengths between 535 and 540 nm (Yamashita and Tateda, 1976). The number of visual cells in the secondary eyes is higher than in the main eyes, 3,000–6,000 cells in an ALE and 8,000–16,000 in a PME (Land, 1972b).

How does a jumping spider use its sophisticated eye structures? The exceptional visual abilities of jumping spiders become most evident during prey capture (Homann, 1928; Land, 1969b; Forster, 1982). Moving prey is usually first detected by the wide-angle lenses of the secondary eyes. These eyes respond to small movements of an object, analogous to our peripheral vision (Duelli, 1978; Komiya et al., 1988). The maximum distance at which a spider can still see something is 30–40 cm. If the object (prey) moves closer than 20 cm, the spider turns to face it, and now the object becomes fixed with the movable retinas of the main eyes. The image of the prey is projected exactly onto the fovea, the central region of the retina. If the prey moves on, the eye muscles shift the retinas appropriately to maintain foveal vision. Behavioral experiments show that a true perception of images occurs at distances of less than 8–10 cm. In this range only about 100 visual cells are stimulaled, yet even this may suffice to allow prey to be distinguished from conspecifics. At this distance the spider will pursue the prey, and at less than 3–4 cm it will stalk it.

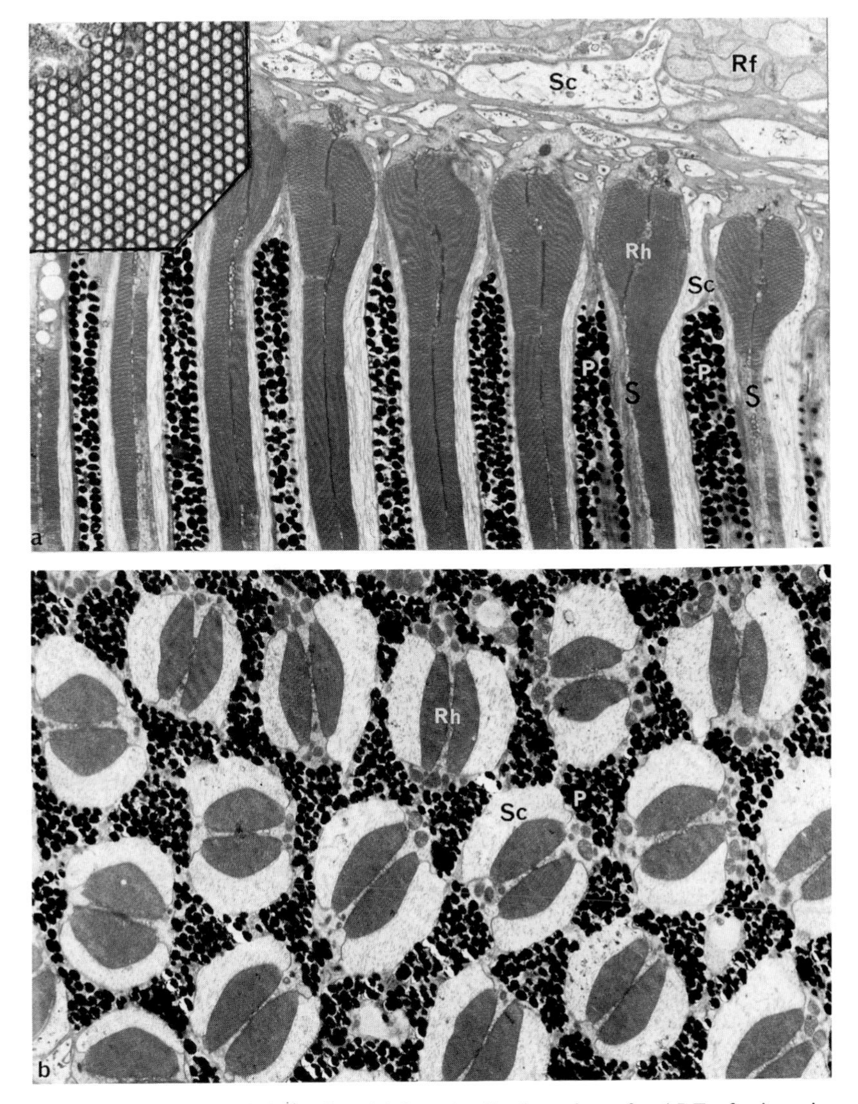

FIG. 77. Fine structure of the retina. (*a*) Longitudinal section of a ALE of a jumping spider. The rhabdomeres (Rh) of the visual cells are surrounded by pigmented (P) and unpigmented supporting cells (Sc). Rf=axons of visual cells. 3,600 ×. *Inset*: The light-sensitive rhabdomeres consist of thousands of tiny, fingerlike processes (microvilli); their regular arrangement is shown here in cross-section. 44,000 ×. (*b*) Cross-section corresponding to (*a*). Note the pronounced isolation of the rhabdomeres by the interspersed pigment cells. 3,000 ×.

During the "recognition" process the muscles of the retinas are in constant motion: the shape of the object is quickly scanned horizontally, and, at the same time, the main eye is rotated along its optical axis. This scanning procedure is apparently necessary to permit identification of an object; only thereafter follows the stalking of the prey. The final leap is performed at a distance of 1–2 cm.

From the foregoing description a definite "division of labor" between the main eyes and the secondary eyes is thus apparent. Although most of the total retinal area belongs to the secondary eyes, it serves only for perceiving and locating *movement*. The small central region of the main eyes retina is the only area suitable for *form perception*. The importance of the main eyes becomes clear when one covers their lenses with opaque wax (Homann, 1928): stalking behavior is completely absent during prey capture, and the final leaps are too long and often unsuccessful. A male with his main eyes blinded will likewise leave out the typical courtship dances that he usually performs in front of the female (Fig. 165). In contrast, eliminating the visual input from the secondary eyes produces less drastic effects. Obviously, the total visual field becomes extremely narrow, and therefore the spider has difficulty in noticing the prey at all. If prey is located, the stalking behavior is initiated much earlier (at a distance of 10 cm) and the final jump is aimed too far.

The perception of images can be impressively demonstrated by setting up a mirror in front of a male jumping spider. The mirror image is then threatened like a real conspecific opponent (Fig. 78; Forster and Forster, 1973). This behavior seems to be unique for jumping spiders; wolf spiders, which also orient visually, do not react to their mirror image (Rovner, 1989). Models of prey will also be attacked if they are moved in front of a jumping spider. It is interesting that three-dimensional models are definitely preferred to two dimensional figures (Drees, 1952).

Whether or not jumping spiders can see colors remains an open question. That jumping spiders are often conspicuously colored has been noted repeatedly

FIG. 78. A male jumping spider (*Trite auricoma*) threatens his own mirror image. (Photo: L. Forster.)

as an indirect argument for their color vision. This seems reasonable, since in fish and birds a colorful phenotype is often correlated with the ability to differentiate colors. Somewhat more convincing evidence comes from behavioral experiments in which jumping spiders could distinguish blue and orange paper strips from 26 shades of gray (Crane, 1949; Kaestner, 1950). The sensitivities of the main eyes to different parts of the visible spectrum further argue for color vision (De Voe, 1975; Yamashita and Tateda, 1976). Since we still lack rigorous proof, we should perhaps be reminded that it took several decades to demonstrate color vision in the domestic cat.

What use jumping spiders make of their ultraviolet (UV) receptors (maximum spectral sensitivity at 360 nm) also remains to be explained. Experiments dealing with prey capture under UV light have given both positive (Young and Wanless, 1967) and negative results (Kaestner, 1950). Most spider eyes seem to be insensitive to red light, as are most insect eyes. Recently, a spectral sensitivity between 330 and 700 nm was found in the jumping spider *Maevia* (Peaslee and Wilson, 1989). However, the sensitivity was much reduced around 700 nm, i.e., in the red part of the spectrum.

It should be emphasized that the visual acuity of jumping spider eyes can easily compete with that of the compound eyes of insects (Homann, 1928). The main eyes of a jumping spider are even superior to the insect's compound eyes in optical resolution. The commonly used term *simple ocelli* does not apply here; on the contrary, the highly developed eyes of jumping spiders demonstrate just how sophisticated ocelli can be.

Temperature Perception

There can be little doubt that spiders can sense changing temperatures. The distal parts of the legs and spinnerets seem to be most sensitive to thermal stimuli, but no specific thermoreceptors have been identified so far (Den Otter, 1974; Pulz, 1986). It has been hypothesized that a change in temperature would deform the shaft of certain hair sensilla, thereby transmitting mechanical forces onto nerve endings. The only proof for temperature-sensitive cells comes from electrophysiological recordings of the tarsal organs (Ehn and Tichy, 1994; see Fig. 68).

Besides perceiving the ambient temperature, spiders can apparently also sense temperatures within their own body. If the body temperature rises above 32°C, tarantulas will move abruptly and display avoidance behavior. Again, it is not known where the internal temperature sensors are located, but the heart ganglion has been suggested as a possible candidate (Pulz, 1987).

Peripheral Nerves

The axons of the various sensory organs form small bundles and join with others to build up sensory nerves. All the nerve fibers passing from the periphery toward the central nervous system (CNS) are called *afferents*. Conversely, all the nerve fibers emanating from the CNS to reach the effector organs (muscles, glands) in the periphery, are called *efferents*. The peripheral nerves may consist of purely afferent (*sensory*) fibers, or purely efferent (*motor*) fibers, or both. Some of the efferent axons may contain small dark granules, which identifies them as neurosecretory fibers.

Only the peripheral nerves of the legs and palps have been studied in some detail. There are three main nerves (A,B,C; Fig. 25) in each extremity, a small nerve A (motor and sensory fibers), a medium-sized nerve B (only motor fibers), and a large nerve C (only sensory fibers) (Rathmayer, 1965a,b). Nerve C is not only the largest, but also contains thousands of small nerve fibers, whereas nerve B has only a few hundred axons, but of large diameter. The actual number of nerve fibers depends on the species and also on the size of the spider. Large spiders, having many more hair sensilla than small ones, also possess many more afferent fibers.

In the wandering spider *Cupiennius*, the course of the three leg nerves has been followed through all the leg segments (Seyfarth et al., 1985). In the distal parts (tarsus, metatarsus) only nerve C is conspicuous, consisting of two separate bundles of sensory fiber. Within the tibia they merge into a large single nerve. In the proximal parts (femur, trochanter, coxa) many more sensory nerves are added (Fig. 79). The large motor fibers of nerve B lie on the dorsal side of nerve C. Although the distribution of the three nerves is the same in all legs, the number of nerve fibers is not. The front legs have more sensory nerve fibers than the hind legs, because the front legs bear more sensory organs. This was

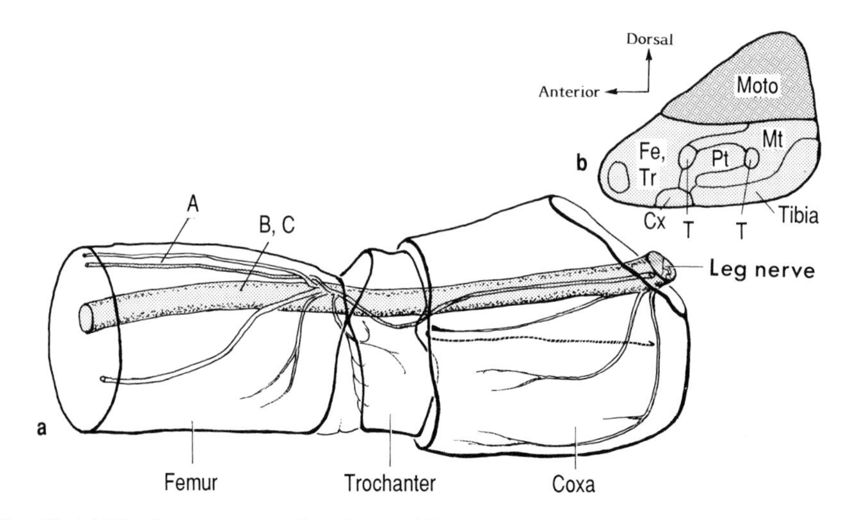

FIG. 79. (*a*) The leg nerves consist of a small branch A and a massive branch B, C (*Cupiennius*; after Seyfarth et al., 1985). (*b*) Cross-section of a leg nerve at the coxa level. The dorsal part (of branch B, C) is made up by large motor axons (Moto), the ventral part by small sensory axons. These sensory fibers are grouped in bundles, according to their origin from different leg segments. Fe = femur, Tr = trochanter, Cx = coxa, Pt = patella, Mt = metatarsus, T = tarsus. (After Brüssel and Gnatzy, 1985.)

nicely demonstrated in the small orb-web spider *Zygiella* where 7,000 sensory axons were counted in leg 1 (coxal level), 4,400 in leg 2, 4,200 in leg 3 and 3,900 in leg 4 (Foelix et al., 1980).

The sensory fibers are of rather small diameter (0.35 μm average) thus allowing for tight packing of thousands of axons within a slender spider leg. Furthermore, there is also a rather orderly packing of the sensory fibers within the nerve C (Brüssel and Gnatzy, 1985). For instance, all of the hundreds of axons coming from tibial sensory organs remain together in one group without intermingling with axons coming from the metatarsus (Fig. 79b). This orderly arrangement of sensory fibers is even maintained after entering the CNS; receptor fibers from distal leg segments terminate in ventral fiber tracts, those from proximal leg segments project into dorsal sensory tracts (Anton and Barth, 1993). A corresponding *somatotopic* organization of sensory fibers is also known in insect leg nerves (Zill et al., 1980).

Microscopically, each leg nerve consists of several axon bundles which are separated by glial cell extensions. Each bundle may contain from a few up to several hundreds of "naked" axons, which are not insulated from each other—as is the case in vertebrates. On the outside the glial cells are covered with a basal lamina and also with collagen fibrils (Foelix et al., 1980; Foelix, 1985). Although one might say that the fine structure of peripheral spider nerves is very similar to that of insect nerves, there is one important difference: within the peripheral nerves of spiders we often find synaptic connections (Foelix, 1975, 1985), which are lacking in insects. Such synaptic contacts may occur between axons or between axons and sensory cells (Fig. 80). Their functional implication

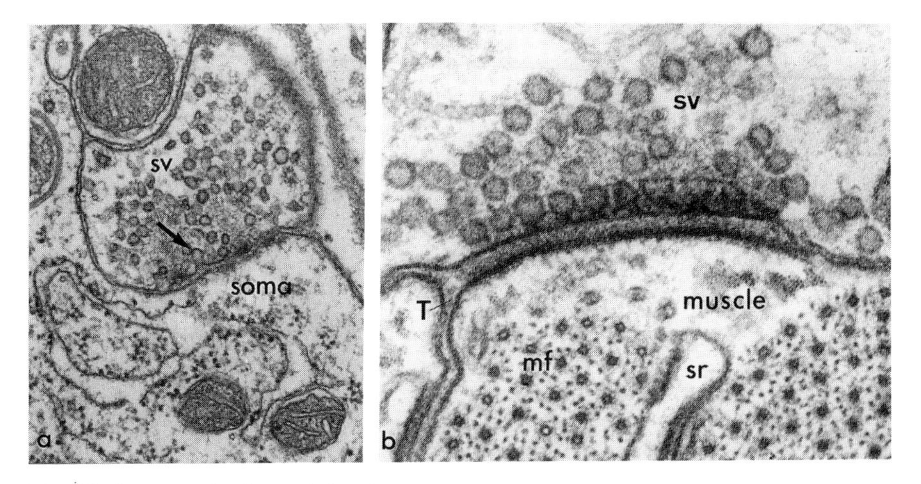

FIG. 80. Synapses in the peripheral nervous system. (*a*) An axosomatic contact on a sensory neuron in the leg of *Amaurobius*. Note one synaptic vesicle (arrow) fusing with the presynaptic membrane. 36,000 ×. (*b*) Motor endplate on a leg muscle of *Zygiella*. The synaptic vesicles (sv) align in a single row at the presynaptic membrane before releasing their neurotransmitter onto the muscle membrane. mf = myofilaments, sr = sarcoplasmic reticulum, T = T-system. 75,000 ×.

(peripheral summation, central inhibition?) is still unclear, yet we must assume that some nervous integration is already taking place at the periphery.

Central Nervous System

Compared with other arthropods, spiders have a highly condensed central nervous system (CNS). Instead of a chain of interconnected ganglia (a ventral nerve cord) extending throughout the body, spiders have only two compact ganglia: the *supra-* and the *subesophageal ganglion*. They are both located in the prosoma (Figs 81, 82). True abdominal ganglia exist only in the embryo. During development they migrate into the prosoma where they fuse with the ganglia of the appendages, giving rise to the large subesophageal ganglion. The much smaller supraesophageal ganglion sits in front and on top of the subesophageal ganglion; it consists of the cheliceral ganglia and the "brain" (the association centers). The boundary between supra- and subesophageal ganglion is marked by the esophagus, which divides the CNS horizontally (Figs 81, 82b). Several nerves emanate from each ganglion and together constitute the peripheral nervous system.

General Structure

That the compact CNS of spiders is composed of individual ganglia can clearly be seen in young spiderlings, where septa of connective tissue, which also

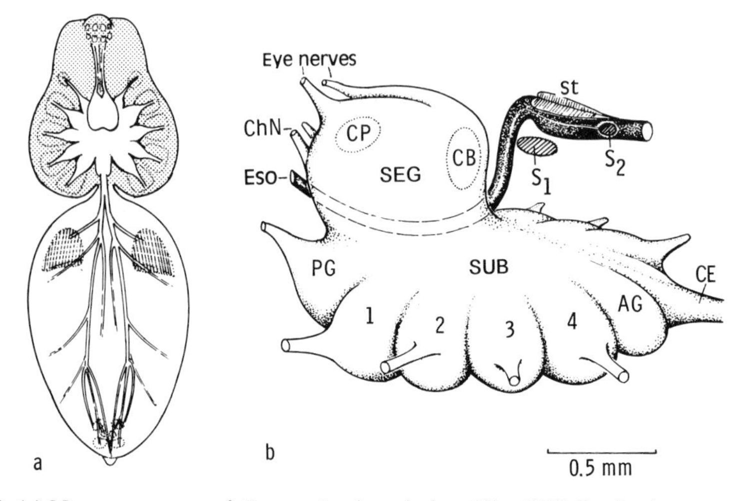

FIG. 81. (*a*) Nervous system of *Tegenaria*, dorsal view. The CNS lies in the prosoma and gives off nerve bundles to the extremities and to the organs of the opisthosoma. (After Gerhardt and Kaestner, 1938.) (*b*) CNS of *Tegenaria* lateral view. The esophagus (Eso) divides the CNS into a supra- (SEG) and a subesophageal ganglion (SUB). ChN = cheliceral nerves, PG = palpal ganglion, 1–4 = leg ganglia, AG = abdominal ganglion, CE = cauda equina, CB = central body, CP = corpora pedunculata, $S_{1,2}$ = Schneider's organs 1 and 2, St = stomach. (After Kühne, 1959.)

contain blood vessels and sometimes tracheoles, delineate the borders of adjacent ganglia (Fig. 82a). In the "ancient" Mesothelae, segmentation of the CNS is still apparent in the adult (Millot, 1949).

The star-shaped subesophageal ganglion consists mainly of the fused ganglia of the ten appendages (Fig. 81a). The different ganglia are interconnected by

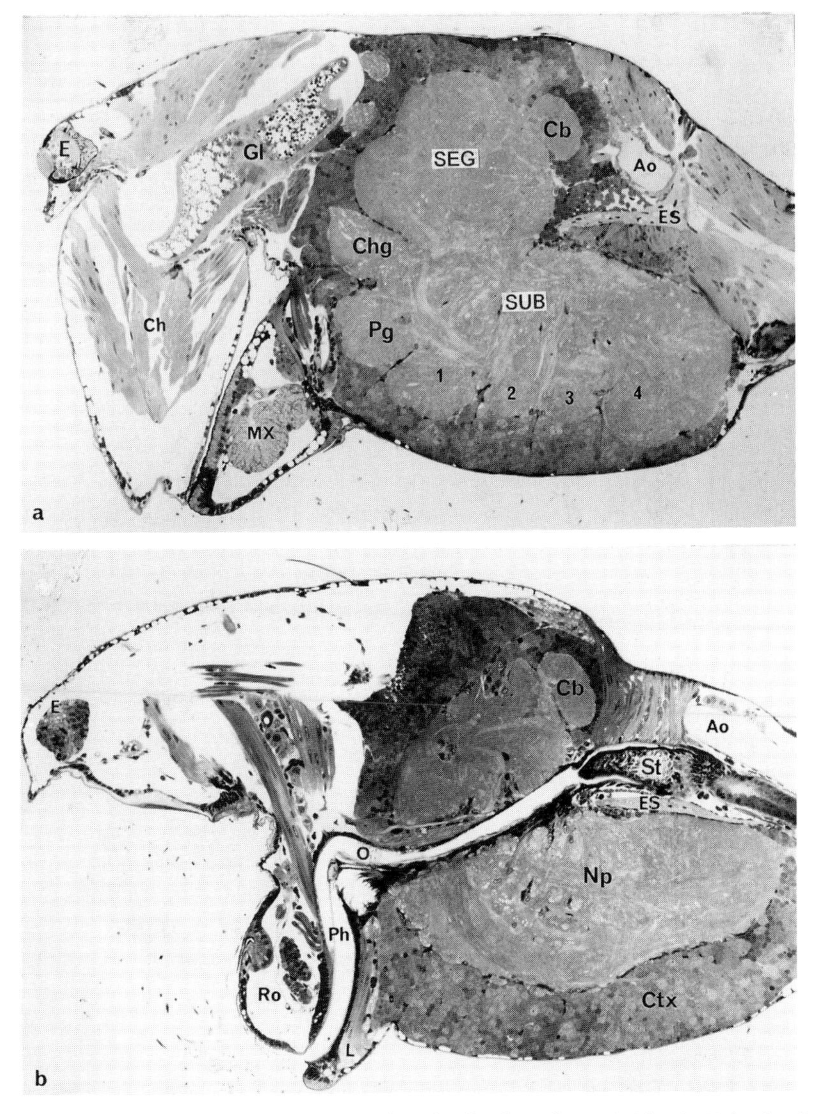

FIG. 82. Prosoma of a young *Tegenaria*, longitudinal section. (*a*) The CNS is clearly divided into separate ganglia (cf. Fig. 81b). Cb = central body, Ch = chelicera, Chg = cheliceral ganglion, E = eye, ES = endosternite, Gl = poison gland, MX = maxilla, Pg = palpal ganglion. (*b*) This sagittal section demonstrates the subdivision of the CNS into a central neuropil (Np) and a peripheral cortex (Ctx). E = esophagus, Ph = pharynx, Ro = rostrum, St = sucking stomach. 220 ×.

many interneurons. Motor nerves for the palps and legs are given off laterally from these ganglia. Caudally, the abdominal ganglia are attached to the subesophageal ganglion, and from them arises a tapered nerve bundle, the *cauda equina*, which innervates the entire opisthosoma. The cheliceral ganglia stem originally from the subesophageal ganglion, but during embryonic development they migrate forward and come to rest on either side *above* the esophagus. Topographically then, they belong to the supraesophageal ganglion, though remarkably, the fibers that connect these ganglia remain behind the esophagus (the *postoral commissures*). The cheliceral ganglia give off the nerves for the musculature of the chelicerae, the pharynx, and the poison glands (Saint Rémy, 1887). The "brain" itself receives only the optic nerves and thus contains only visual and association centers. The combination of the cheliceral ganglia and the brain is called the syncerebrum, or supraesophageal ganglion.

Cellular Structure

Histological sections show a distinct division of the CNS into a marginal layer of neurons (the *cortex*) and a central mass of nerve fibers (the *neuropil*) (Fig. 83). The surface of the CNS is covered by a connective tissue layer, which is rather thick in Orthognatha (Legendre, 1958, 1961b; Babu, 1965). The cell bodies of the neurons are found mostly on the ventral side of the CNS, and most of them give off only a single axon (that is, they are pseudo-unipolar; see Fig. 88). Several different types of neurons can be distinguished, the largest of which are thought to be motoneurons; others are interneurons, which connect different ganglia, and some can be identified as neurosecretory cells (Babu, 1965, 1969). The total number of motoneurons has been counted as 900 for a female orb-web spider *Argiope* but only 60 for the male (Babu, 1975). The entire CNS apparently contains 30,000 neurons, 13,000 of which belong to the supraesophageal ganglion. In the larger wandering spider *Cupiennius*, 100,000 neurons were counted for the whole CNS: 51,000 in the supraesophageal ganglion and 49,000 in the subesophageal ganglion (Babu and Barth, 1984).

The typical synapse is a so-called *dyad synapse*, in which one presynaptic fiber containing synaptic vesicles contacts two postsynaptic nerve fibers (Fig. 84). This divergent arrangement of synapses is commonly seen in the CNS of arthropods (Fahrenbach, 1979; Foelix, 1985). Often, the connectivity is much more complex, e.g., by forming *serial* or *reciprocal* synapses between several different nerve fibers.

The neuropil is a dense network of nerve fibers, and most of the synaptic connections occur there (Fig. 83b). During postembryonic development the number of nerve fibers forming the neuropil increases markedly (Meier, 1967), because of the repeated branching of the dendritic processes. The number of neurons within the cortex remains constant, however (Babu, 1975). Thus the apparent growth of the CNS is caused mostly by an increase of the neuropil. In very young spiderlings the CNS takes up almost 50% of the volume of the prosoma; in adult spiders the volume has dropped to 5–10%, even though the adult CNS is 10–20 times larger than in early nymphal stages (Babu, 1975). In

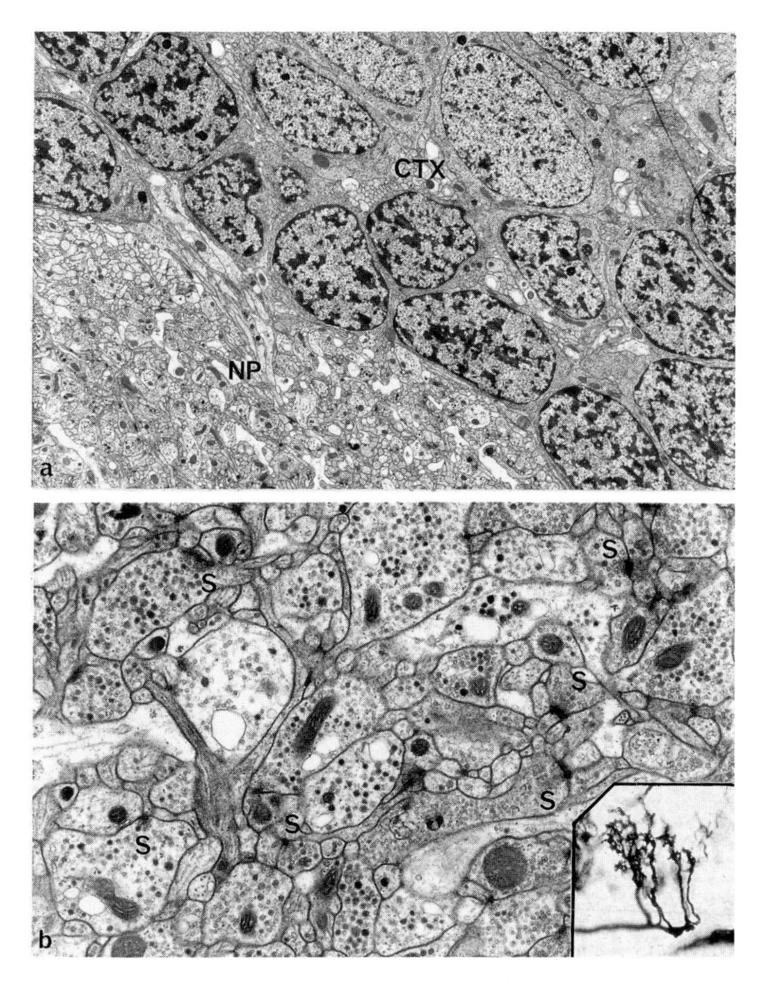

FIG. 83. The complexity of the CNS is apparent in electron micrographs. (*a*) Border region of cortex (CTX) and neuropil (NP) of the central body. 2,700 ×. (*b*) Detail from the neuropil. Innumerable small nerve fibers intertwine and make synaptic contacts (S). 11,500 ×. Inset: The delicate ramifications result from repeated branching of one dendrite (Golgi preparation, 390 ×).

terms of weight the spider's CNS makes up only 0.1% of the total body weight in a tarantula, but 2.5% in wolf spiders and even 5% in jumping spiders (Meyer et al., 1984). These differences are most likely due to the optical centers, which are much more pronounced in the visually oriented wolf and jumping spiders.

The neurotransmitters in the spider CNS are: acetylcholine, about ten different amino acids (e.g., *glutamine*, *glycine*, *GABA*, *serine*, *taurine*, etc.), and several biogenic amines (e.g., *histamine*, *serotonin*, *dopamine*, *norepinephrine*; Meyer et al., 1984; Schmid et al., 1992). Analogous to the insect CNS (Walker, 1982), it is believed that some of these neuroactive substances are excitatory (e.g., *aspartate* and *glutamic acid*), while others are inhibitory (e.g., *glycine*

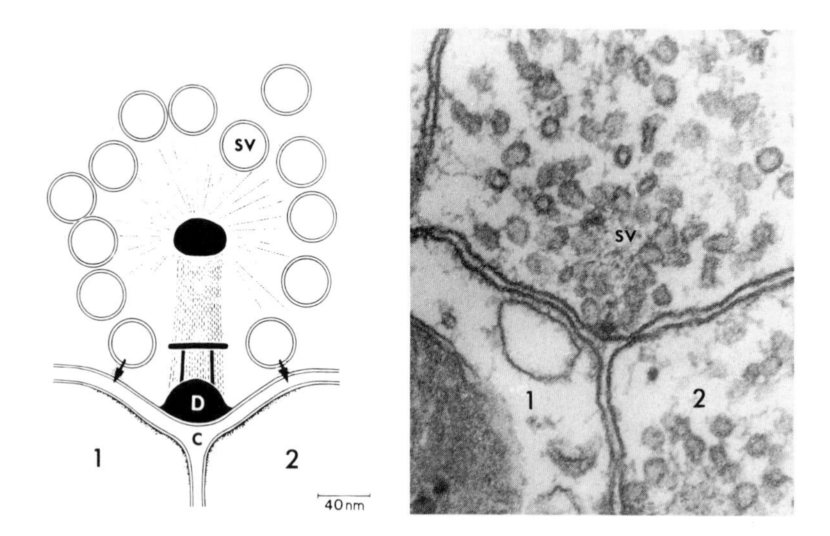

FIG. 84. (*a*) Diagram of the typical *dyad* synapse found in the spider's nervous system, in which one presynaptic fiber contacts two postsynaptic fibers (1, 2). Synaptic vesicles (sv) aggregate around a presynaptic density (D), fuse with the cell membrane (arrows) and release their neurotransmitter into the synaptic cleft (*c*). (After Foelix and Choms, 1979.) (*b*) Dyad synapse from the peripheral nervous system in *Zygiella*, as seen at high magnification in the electron microscope (72,000 ×).

and *GABA*). Catecholamines can modulate certain behaviors or may cause unspecific behavioral arousal. *Histamine* was found in the visual system and seems mainly a neurotransmitter of photoreceptors. It also occurs in six giant neurons of the brain. Since their dendrites branch into specific areas of the subesophageal ganglion, it is likely that they play a modulatory role on both the sensory inputs and the motor outputs (Schmid and Duncker, 1993). Recently, *octopamine*—the invertebrate counterpart of noradrenaline—was identified in the spider CNS. Its distribution suggests a role as a neurotransmitter/ modulator or even as a hormone at peripheral sites (Seyfarth et al., 1993).

Supraesophageal Ganglion ("Brain")

The dorsal part of the brain, the *lobus opticus*, receives information from all eight eyes and is a particularly complex part of the CNS. The visual fibers of the optic nerves project into specific areas of the optic lobe, where they make synapses with ganglia cells (Trujillo-Cenóz, 1965). If a fluorescent dye is injected into a specific eye, the dye will slowly spread from the retina through the optic nerve into a well-defined area within the optic lobe, the *first optic neuropil* (ON_1). Using this method on each eye separately, one can draw a precise map of each ON_1 for each eye (Fig. 85). It becomes readily apparent that neighboring eyes also have neighboring neuropils and that there is no overlap between those areas. In other words, the distribution of these optic neuropils in the brain

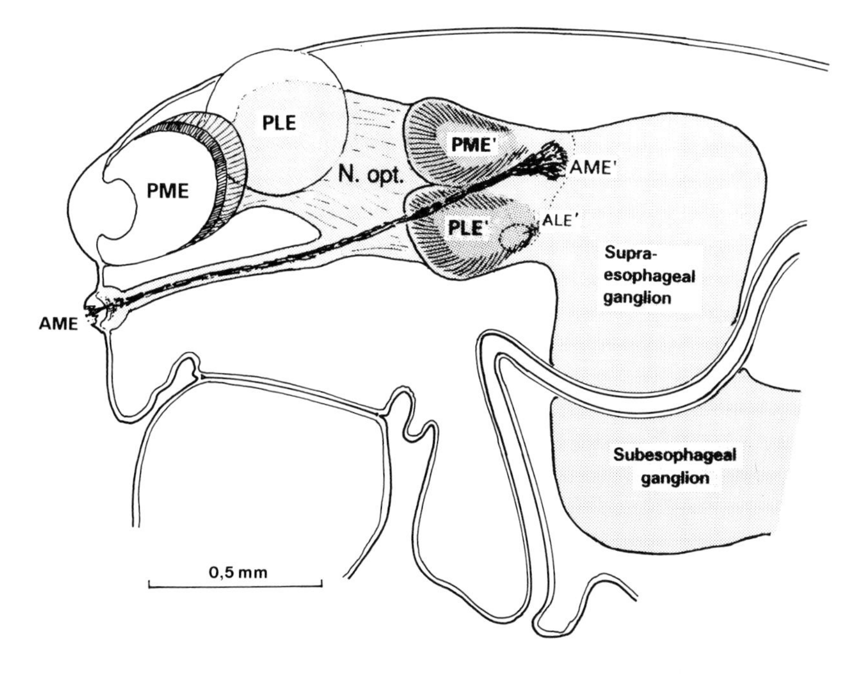

FIG. 85. Visual pathway from the eyes to the brain in a wolf spider (*Pardosa*). A small main eye (AME) is shown after injection of a fluorescent dye: the dye has spread through the 300 visual fibers into the first optic neuropil located in the dorsal brain (AME′). From similar experiments with the secondary eyes, a precise mapping of each eye and its target in the brain was obtained. The large PME and PLE (4,000 visual fibers each) project through the optic nerve (N. opt.) into specific dorsal and ventral regions (PME′, PLE′) of the brain. The tiny ALE (100 visual fibers) terminate in a small ventral region (ALE′). (After Foelix, 1989, unpubl.)

corresponds to the arrangement of visual fields of the eyes (see Fig. 76). Thus we can speak of a similar order in the visual pathway of spiders as we know it also from the visual system of vertebrates (*retinotopy*). The visual pathway, however, does not end in the first optic neuropil (ON_1) but continues into two further relay stations (ON_2, ON_3). Furthermore, this pathway differs between main eyes and secondary eyes (Fig. 86; Strausfeld and Barth, 1993). Generally speaking, the pathways coming from the main eyes seem to be involved in the perception of form and texture, whereas those originating from the secondary eyes are specialized for detecting (horizontal) motion (Fig. 87; Strausfeld et al., 1993).

The three secondary eyes (ALE, PLE, PME) supply three successive neuropils (ON_1, ON_2, ON_3); the main eyes (AME) have their first and second optic neuropil (ON_1, ON_2) in a similar location, but their third optic neuropil (ON_3, *central body*) lies at the rear of the brain. The ON_3 of the secondary eyes lies right behind ON_2 and was termed *corpora pedunculata* or *mushroom body* (Hanström, 1921, 1928, 1935) in analogy to the insect brain. However, it is probably not homologous nor functionally analogous to the mushroom

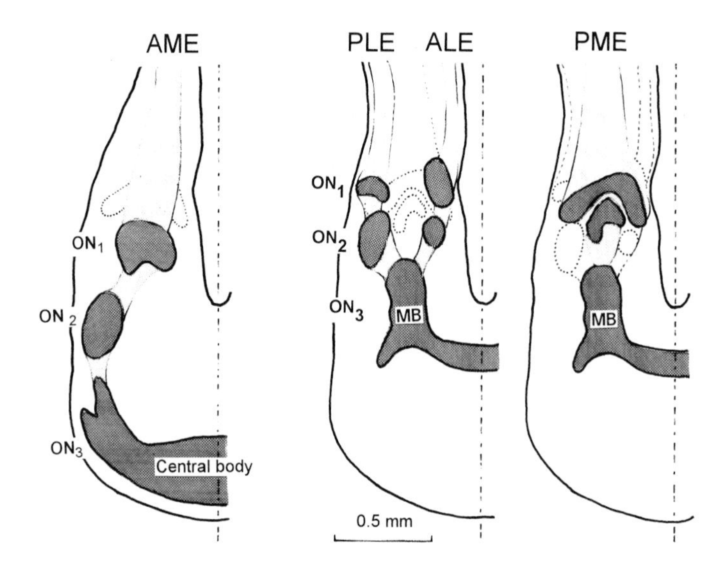

FIG. 86. Visual pathways of main and secondary eyes differ in the spider brain (*Cupiennius*). In the main eyes (AME) the visual information passes through two optic neuropils (ON_1, ON_2) and projects into a third one (ON_3) situated in the rear of the brain, called the central body. Each of the secondary eyes (PLE, ALE, PME) has its own set of three successive neuropils; the last optic neuropil (ON_3) lies more anteriorly than the central body and is termed the *mushroom body* (MB). (After Strausfeld and Barth, 1993.)

bodies of insects (Strausfeld et al., 1993). The central body is not only integrating visual information, but is probably also a motor and association center. For a long time it was claimed to be especially well developed in orb-web spiders (and supposedly related to the construction of their elaborate webs). Recent studies comparing the relative volume of the central bodies in web spiders and wandering spiders did not show any differences, however (Weltzien and Barth, 1991).

Subesophageal Ganglion

As mentioned earlier, the subesophageal ganglion lies ventrally, below the esophagus (Fig. 81) and consists of the fused ganglia of the appendages. Each ganglion receives incoming sensory fibers, which may terminate there, or they may continue to neighboring ganglia—even onto the other side (*contralateral*) of the subesophageal ganglion (Fig. 88). The outgoing fibers stem from large motoneurons, situated in the peripheral cortex. The various ganglia are connected by many *interneurons*. Functionally, they link the sensory input to the motor output (Milde and Seyfarth, 1988; Gronenberg, 1989, 1990). Morphologically, they may be simple and restricted to a small area, or they may be more complex, connecting all ganglia on both sides. About half of these *plurisegmental* interneurons combine the information of specific

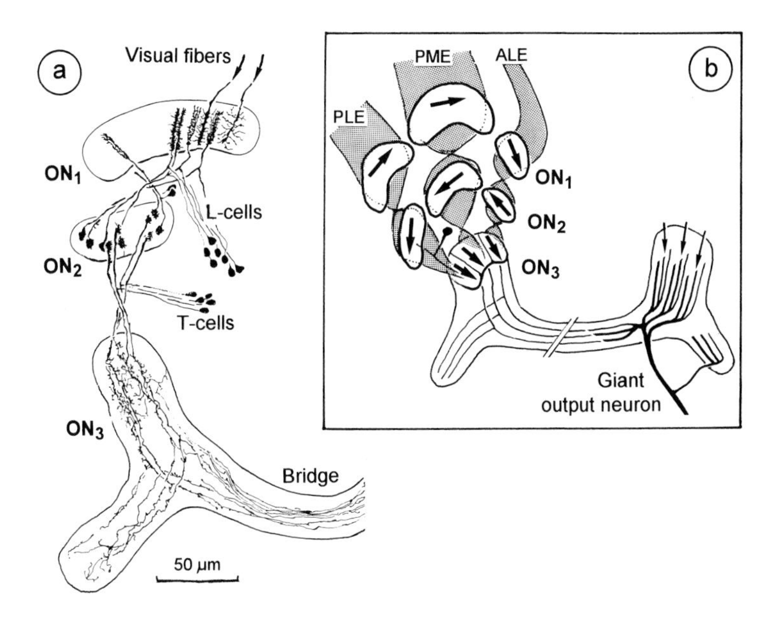

FIG. 87. Organization of the optic neuropils (ON_{1-3}) of the secondary eyes (*Cupiennius*). (*a*) The visual fibers from a PME terminate in a ON_1. L-cells pass the visual information from ON_1 to ON_2, and T-cells further from ON_2 to ON_3. Note the crossing of fibers (*chiasmata*) between optic neuropils. (*b*) Diagram of visual pathways of secondary eyes. The projections change their direction but seem to maintain a certain order while passing from one neuropil to the next. The horizontal direction (arrows in ON_1) of the visual world becomes represented as three parallel strips in ON_3. T-cell axons project across the bridge onto giant output neurons, which lead to thoracic muscles. (After Strausfeld and Barth, 1993.)

mechanoreceptors (tactile hairs, trichobothria, slit sensilla), and the other half processes information of various other sensory organs. Knowing the intersegmental connections between the sensory system and the motor system allows us to explain certain aspects of spider behavior, e.g., the inhibition observed between neighboring legs (Hergenröder and Barth, 1983b; see below).

The dorsal neuropil is traversed by six pairs of longitudinal tracts that enter the supraesophageal ganglion on the anterior side; they are supposed to consist mainly of motor fibers (Babu and Barth, 1984). Additionally, five pairs of sensory tracts pass through the medial part of the neuropil. There are also well developed cross-commissures, connecting the left and the right side of the subesophageal ganglion (Babu et al., 1985).

Neurosecretion (Stomatogastric System)

As their name implies, neurosecretory cells exhibit characteristics of both nerve cells and gland cells for they have axons and produce secretions. Functionally these cells are a link between the nervous system and the hormonal system. The activity of neurosecretory cells in spiders is probably correlated within

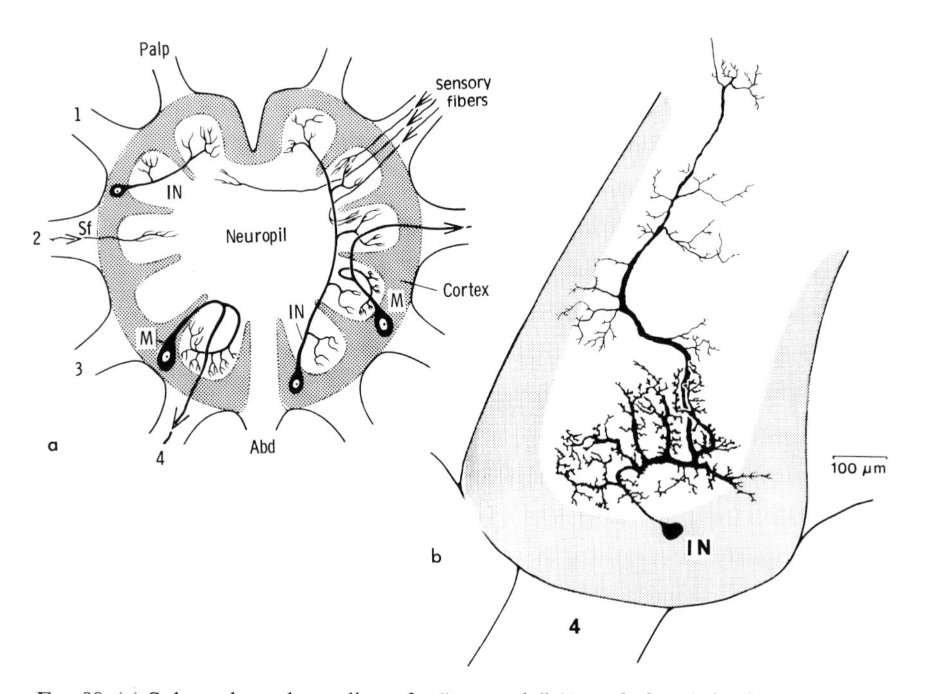

Fig. 88. (*a*) Subesophageal ganglion of a "tarantula" (*Poecilotheria*), horizontal section. Simple connections are pictured on the left: an interneuron (IN) connecting the pedipalp ganglion with the ganglion of leg 1; an afferent fiber (Sf) coming from leg 2; a motoneuron (M) exiting into leg 4. Connections of greater complexity are shown on the right: an interneuron (IN) sending branches to all extremities; a motoneuron (M) of the third leg ganglion, the axon of which exits into the second leg; three sensory fibers from the first leg, which either remain ipsilateral (near leg 2) or continue contralaterally to the palpal ganglion. Abd = abdominal nerve. (After Babu, 1969.) (*b*) A local interneuron in the fourth ganglion of *Cupiennius*, which transmits tactile input from leg 4 to a motoneuron. The fine dendritic branches were made visible by injecting a dye intracellularly. (After Milde and Seyfarth, 1988.)

certain events in their life cycle such as molting or sexual maturation. (That hormones control development in insects has long been established; the respective endocrine organs are the corpora cardiaca and the corpora allata.)

The neuroendocrine system of spiders consists of several groups of neurosecretory cells in the CNS and several "retrocerebral" organs which act as reservoirs or even as production sites of neurosecretions (Gabe, 1955). Clusters of neurosecretory cells are present in the supra- as well as in the subesophageal ganglion (Fig. 89; Juberthie, 1983). In the subesophageal ganglion a serial (metameric) distribution of these cells is evident. Neurosecretory cells can be identified histologically by specific stains such as paraldehyde–fuchsine or chrome–hematoxylin–phloxine. Behind the brain ("retrocerebral") lie several neuroendocrine organs: *Schneider's organs* (also known as stomatogastric ganglia or sympathetic ganglia; Legendre, 1953, 1985), the *Tropfenkomplex*, the

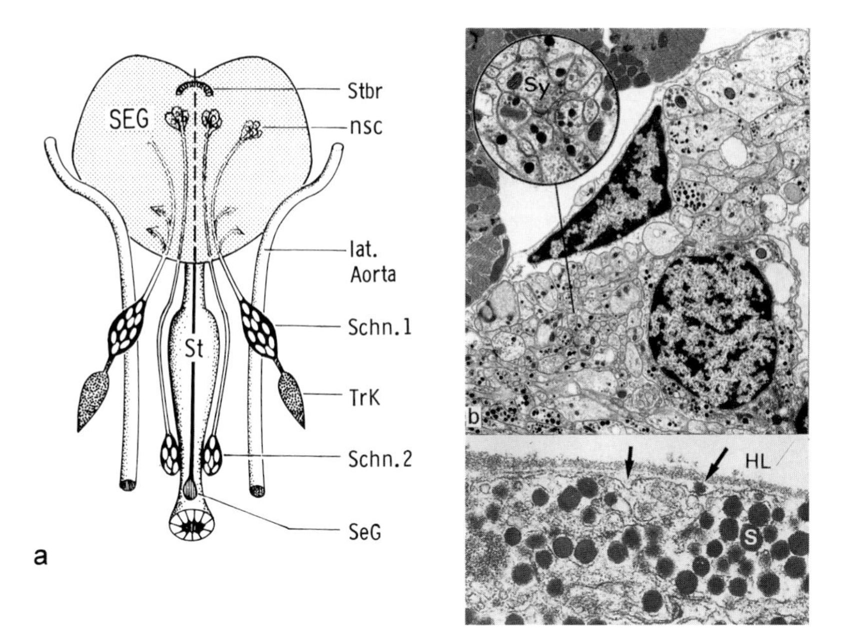

FIG. 89. Neuroendocrine and stomatogastric systems. (*a*) Dorsal view of the supraesophageal ganglion (SEG). Stbr = stomadeal bridge, nsc = neurosecretory cells, Schn. 1, 2 = Schneider's organs 1 and 2, TrK = Tropfenkomplex, SeG = esophageal ganglion with nervus recurrens, St = sucking stomach. (After various authors.) (*b*) Detail from Schneider's organ 1 of *Tegenaria.* Many nerve fibers possess dark neurosecretory granules and also make synaptic contacts (Sy). 3,800 ×. *Inset*: 15,000 ×. (*c*) The actual release (arrows) of neurosecretory droplets (S) into the hemolymph (HL) is pictured here at high magnification. 26,000 ×. (Photo: Bonaric.)

stomadeal bridge, and the *esophageal ganglion* with the nervus recurrens. The arrangement of this complex is shown in Fig. 89a.

Schneider's organs 1 lie directly behind the supraesophageal ganglion and are continuous with the Tropfenkomplex, where the neurosecretions are released into the hemolymph. Schneider's organs 2 are situated along the lateral walls of the sucking stomach, close to the aorta. The "stomatogastric system" consists of the stomadeal bridge, the anterior rostral nerve, and the posterior nervus recurrens. The nervus recurrens terminates in the small esophageal ganglion, which has not yet been shown to be a source of neurosecretion (Kühne, 1959). Further endocrine storage organs (*paraganglionic plates*) have been described in the subesophageal ganglion and in the posterior part of the brain (Bonaric, 1980).

The specific functions of the neuroendocrine organs are largely unknown. Definite neurosecretory cells appear only shortly before the spider reaches maturity, and it is believed that maturation of the reproductive cells is closely related to neurosecretion (Babu, 1973). Whether or not the molting process is controlled by neurosecretions is still very uncertain (Kühne, 1959; Streble, 1966; Eckert, 1967).

The CNS and Behavior

We want to examine here a few examples of the ways that a spider's CNS works to integrate a variety of sensory information, enabling the spider to respond "appropriately" to environmental stimuli, and, especially, to orient itself. External sensory input as well as proprioceptive input, can be stored in the memory and be recalled later. Actually, the process of learning may occur within seconds: a single run out of the retreat is often sufficient for the sheet-web spider *Agelena* to return to it in a direct course (Bartels, 1929; Görner, 1958). Such "engrams" can be formed either through visual or kinesthetic input. Usually the behavior of a spider is the result of the integration of several kinds of sensory input.

When *Agelena* catches prey on its sheet web and then carries it back to its retreat, it uses visual and kinesthetic cues to make the return run. Additionally, the inclination of the web (gravity) and the pattern of elasticity of the web may play a role in the spider's navigation (Bartels and Baltzer, 1928; Baltzer, 1930; Holzapfel, 1933, 1934; Görner, 1966, 1973, 1988). Even when all external cues are eliminated, *Agelena* can still return to its retreat in a straight line; in this case the spider's orientation has to be purely kinesthetic (idiothetic). If one designs an experiment in which visual and kinesthetic cues contradict each other, the spider reacts with a compromise. For instance, if the position of a light source is shifted immediately after prey is captured, *Agelena* initially turns in the correct direction, toward the retreat. While running back, however, the spider gradually drifts off to one side in an attempt to correct its course according to the new position of the light (Fig. 90a) This deviation from the correct course shows that a comparison must take place in the CNS between the course as judged visually and that based on kinesthetic cues. Indeed, later

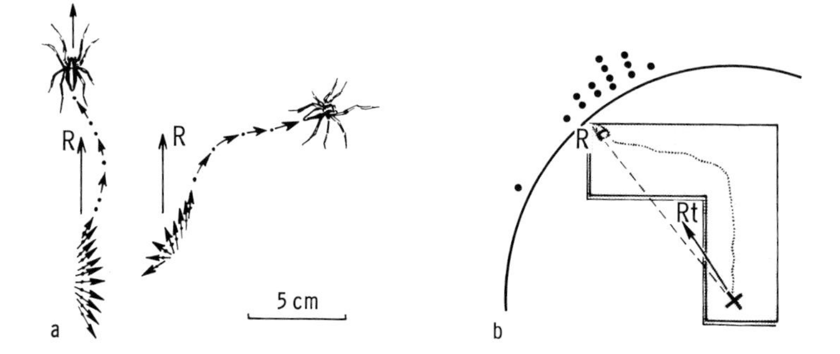

FIG. 90. Direction of return runs to the retreat by *Agelena*. (*a*) On the left, a control run, on the right, a run after the light source had been displaced clockwise by 110°. The spider is influenced by the visual cue; shifting the light causes it to deviate markedly from the straight course of return (R). (After Görner, 1972.) (*b*) Choice of starting direction on return runs of *Agelena* in a right-angle frame under polarized light. The spider was lured to prey (X) via a detour (stippled line). The black dots give the initial direction of each individual return run. Note that the average starting direction of all the returns (Rt) is very close to the ideal return route XR (dashed line). (After Görner, 1958.)

investigations showed that visual and kinesthetic information seem to enter a common storage site, where they are integrated to give a compromise solution to the problem of direction (Görner, 1966; Dornfeldt, 1972). Apparently one of the two means of orientation can also be "switched off:" some spiders orient themselves purely kinesthetically despite existing visual cues (Moller, 1970).

The ability of spiders to evaluate sensory information to alter their behavior has been seen in many experiments. For instance, *Agelena* returns straight back to its retreat even if it had been forced to make a detour on the way out (Fig. 90b). This means that during the run out onto the web the spider processes visual and kinesthetic information in the CNS and then "calculates" the shortest return. Besides the ability to orient themselves, *Agelena* and *Cupiennius* can also estimate distance (Seyfarth and Barth, 1972; Dornfeldt, 1975). This is evident when the spider searches for lost prey: at first the spider returns in a straight line to the capture site; when it nears the site it changes its course into irregular "searching" loops. The same kind of searching movements are performed by female wolf spiders, if their cocoon has been taken away (Görner and Zeppenfeld, 1980).

Integration of information by the CNS also occurs in the visual system of salticids (Land, 1971). A jumping spider adjusts the long axis of its body to be perpendicular to the path of a moving object so that the object can be fixed with the main eyes. The turn is triggered by input from the secondary eyes and is executed without any visual feedback. Thus when an object appears 80° to the left, the information coming from the retinas is sent to the legs as the command "turn 80° to the left, then stop." Each retinal area receives a slightly different image and therefore conveys slightly different information to the leg muscles. The CNS has to transform the information about the position of an object provided by the secondary eyes into a complex motor pattern that results in a precise and "precalculated" turn. When a moving object enters the visual field of the secondary eyes, a jumping spider will not always react by turning towards it. Sometimes the spider simply sits there and shows no apparent reaction at all. However, a closer examination reveals that the muscles of the main eyes are active, although the legs do not move. Apparently the CNS receives the stimulus from the secondary eyes, but transmits the information only to the main eyes and not to the leg muscles (Land, 1971).

The processing of vibration stimuli could be analyzed indirectly in behavioral experiments on the wandering spider *Cupiennius* (Hergenröder and Barth, 1983a,b). Apparently, information from the front legs is more important than that coming from the hind legs. Furthermore, a neural network seems to exist that is graded in the same direction: if a front leg is stimulated, then an inhibitory effect is observed on the posterior legs—not only on the same side (*ipsilateral*) but also on the other side (*contralateral*). It could also be shown that input from the trichobothria (air vibration) and from the slit sensilla (substrate vibration) is processed together in the CNS. If both systems signal "prey," the stimuli are added and attack behavior is elicited. However, if the trichobothria signal "no prey," they will have an inhibitory effect and any reaction is suppressed. One can eliminate the trichobothrial input by plucking off all trichobothria with fine forceps. This will leave only the slit sensilla as a source of vibrational input and, as a result, the spider will turn toward the prey.

To attribute specific functions to specific areas of the brain is not yet possible. Short exposures of the supraesophageal ganglion to a laser beam have produced several irreversible changes in the web pattern of Araneus, but such experiments have not led to a mapping of the centers responsible for web building (Witt, 1969).

We can speak of *learning* as the alteration of behavior by experience. Although spiders are better known for their stereotypical behavior, many experiments have demonstrated their capacity to learn and their behavioral plasticity. In one such experiment, dead flies that had been soaked in either a 6% glucose solution (sweet) or a 0.5% quinine solution (bitter) were thrown into the orb webs of *Araneus diadematus*. The spiders were then lured to the quinine flies with a tuning fork of the frequency of middle C and to the glucose flies with a fork of the frequency C_1 (one octave above C). Glucose flies were always bitten and eaten, but the quinine flies were rejected. If the flies were replaced by glass beads and touched with a C_1-tuning fork, the spiders still tried to bite into them. In contrast, glass beads vibrating at the frequency C were rejected, indicating that the spider associated the frequency C with the bitter taste of the quinine flies (Bays, 1962).

Young jumping spiders (*Phidippus*) seem to learn to avoid ants. At first, inexperienced spiderlings tackle ants but are quickly repelled by the ant's defence (bites, stings, formic acid). During later encounters with ants, the spiders always back away, probably because they remember their "bad experiences" (Edwards and Jackson, 1994).

Another experiment showing a spider's ability to learn from experience involved the araneid *Zygiella*. When prey hits the web, the spider will rush down along the signal thread; at the hub it chooses the radial thread that leads to the prey. When returning to its retreat it uses exactly the same path, in reverse. If the entire web is turned upside down, and the spider is then lured onto the web by a fly or tuning fork, the spider can still quickly reach the source of vibration. However, it has great difficulties on its way back. At first the spider runs in the original direction, which leads it to the periphery of the web. Only after some searching and several periods of rest does the spider find its way back to the retreat. If this experiment is repeated many times, the spider returns much faster (LeGuelte, 1969). Such improvement depends on the number of trials and also on the amount of time that elapses between trials. Definite improvement occurs only if trials are separated by less than 4 hours, suggesting that spiders have something akin to a "short-term memory." Long-term effects can also be demonstrated. Spiders tested after 24 hours do not return to the retreat quickly on their first attempt, yet in subsequent trials they learn the way faster than inexperienced control spiders do.

A clear example of memory in spiders is provided by web spiders, which can relocate prey previously captured. If a spider is busy feeding on a fly when a second fly gets stuck in the web, she leaves the first prey at the hub and rushes to the new one. After wrapping it up, she will leave it at the capture site and return to the first fly. If that fly has been removed in the meantime, the spider makes typical searching movements. The same is true if the "storage prey" is taken out of the web (Grünbaum, 1927; H. M. Peters, 1951). In both cases the spider seems to remember the existence and location of a previously captured fly. Jumping spiders can even remember uncaptured prey. They will often make detours while pursuing prey they have spotted at a site (for example, on a leaf) that is not directly accessible, but they retain a memory of the relative position of the prey at all times. This is expressed in the form of quick turns that reorient the spider to face the expected position of the prey (Hill, 1979; Tarsitano and Jackson, 1992).

From these descriptions it is clear that spiders are not simply reflex machines; they can react quite differently to the same stimulus. Their behavioral plasticity was highlighted when two *Araneus diadematus* built their orb webs under condition of weightlessness in the space capsule Skylab (Witt et al., 1977). This is particularly remarkable because gravity is normally an essential factor in the orientation of *Araneus* (H. M. Peters, 1932).

5

Spider Webs

> The evolution of spider silk has been an event comparable in importance to the evolution of flight in the insects, or warmbloodedness in the vertebrates.
>
> M. R. Gray, 1978

The most characteristic feature of spiders is their ability to produce silken threads. Although certain insects (for example, some lepidoptera, hymenoptera, and neuroptera) are also capable of silk production, this ability is usually restricted to a single stage in their life span, such as building a cocoon prior to pupation. In contrast, *all* spiders possess spinning glands, which they use not only for making egg sacs (cocoons) and draglines but also for building snares.

Spider Silk

Silk is the secretory product of the spinning glands. Although different spinning glands produce different kinds of silk, all silks are proteinaceous and belong to the fibroins. The molecular weight of the fibroin from the orb weaver *Nephila* has been determined to be 30,000 (Braunitzer and Wolff, 1955). This value refers to liquid silk inside the spinning gland; the molecular weight of a solidified silk thread is 200,000–300,000, or about ten times higher. The transition from the water-soluble liquid form into the insoluble, solid silk thread is not yet fully understood, but apparently the molecules of the polypeptide chain change their orientation from an α-configuration that is soluble (with predominantly intramolecular hydrogen bonds) to the insoluble β-configuration (β-pleated sheet form with intermolecular bonds). From the silk moth it is known that the liquid α-fibroin can be transformed into the β-configuration simply by pulling it.

A closer look at the spider silk reveals that each thread is actually a composite of α-chains and β-pleated sheets (Fig. 91). These sheets are stacked up in an accordionlike fashion, thereby forming protein crystals; they are embedded into a matrix of loosely arranged amino acids (Vollrath, 1992). The protein crystals are responsible for the strength of a silk thread, while the amorphous matrix has rubberlike properties (Gosline et al., 1984, 1986; Xu and

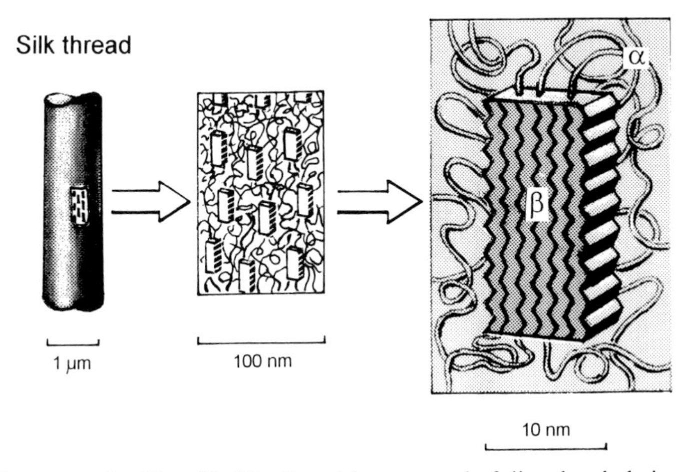

FIG. 91. Structure of spider silk. The thread is composed of disordered chains of amino acids (α-configuration) and ordered crystals of amino acids (β-configuration). The crystalline regions give the silk thread its stiffness, whereas elasticity is provided by the loose α-helices. (After Vollrath, 1992.)

Lewis, 1990). How strong or how elastic a thread is depends also on its water content (Work, 1977; Work and Morosoff, 1982; Edmonds and Vollrath, 1992). A dry silk thread is rather stiff and breaks if extended beyond 30% of its length. Wet threads are highly viscoelastic and can be stretched more than 300% before snapping. In many spider webs we find dry, stiff threads for the basic framework and moist elastic threads for catching prey (Tillinghast et al., 1984; Tillinghast, 1987).

The tenacity of spider silk is slightly less than that of Nylon (Lucas, 1964), yet its elasticity is twice as high (31% vs. 16%). Compared to many other natural and technical materials, spider silk is quite remarkable (Fig. 92). In terms of tensile strength it is clearly superior to bone, tendon or cellulose, and it is still half as strong as the best steel. Another vivid illustration would be to

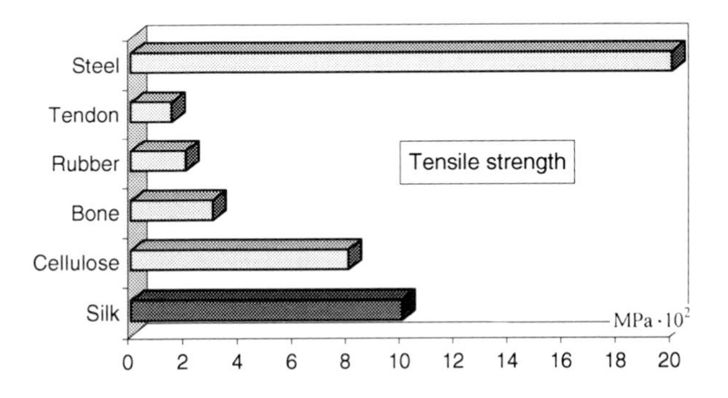

FIG. 92. Tensile strength is a measure of the greatest longitudinal stress a material will take before tearing apart. It is evident that spider silk is stronger than many other natural products and about half as strong as steel. (After data in Vollrath, 1993.)

TABLE 2. Amino-acid Composition of the Silk of *Araneus diadematus*. (After Witt et al., 1968.) The Amino Acids Arginine, Aspartic, Isoleucine, Lysine, Phenylalanine, Threonine, Tyrosine, and Valine Make Up Only 1–6% of the Total Fibroins.

Amino Acid	*Total Web*	*Frame* (Ampullate gl.)	*Cocoon* (Tubular gl.)	*Attachment Disc* (Piriform gl.)
Alanine	27%	33%	25%	29%
Glycine	20%	24%	12%	25%
Serine	5%	6%	19%	5%
Glutamic	9%	18%	14%	15%
Proline	13%	2%	4%	5%

calculate the so-called breaking length: a dragline would have to be 80 km long before rupturing under its own weight.

Spider silk is a highly unusual protein in its amino-acid composition (Table 2). Amino acids with short side-chains (alanine, glycine, serine) constitute 50–60% of the total fibroin (Peakall, 1969; Work and Young, 1987; Lombardi and Kaplan, 1990). It seems that formation and disruption of the α-helical alanine-rich regions account for the elasticity of the silk thread (Dong et al., 1991). Recently, biochemical investigations have shown that the dragline silk (*Nephila*) is actually composed of multiple proteins and first attempts of analyzing their genes are underway (Hinman and Lewis, 1992).

It is quite remarkable that the silk proteins of the spider's web are largely conserved during successive web building (Peakall, 1971). Before building a new web, the spider (*Araneus*) eats its old web. When the old web is labeled radioactively with [^{3}H]alanine, 80–90% of the initial radioactivity appears in the new web. Since this occurs after only 30 minutes, either the silk of the old web has been incompletely broken down by digestive enzymes, or the amino acids are so rapidly absorbed by the glandular epithelium that little is diverted into making body proteins. In any case, the old silk is apparently quickly recycled and hardly any body proteins are needed for its synthesis.

Silk Glands

All silk glands are located inside the abdomen. Each gland leads to a specific spinneret, which opens to the outside in the form of a tiny spigot. As one might expect, the most simple spinning glands occur among the ancient spiders (Orthognata), which often exhibit only a single gland type (Palmer et al., 1982; Palmer, 1985). In contrast, at least four different kinds of silk glands are found in wandering spiders and even more (7–8) in the highly developed orb weavers (H. M. Peters, 1955; Mullen, 1969; Richter, 1970). The following glands can be distinguished morphologically and histologically: ampullate glands, aciniform glands, tubuliform glands, aggregate glands, piriform glands, and flagelliform (or coronate) glands. Each type of gland secretes a different kind of silk with its own specific characteristics (H. M. Peters, 1982). Figure 93 surveys the

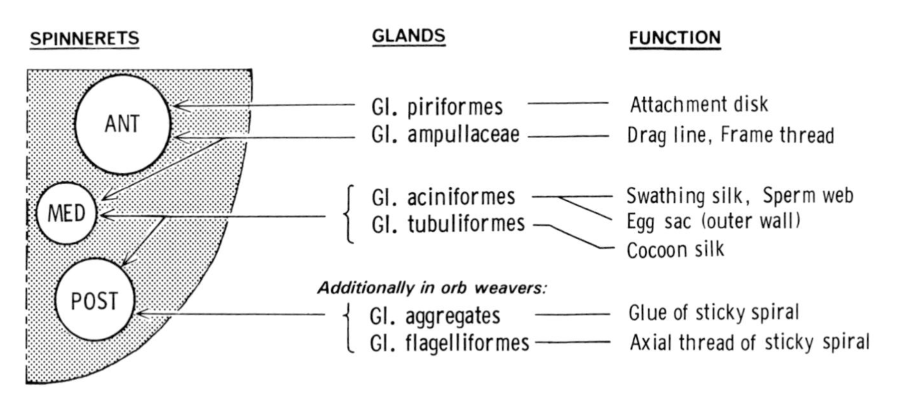

FIG. 93. Location and function of the various silk glands.

location as well as the function of the various silk glands. The tubuliform glands provide the silk thread for the egg sac and are thus found only in female spiders. The aggregate glands are typical of the araneids; they produce the glue substance for the web's catching spiral. The entire set of spinning glands of a tropical orb weaver is depicted in Fig. 94.

Structure

Each spinning gland comprises the gland proper and a thin duct. Most glands are pear-shaped and always consist of a single layer of epithelial cells (Fig. 95).

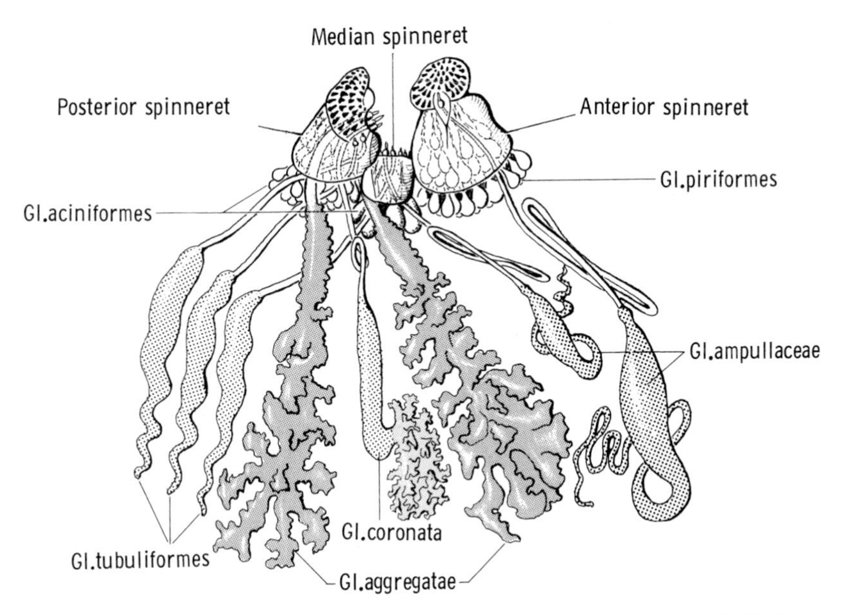

FIG. 94. Spinning apparatus of the golden silk spider *Nephila*. Only one half of the paired set of silk glands is depicted. (After H. M. Peters, 1955.)

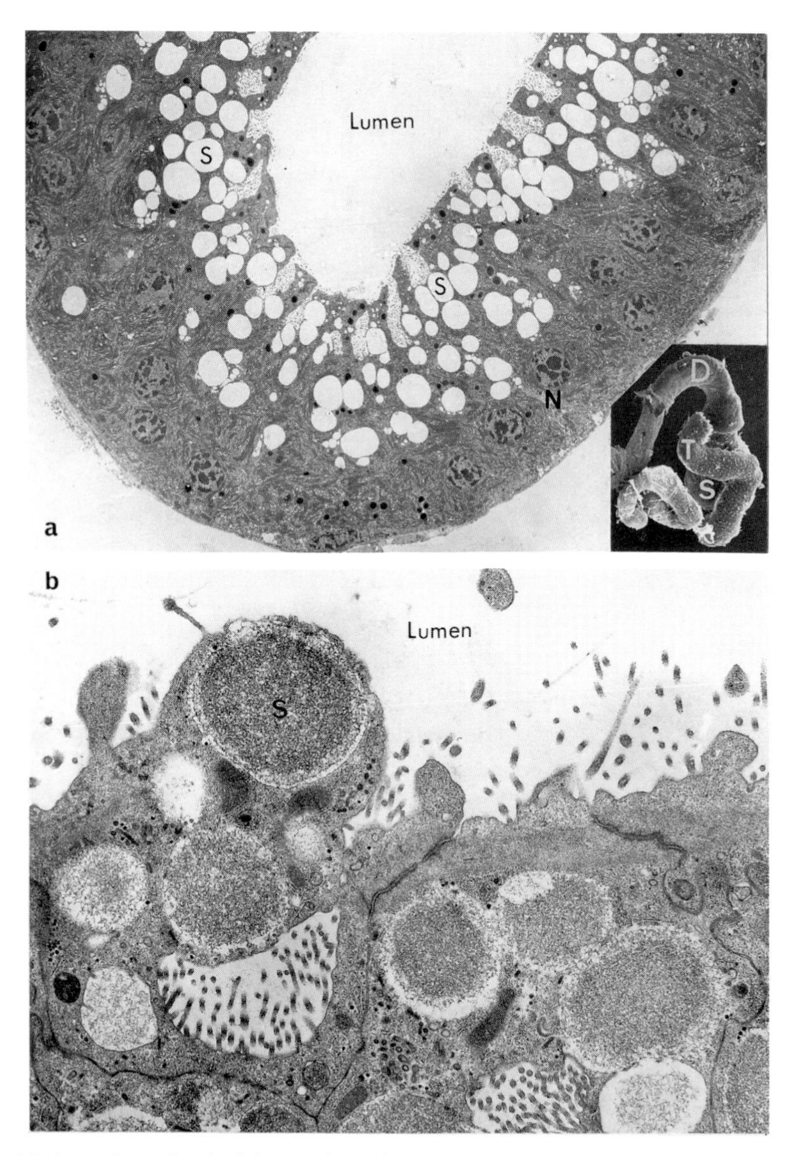

FIG. 95. Ampullate gland of the garden spider *Nuctenea cornuta*. (*a*) *Inset*: the gland consists of a coiled tail (T), a sacklike midpiece (S), and a looped duct (D). 70 ×. The glandular epithelium of the "sac" exhibits basally located nuclei (N) and apically concentrated secretory droplets (S). 1,300 ×. (*b*) Secretory droplets (S) just prior to discharge into the glandular lumen. 7,600 ×.

Silk is secreted into the glandular lumen in the form of tiny protein droplets and can be stored in the sacklike middle part of the gland. The extrusion via the glandular duct is most likely caused by an increase of the spider's abdominal hemolymph pressure and sometimes also by the active extraction of silk from the spigots with the hind legs. The transition from liquid silk into a solid thread

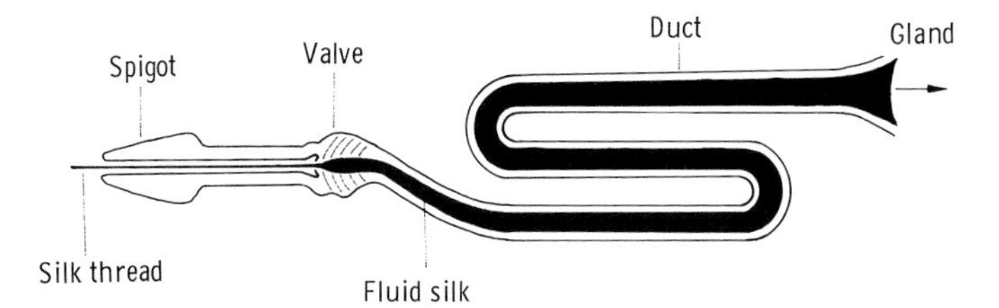

FIG. 96. Diagrammatic representation of dragline formation in *Araneus*. If the valve is almost closed, as shown here, the transition of liquid to solidified silk takes place directly behind the valve. When the valve is in the "open" position, the lumen of the spigot is filled with liquid silk and a solid thread is first formed at the tip of the spigot. (After R. S. Wilson, 1962b.)

is irreversible. It has nothing to do with exposure to air but is simply brought about by tension that orients molecules of the fibroin so that they are parallel to each other.

Unlike poison glands, spinning glands do not possess any musculature for expelling their secretions. In orb-web spiders, though, the amount of silk leaving the gland can be controlled by valves situated just in front of the spigots (Fig. 96). The diameter of the valves can be regulated by muscular action, thereby varying the thickness of the thread (R. S. Wilson, 1962a,b). Most silk threads have a diameter of only a few micrometers (Fig. 97), and many are even thinner than 1 μm. The dragline of orb-web spiders is less than 2 μm thick (Denny, 1976; Brandwood, 1985); its diameter depends on the weight of the spider, i.e., it is larger in heavy spiders.

A well-known example of extremely fine silk threads is found in the catching silk ("hackle band") of cribellate spiders, where the individual fibers measure only 0.01–0.02 μm (Fig. 99c).

Physiology

The most thoroughly investigated silk glands are the ampullate glands (*gl. ampullaceae*); they will therefore serve as a representative example for the other types of glands. The ampullate gland consists of a tubular tail portion, a sacklike midpiece, and a thin looped duct (Figs 94, 95). Experiments using radioactive marker substances have shown that the synthesis of silk is restricted mostly to the tail portion, while the "sac" functions mainly as a reservoir (Peakall, 1969). The secretory cells are typical columnar epithelial cells with basally located nuclei and apically concentrated secretory granules (Fig. 95a). The granules consist of protein and are a product of the endoplasmatic reticulum; Golgi bodies are lacking in these gland cells (Bell and Peakall, 1969). The secretory granules move toward the apical cell membrane and empty into the glandular lumen (Fig. 95b). Resorption of water presumably takes place within the lumen of the long duct (Kovoor and Zylberberg, 1980).

The silk thread that exits from the ampullate gland serves as a dragline. In many spiders such a dragline is continually left behind as the spider walks

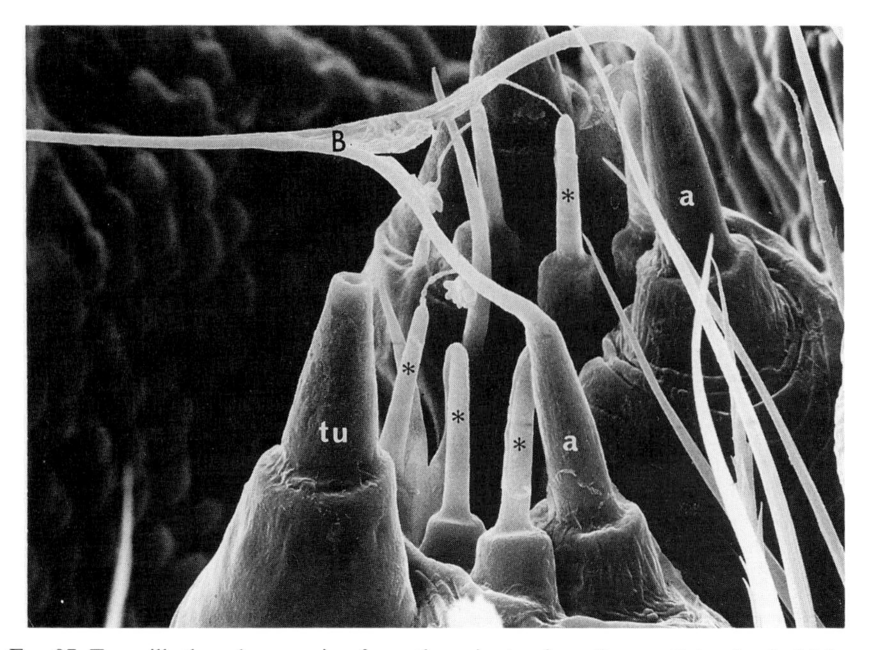

FIG. 97. Two silk threads emerging from the spigots of small ampullate glands (a) fuse to form a "bridging thread" (B). A large spigot of a tubuliform gland (tu) and several smaller spigots of aciniform glands (*) are located on the same spinneret (*Linyphia triangularis*). 1,900 ×. (Photo: Peters and Hüttemann.)

around. It is anchored at intervals to the substrate by the action of the piriform glands. This results in distinct "attachment discs" (Fig. 98), which are barely perceptible as tiny white dots. Aside from draglines the ampullate glands also provide the material for the frame threads of orb webs and for the gossamer threads of young spiderlings.

When the reservoir of a silk gland becomes depleted, new synthesis of fibroin starts within a few minutes in the tail portion of the gland. Many spinning glands seem to have two different cell types which contribute chemically different secretory products (Kovoor, 1972, 1987). While one type secretes only fibroin, the other apparently synthesizes mucopolysaccharides. which may contribute to the hygroscopic properties of the sticky spiral of orb webs.

These sticky capture threads have been closely investigated over the past years (Tillinghast, 1981, 1984; Vollrath and Tillinghast, 1991; Vollrath et al., 1990). In short, the viscous coating has not only an adhesive function but is also a water reservoir to maintain the high elasticity of the catching thread. When this thread is produced, the two core fibers coming out of the ampullate gland spigots are immediately covered by a viscous secretion of the neighboring aggregate glands. This coating is hygroscopic, i.e., it takes up water from the atmosphere, and quickly separates into small droplets, which sit astride on the core fibers (Fig. 109). A chemical analysis of those droplets has shown glycoproteins, covered by an aqueous solution of organic substances (mostly

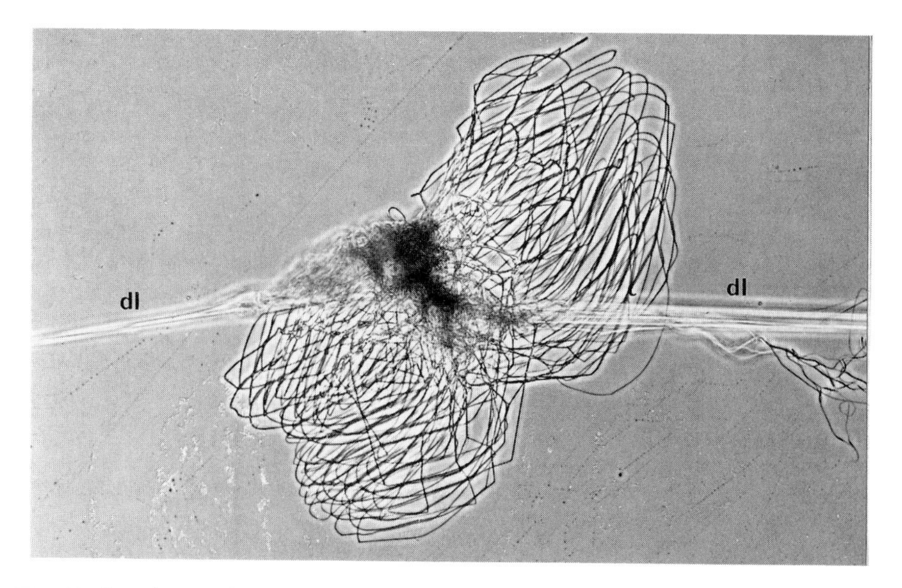

FIG. 98. Attachment disc of the crab spider *Xysticus* on a glass slide. The horizontal dragline (dl) can be clearly distinguished from the many delicate threads of the attachment disc itself. The symmetrical arrangement of these threads is due to their origin from the paired anterior spinnerets. 180×.

amino acids) and inorganic salts (KNO_3, KH_2PO_4). The function of the glycoproteins is to act as a glue, while the surrounding salt solution attracts water. When water molecules enter the core fibers, they become plasticized, giving the capture thread its enormous elasticity. It has been claimed that the sticky droplets can act as tiny windlasses, being able to reel in the core fibers and thus keeping the capture thread taut (Vollrath and Edmonds, 1989). However, this interpretation has been questioned recently (Schneider, 1995).

It is remarkable that old, abandoned spider webs are not attacked by fungi or bacteria. Earlier chemical investigations had shown that the sticky substance of the orb web's catching thread is acidic (pH 4) and hence resistant to decomposition by bacteria (Schildknecht et al., 1972). The axial threads of the web are apparently protected from bacteria by the hydrogen ions that arise from the dissociation of the phosphate. On the other hand, the acidity produced by the hydrogen ions would denature the proteins and the stickiness would be lost. Since this is not the case, the presence of nitrate ions probably prevents denaturation of the proteins.

Spinnerets

A spider has three pairs of spinnerets on its abdomen; phylogenetically they represent modified extremities (Shultz, 1987b). The spinning glands terminate in "spigots" on the surface of each spinneret (Figs 97, 99). All three pairs of

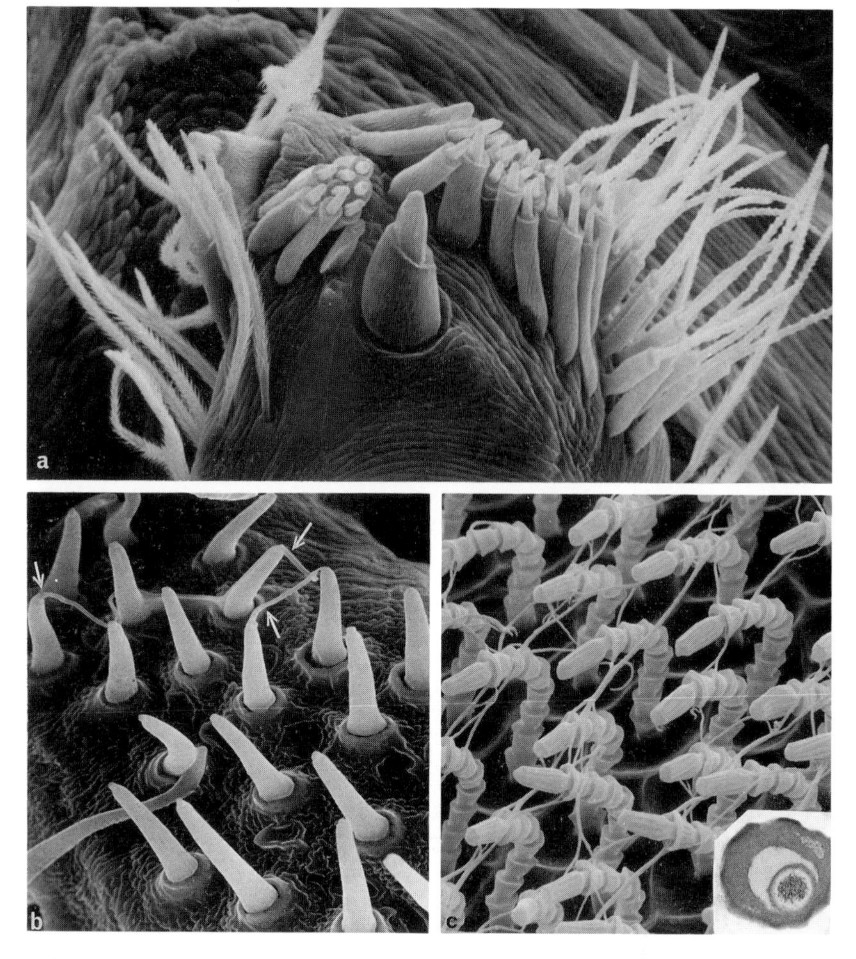

Fig. 99. (*a*) Median spinneret of the cribellate orb weaver *Uloborus* showing three different types of spigots. 570 ×. (*b*) Spigots of the piriform glands on the anterior spinneret of *Zygiella*. The emitted silk threads (arrows) are the basic elements of the attachment discs (see Fig. 98). 850 ×. (*c*) Tiny spigots from the cribellum of *Uloborus plumipes*. Note the pliable spigot shafts which exude the fine threads of the "catching wool". 8,500 ×. (Photo: Peters and Hüttemann.) *Inset*: Cross-section of a spigot from the cribellum of *Amaurobius*. The silk thread lies within the circular lumen and measures only 0.08 μm in diameter. 57,000 ×.

spinnerets—anterior, median, and posterior (Figs 100, 101)—are extremely mobile because they are equipped with a well-developed musculature. In some "tarantulas" the spinnerets can be seen moving in step with the legs, alternating as the spiders walk (Seyfarth, pers. comm.). Like the other extremities, the spinnerets have more flexors than extensors. If the muscles are attached at the joint membrane between two segments of a spinneret, then both segments are moved with respect to each other. Some muscles traverse the spinneret and can

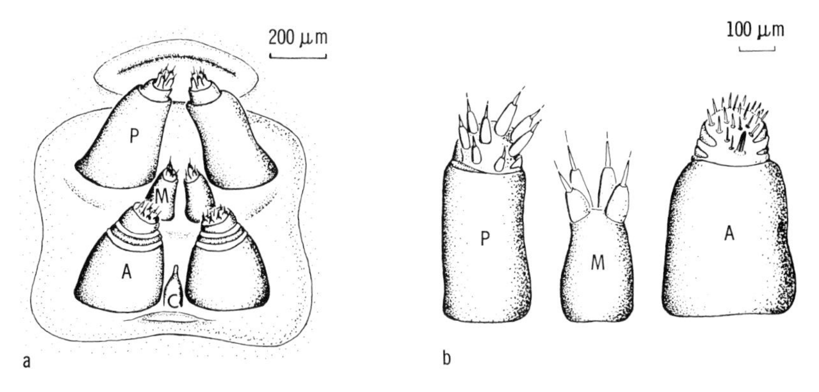

FIG. 100. Spinning apparatus of *Segestria senoculata.* (*a*) Arrangement of the three pairs of spinnerets and the colulus (C). (*b*) The anterior spinnerets (A) consist of three segments, the posterior (P) of two, and the median (M) of only a single segment. The different spigots belong to four different kinds of silk glands. (After Glatz, 1972.)

shift the entire spigot-studded terminal area (R. Peters, 1967). During the construction of a web or cocoon the spinnerets must move independently, yet must also be able to work together in a highly coordinated manner. The spinnerets can move in several ways: lifting, lowering, twisting, and also a synchronous spreading of all spinnerets can be observed. The effective working range of the spinnerets is enhanced to a great extent by the movements of the abdomen itself.

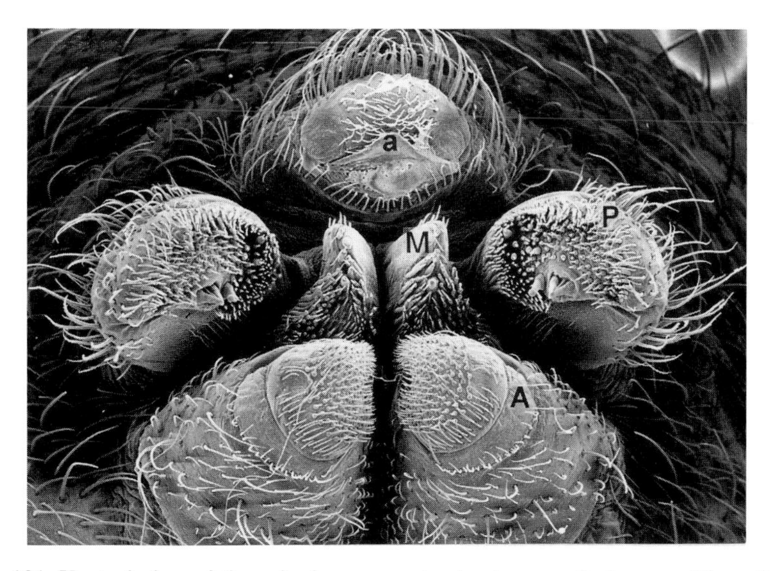

FIG. 101. Ventral view of the spinning apparatus in *Araneus diadematus.* More than 1,300 spigots of six different silk glands terminate on the three pairs of silk glands. A = anterior, M = median, P = posterior spinnerets, a = anus. 30 ×. (Photo: Peters and Hüttemann.)

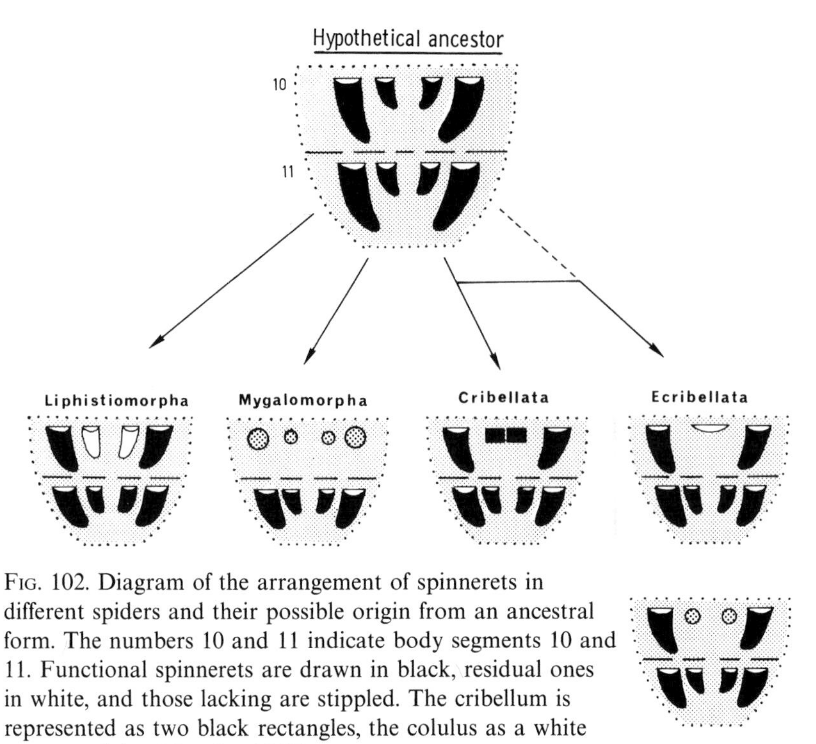

FIG. 102. Diagram of the arrangement of spinnerets in different spiders and their possible origin from an ancestral form. The numbers 10 and 11 indicate body segments 10 and 11. Functional spinnerets are drawn in black, residual ones in white, and those lacking are stippled. The cribellum is represented as two black rectangles, the colulus as a white crescent. (After Marples, 1967.)

Phylogenetically, spiders originally had four pairs of spinnerets (Fig. 102), but among extant spiders this ancient trait is retained only in some Mesothelae, that is, those spiders which also have other "primitive" characteristics. In the less "primitive" Orthognatha we find two to three pairs of spinnerets, and sometimes even a reduction to one pair (Glatz, 1972, 1973). For instance, in *Nemesia* the anterior spinnerets are lacking, the median pair is rudimentary, and only the posterior spinnerets carry functional spigots. Most extant spiders possess three pairs of functional spinnerets. They develop embryologically from the *anlagen*, or primordia, of the extremities of body segments 10 and 11. Each anlage splits in half so that two further pairs of spinnerets originate medially, resulting in eight spinnerets altogether. However, the anterior median pair (on body segment 10) is often extremely reduced, and many ecribellate spiders (such as Araneidae, Linyphiidae, Theridiidae, and Thomisidae) have only a vestigial bump, which is referred to as the *colulus*. This rudimentary structure is probably now without any function (Marples, 1967). In many spiders the colulus is absent altogether.

Cribellate spiders possess an additional spinning organ, the cribellum, a small plate located in front of the three pairs of spinnerets (Fig. 103a). It is assumed that the cribellum and the colulus are homologous organs, since both develop from the primordia of the anterior median spinnerets. The cribellum can take the form of a single small plate (as it does in *Hypochilus*), or it may

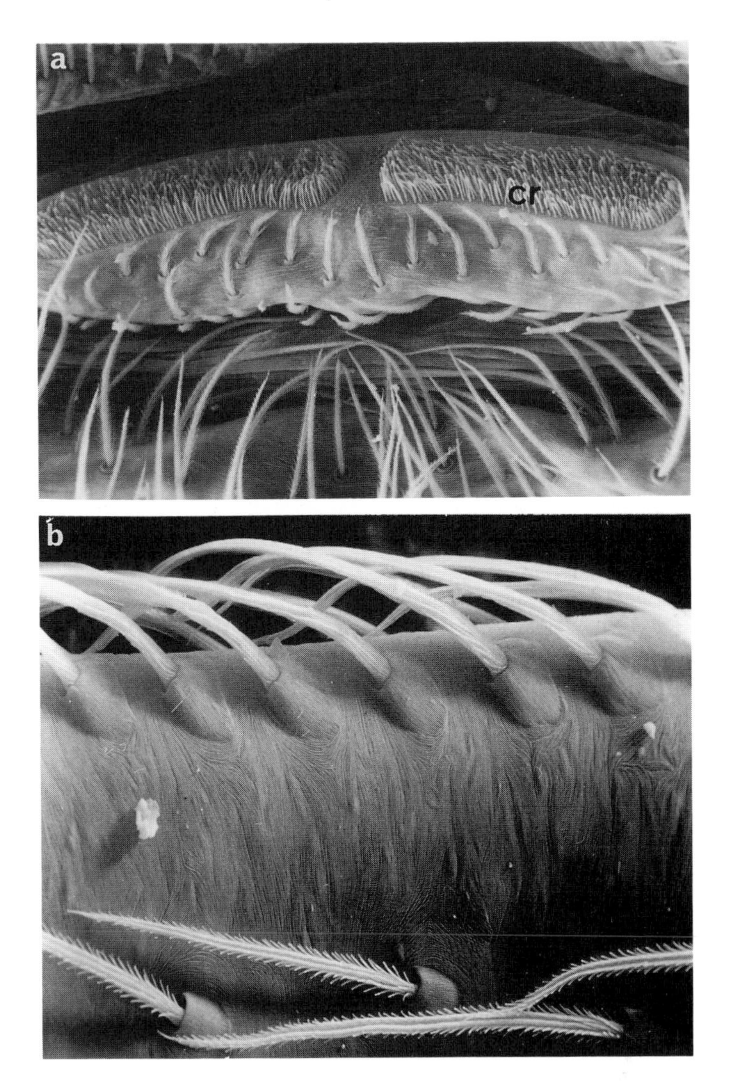

FIG. 103. Cribellum and calamistrum of *Amaurobius*. (*a*) The cribellum (cr) consists of a divided platelet, which is normally hidden in front of the spinnerets. 380×. (*b*) The calamistrum is formed by a row of strong bristles on the metatarsus of leg 4. Each bristle is equipped with fine cuticular teeth (see Fig. 61b), which comb out the cribellate silk from the cribellum. 600×.

be divided into two (as it is in *Amaurobius*), or, rarely, even into four parts (as in *Dresserus*). The cribellar area is densely covered with many tiny spigots (cf. Fig. 99c). In an adult *Stegodyphus pacificus* (Eresidae) the cribellum contains over 40,000 spigots (Kullmann, 1969); young spiders have fewer spigots, but the number increases with each molt. These delicate spigots produce the extremely thin silk threads (0.01 μm) of the "hackle band" (Lehmensick and Kullmann, 1956; Friedrich and Langer, 1969); they are combed out of the cribellum by rhythmic movements of the *calamistrum*, a row of comb-shaped hairs situated on the metatarsi of the fourth legs (Fig. 103b).

Webs

Ecologically one can divide spiders into two groups: the wandering spiders and the more sedentary web spiders. Only web spiders use their ability to produce silk for the construction of snares. The various web spiders have developed quite different webs and prey-catching strategies. The well-known orb-shaped web is one of the many kinds of webs built by different spider families. The funnel weavers (Agelenidae) and the sheet-web weavers (Linyphiidae) construct horizontal sheets, whereas the cobweb spiders (Theridiidae) and the daddy-longlegs spiders (Pholcidae) build somewhat irregular meshes (Kirchner, 1986; Eberhard, 1992). "Primitive" web builders, such as the cribellate *Amaurobius* or the ecribellate *Segestria*, weave only a tubular retreat with simple signal or catching threads radiating from its entrance. The greenish cribellate spider *Dictyna* usually lives on leaves on which it lays down its catching threads in a zigzag fashion on top of underlying radial supporting threads (Fig. 123).

The catching threads ("hackle bands") of cribellate spiders are especially interesting because they function without any gluey substance—unlike the sticky threads of ecribellate spiders. Cribellate catching threads work mainly because of their structure: one or two straight axial threads are surrounded by sinuous support threads and a dense network of cribellate silk (Fig. 104; H. M. Peters, 1992; Eberhard and Pereira, 1993). Prey is easily entangled in this "wool." More importantly, the cribellate silk has some adhesive properties, although these properties are not well understood. Because these threads do not contain any glue (which deteriorates quickly), they are effective much longer than are the sticky threads of ecribellate spiders. However, within a few weeks they also age, gradually losing their adhesiveness (Foelix and Jung, 1978). Cribellate silk, incidentally, is found not only in the rather simple snares of *Amaurobius* and *Filistata*, but also in the orb webs of *Uloborus* and in the reduced orb of *Hyptiotes*.

Sheet Webs and Frame Webs

Sheet webs are built by Agelenidae and by Linyphiidae, yet those sheets differ as much in their structural details as does the behavior of their respective inhabitants. The Agelenidae weave a flat, slightly concave silk mat with a funnel-shaped retreat at one end. There the spider hides and rushes out only when prey has blundered into the web. The webs of *Tegenaria* commonly bridge the corners of rooms in European houses; the corresponding agelenid in North America belongs to the genus Agelenopsis. The much smaller Agelena prefers the outdoors and is abundant in grass and low bushes. Its web consists of a sheet measuring, at most, 40–60 cm, suspended by oblique and vertical threads. Flying insects hit these vertical threads and drop onto the sheet.

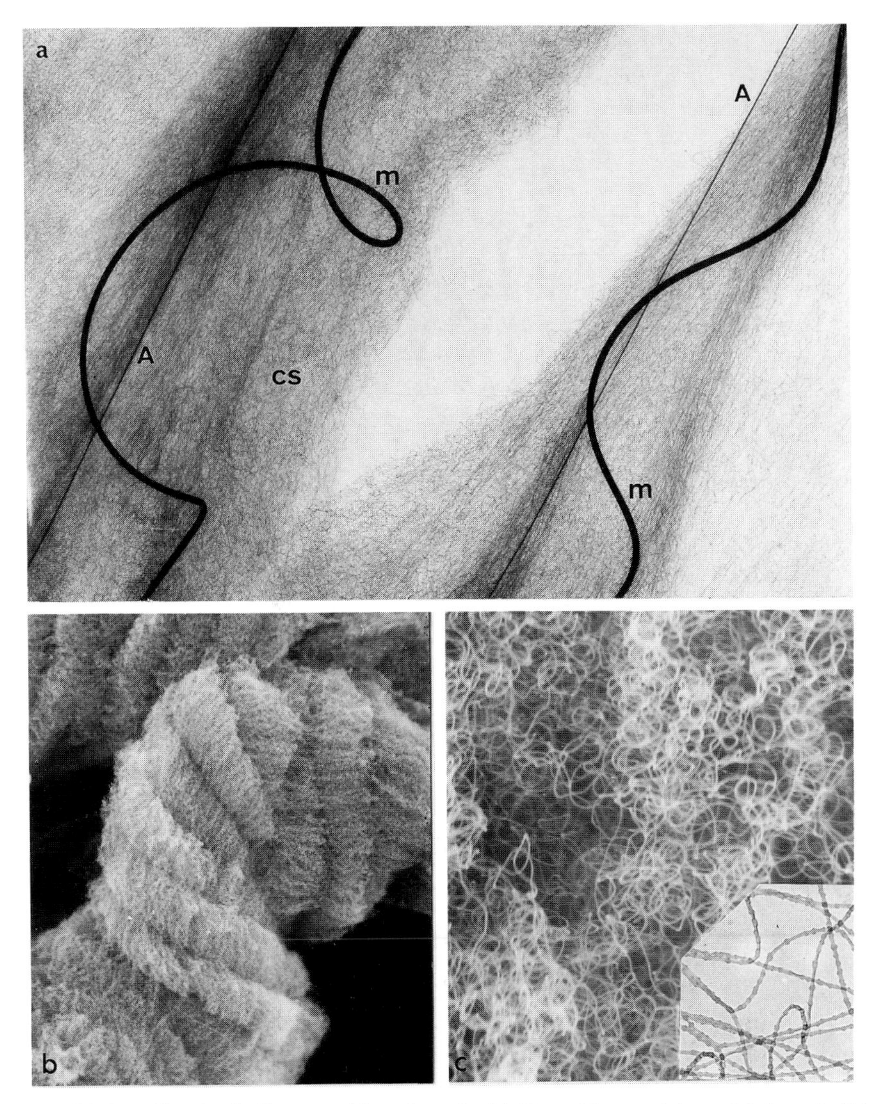

FIG. 104. Details of cribellate catching threads. (*a*) Two thin, straight axial threads (A) support two coiled threads (m) and the surrounding "catching wool" of cribellate silk (cs) (*Amaurobius*) 1,200 ×. (b) The cribellate silk is brushed in regular bundles onto the axial threads by means of the calamistrum (*Mallos*). 5,000 ×. (*c*) The extremely fine threads of cribellate silk measure only 0.015 μm in diameter. 8,000 ×. *Inset*: In some species these threads appear beaded (*Amaurobius*). 35,000 ×.

The very delicate webs of the linyphiids are also horizontal, but convex, sheets with similar vertical threads that serve as tripping lines for insects (Fig. 105). Some silk threads of the linyphiid web contain sticky droplets, but these play an insignificant role in catching prey. In most cases an insect simply becomes trapped between the vertical suspension threads. In contrast to agelenid spiders, linyphiids always hang beneath their dome. When prey is

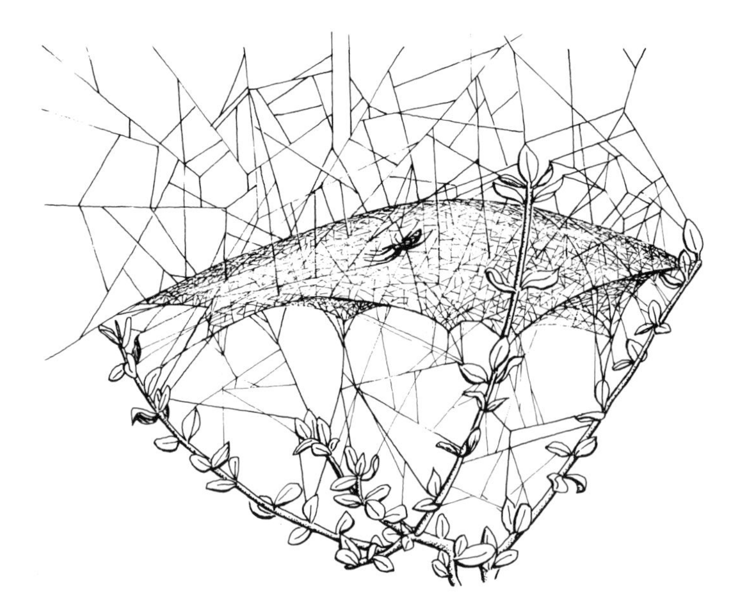

FIG. 105. Web of *Linyphia triangularis*. The spider hangs suspended upside down under the dome of the horizontal sheet web. (After Bristowe, 1958.)

trapped by the threads, the spider shakes the web until the victim falls onto the sheet, then bites through the web and pulls the victim down. The damage done to the web is mended after feeding.

To a certain extent the frame webs of the theridiids resemble those of linyphiids. The structure of the sheet, however, is very loose and irregular, and the trapping threads are studded with glue droplets (Fig. 106; Wiehle, 1949). These threads are attached tightly to the substrate and break easily when an insect (often an ant) touches them. The insect is then glued to the broken thread and suspended helplessly in the air. While trying to free itself, the animal contacts neighboring catching threads and becomes progressively more entangled. Meanwhile the alerted spider quickly climbs down and throws further sticky threads over the victim before biting it. Some theridiids build special retreats inside or outside the frame web. A well-known example is *Achaearanea saxatile*, which constructs an open silk tube that hangs vertically in the middle of the web. The spider camouflages the tube by adding plant and soil material to its surface. The South African *Achaearanea globispira* builds an even more elaborate retreat, which contains a helical silken tube (Fig. 106b; Henschel and Jocquet, 1994). A single thread suspends the retreat, while several glue-studded catching threads extend from the bottom of the retreat to the ground. This remarkable construction serves probably as protection from the sun as well as from predators (wasps and ants).

The more primitive webs, which consist only of a tubular retreat with radial catching threads, have already been mentioned, but one of the most extraordinary variations on such a "retreat" is the silken tube of the purse-web

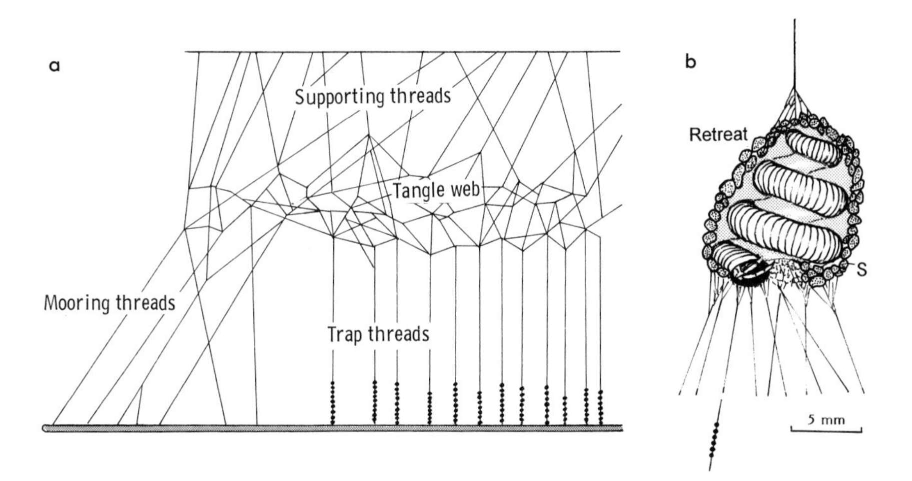

FIG. 106. (*a*) Diagram of the web of *Steatoda castanea* (Theridiidae). The vertical trapping threads are studded with glue droplets. (After Wiehle, 1949.) (*b*) The retreat of *Achaearanea globispira* consists of a silken case covered with sand grains (S) and a spiral silk tube inside that houses the spider. (After Henschel and Jocqué, 1994.)

spider *Atypus*. This catching device is about 15–30 cm long and 1 cm in diameter; most of it is buried vertically in the soil and only a short, fingerlike end extends horizontally on the ground. The spider lives inside the silken tube and attacks insects that happen to crawl over the tube (Fig. 107). The long chelicerae pierce through the walls of the tube into the victim, which is then dragged inside. The two slits cut into the wall of the tube by the action of the cheliceral teeth are quickly repaired before feeding.

Orb Webs

The orb web is certainly the best known of all webs. Essentially it is made up of three elements (Fig. 108): (1) *radial threads*, which converge in a central spot,

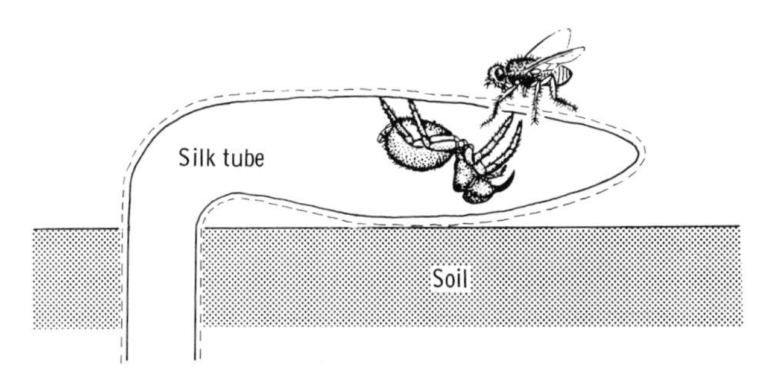

FIG. 107. The silken tube of the purse-web spider *Atypus*. The spider bites the prey through the wall of the silk tube and then pulls it inside. (After Bristowe, 1958.)

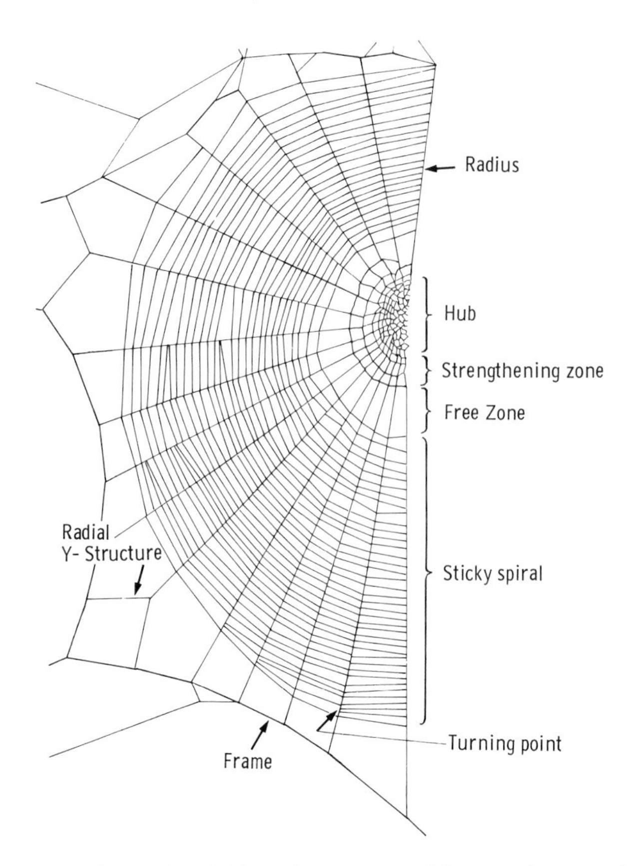

FIG. 108. Structure of an orb web. Drawing was traced from a photograph of a web of *Araneus diadematus*.

the hub; (2) *frame threads*, which delineate the web and serve as insertion sites for the radial threads; and (3) the *catching spiral* (Wiehle, 1927, 1928, 1929, 1931). The scaffolding of the orb web is provided by the radial and frame threads; neither type is sticky. The catching spiral, in contrast, consists of a thread studded with glue droplets (Fig. 109). The microscope reveals that there are in fact two axial threads (1–2 μm in thickness), which are embedded in a gluey substance. The coating is hygroscopic and this is why the catching thread is so elastic (see p. 110). If it is experimentally removed, then the catching spiral becomes as stiff and inelastic as a radial thread (Vollrath and Edmonds, 1989). The orb web as a whole is thus a combination of strong frame and radial threads with a very elastic catching spiral. Frame and radii provide a mechanically stabile construction, which is also well suited for signal transmission (courtship plucking, prey vibrations). The elastic capture thread is an adaptation to struggling prey, i.e., the impact and movements of an insect cause little damage, because the kinetic energy is largely absorbed by the yielding catching spiral (Denny, 1976; Lin and Vollrath, 1995, pers. comm.).

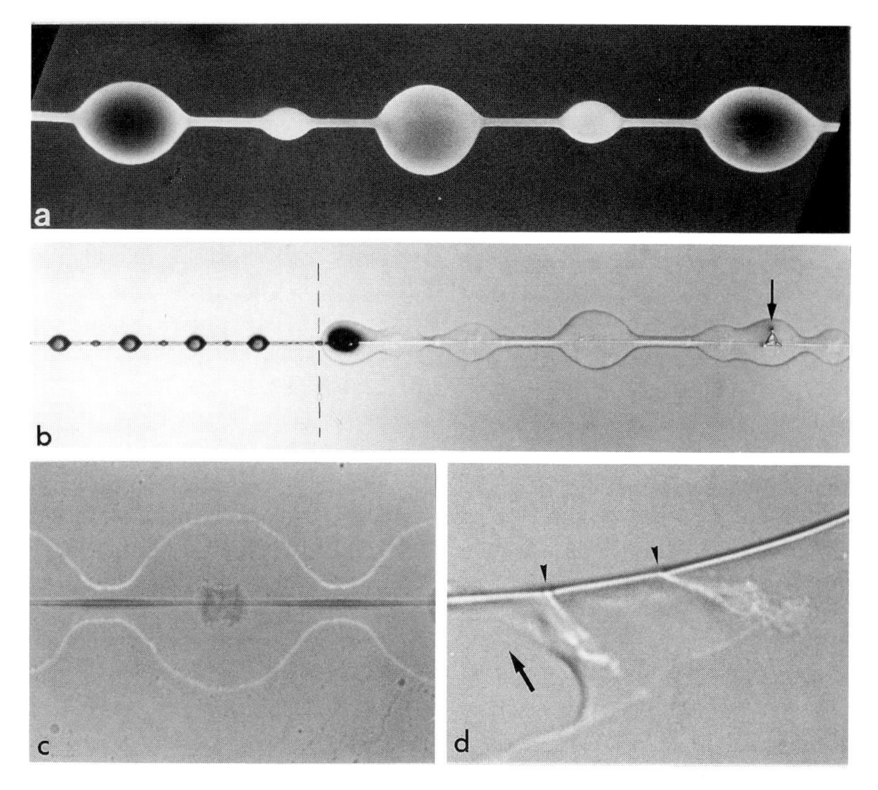

FIG. 109. Fine structure of the sticky spiral in *Araneus diadematus*. (*a*) The viscous coating forms regular droplets of 5–10 μm along the axial threads. 1,100 ×. (Photo: Foelix and Kaufmann.) (*b*) Sticky spiral thread suspended in the air on the left, touching a glass plate on the right. After making contact the droplets spread on the surface and a dense core becomes visible; some droplets exhibit triangular crystals (arrow). 220 ×. (*c*) Higher magnification shows that the dense core consists of a fibrous material (glycoproteins), that is attached to the two axial fibers. 520 ×. (*d*) The attachment points (arrow heads) can be seen when the axial threads are pulled sideways (arrow). Note that most of the core material adheres to the substrate. 300 ×.

The hub usually consists of irregularly interconnected threads. In some orb weavers (such as *Tetragnatha*) the hub is removed after its construction and the web then appears with a large hole in the center (an "open hub"); in other genera (such as *Argiope*) the hub may be covered with a fine sheet of silk. Immediately surrounding the hub is the so-called *strengthening zone*, but this part of the web cannot always be determined as separate from the hub itself. More important is the next part, the "*free zone*," which is crossed only by the radial threads. Here the spider can easily change from one face of the web to the other. The actual catching area is that part of the web covered by the sticky spiral.

Wherever two threads cross in the web, they form a solid connection. It is still a matter of debate, however, whether these connections represent a genuine fusion of two threads or whether another substance cements them together. Several studies have demonstrated that spiders do use a cement that can form either stiff or flexible connections (Eberhard, 1976; Kavanagh and Tillinghast, 1979). Altogether, an orb web contains 1,000–1,500 connection points, most of which occur between the radii and the spiral thread (Jackson, 1971).

The number of radii varies little within a particular species of orb weaver, and is often characteristic of that species. The web of the garden spider *Araneus diadematus*, for instance, usually has 25–30 radial threads forming fairly regular angles of 12–15°. Webs of very young *A. diadematus* often have many more radii than those of adults (Witt et al., 1972). In the horizontally inclined orbs of *Tetragnatha* one finds only few radii (about 18), whereas 50–60 are present in the webs of the small araneid *Mangora*. These numbers of radii imply that many orb weavers show a species-specific geometry in their webs. It is thus often possible to identify a certain spider solely by its characteristic web structure.

Zygiella, for instance, typically builds a web with a "free sector" that is crossed only by a sturdy signal thread (Fig. 110a); this thread connects the hub with a silken retreat outside the web. To observe how this free sector comes about is quite interesting. While laying down the sticky spiral, the spider always stops before reaching the signal thread and reverses its course from a clockwise to a counter clockwise direction, and vice versa. Thus the sticky spiral actually

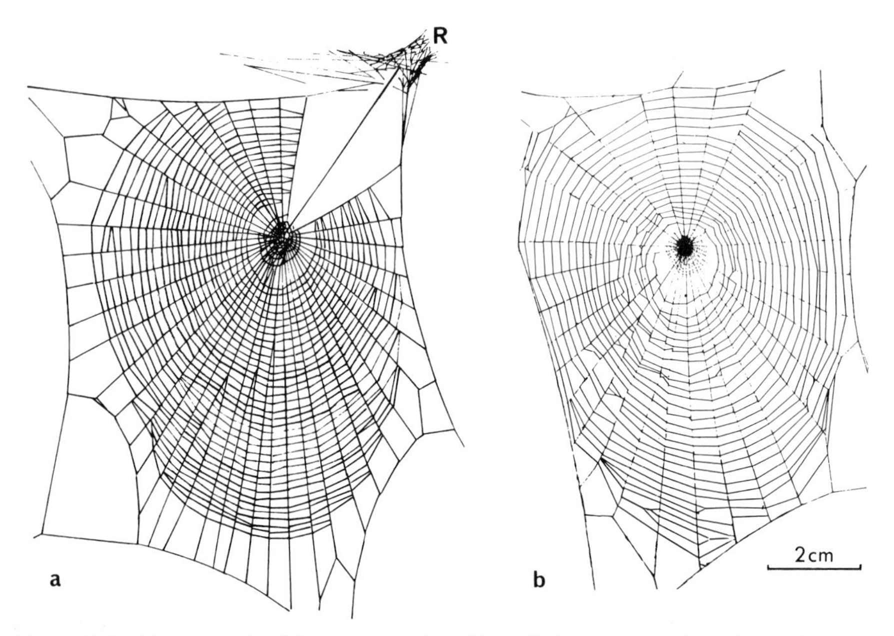

Fig. 110. (*a*) Typical web of the sector spider *Zygiella.* During daytime the spider hides in the retreat (R), but keeps contact with the hub by means of a signal thread (tracing of a photograph by Witt). (*b*) First web built by a cross spider (*Araneus diadematus*). (Photo: Witt.)

results from the spider's moving like a pendulum. During the day *Zygiella* hides in a retreat, keeping one leg touching the signal thread to detect any vibrations occurring in the web. Although *Zygiella* typically builds free-sector webs, closed orb webs are often built by the juveniles. Even adults occasionally produce orbs without the free sector; this happens if the signal thread deviates from the plane of the web by more than 40°. Signal threads leading to a safe retreat, a curled leaf, for instance, occur in many other araneid spider webs. Having such a signal thread means that the spider is less exposed to predators (such as birds or wasps), for it needs to be present on the web only when actually catching prey.

Another common structural variation in the orb web are the so-called *stabilimenta* (Figs 111, 146). These silken bands leading to the hub are added to neighboring radii in a zigzag manner; sometimes they are placed in a circular fashion around the hub. Whether these structures have any stabilizing effect, as their name would suggest, is dubious. More likely the stabilimenta serve as some kind of camouflage, because they are built only by those spiders that always sit in the hubs of their webs. *Cyclosa* is one of the best examples of such a spider. In its web the stabilimenta extend vertically from the hub to the catching area; additionally, they are adorned with remnants of prey and old shed skins (Rovner, 1976; Neet, 1990). The spider itself, sitting in the hub with its legs pressed tightly to its body, blends perfectly with the brownish

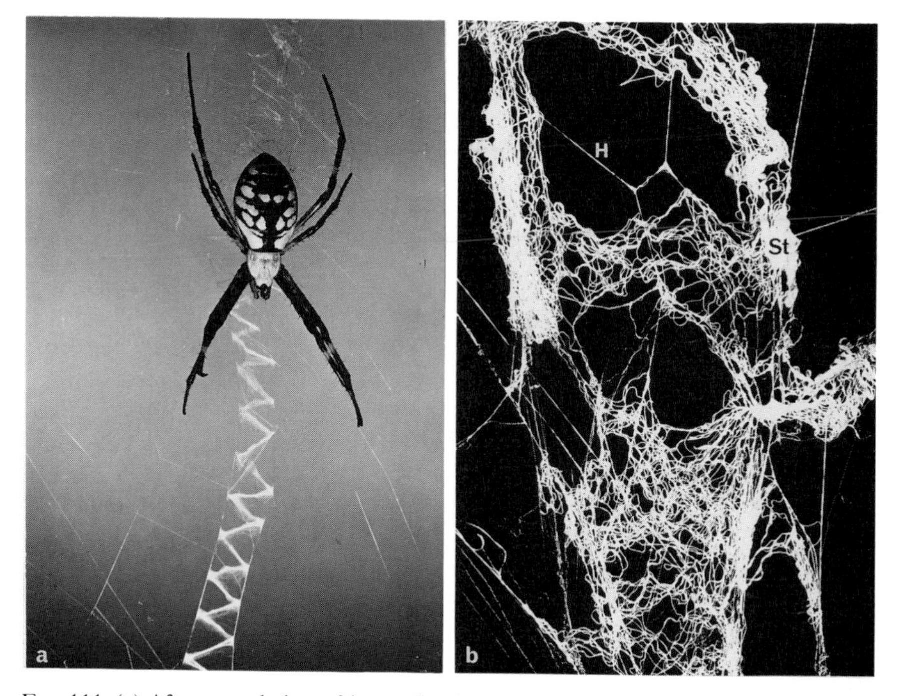

FIG. 111. (*a*) After completion of her web, *Argiope aurantia* adds a zigzag pattern, the *stabilimentum*, radiating from the hub. (*b*) Stabilimentum threads (St) decorating the hub (H) of a *Uloborus* web. The silk stems from aciniform and piriform glands.

stabilimentum. However, the stabilimenta in the webs of the brightly colored *Argiope* species hardly render the species less conspicuous. On the contrary, they may even have a signal function, e.g., to warn birds not to fly into their webs (Eisner and Nowicki, 1983; Nentwig and Heimer, 1987). Perhaps the most important function of stabilimenta is to attract insects. The silk of the stabilimentum reflects UV light and pollinating insects are positively lured to these threads (Craig and Bernard, 1990).

Quite probably, different functions must be attributed to the stabilimenta of different types of orb webs (Robinson and Robinson, 1973). Aside from camouflaging the spider (such as *Cyclosa* or *Uloborus*), stabilimenta may also serve as a molting platform (as for *Nephila*) or as a protective shield against heat radiation from the sun (Humphreys, 1992). Stabilimenta could also be merely deposits of surplus silk (Wiehle, 1928) and thereby might regulate silk production (H. M. Peters, 1993a). Argiopids and uloborids both use large amounts of silk while wrapping their prey; if too much silk has accumulated inside the glands, it could be deposited in form of stabilimenta.

The silk glands producing the stabilimenta were not known until quite recently. By dripping hot paraffin onto the spinnerets (H. M. Peters, 1982) while the spider (*Argiope*) was weaving its stabilimentum, fine silk threads could be preserved that were just exiting the spigots of the aciniform glands. In *Uloborus* the piriform glands are additionally involved (H. M. Peters, 1993a). Exactly the same glands also provide the threads for wrapping prey (*swathing silk*, Fig. 142), and this would support the "surplus theory" cited above. Other arachnologists see hardly any advantage for a stabilimentum and consider it as a "nonspecific reaction to any form of stress" (Nentwig and Rogg, 1988; Eberhard, 1990). In other words: Stabilimenta are still puzzling structures.

Most orb webs are built vertically, but Uloborus builds an orb web that is usually oriented in a horizontal plane. With the exception of cribellate catching silk it closely resembles the araneid orb web, even in details. This remarkable example of convergence is interesting from a phylogenetic viewpoint and will be discussed later.

All orb webs presented so far have been basically two dimensional. Several orb weavers, however, add further silk lines to the orb and thus make it three dimensional. Many tropical uloborids make quite elaborate variations, e.g., by adding a cone-shaped orb onto a regular orb web (Fig. 112; Lubin et al., 1982; Lubin, 1986). Another example is *Theridiosoma,* which builds a complete orb web but attaches a thread laterally to the hub. When this thread is tightened, the web bulges, resembling an inside-out umbrella (Fig. 113). The spider itself sits as a living bridge: its hind legs grasp the radii near the hub while its front legs pull on the tension thread. When an insect flies into the web, *Theridiosoma* at once releases its front legs so that the web jerks back into the two-dimensional plane and entangles the hapless victim.

A similar spring trap is built by the cribellate spider *Hyptiotes* (Uloboridae). Its triangular web is most likely a reduction from an orb (Fig. 114). Four radial threads, which enclose three sectors, converge into one mooring thread. Again, the spider forms a living link by holding the mooring thread tightly with its

FIG. 112. *Uloborus bispiralis* from New Guinea exhibits an orb plus cone construction. The spider sits well protected on the inside of this "cage web." (Photo: Lubin.)

front legs. When prey is encountered, the tension is released and the web collapses. Obviously, the web becomes so damaged during the capture that a new one has to be constructed after each catch.

Since there are more than 2,000 orb-weaver species, I cannot touch upon all the structural variations that orb webs may exhibit, but in many cases the differences are relatively minor and concern only the symmetry of the web. *Nephila,* for instance, builds a broad orb web in which the lower half is much larger than the upper. Consequently, the sticky spiral does not really spiral around the center but only crosses the lower half of the web in a "pendulum" fashion. A further characteristic of the *Nephila* web is that an auxiliary spiral persists in the finished web. *Nephila* also constructs a three-dimensional mesh above the hub, called a *barrier web.*

The rather different web of the "orb weaver" *Cyrtophora* also consists of a three-dimensional network but has a central tentlike construction (Fig. 115). This "canopy" resembles an orb web inasmuch as it also contains radial threads and a spiral thread. However, the number of radii is much higher (around 200 split radii), and the spiral is not sticky. Perhaps the *Cyrtophora* web represents an intermediate step between the sheet web of the linyphiids and the orb web of the araneids: the transition to the orb could have evolved by a reduction of the irregular network and a tilting of the central sheet (Kullmann, 1972a). However, the opposite opinion has also been expressed, namely, that the *Cyrtophora* web is derived from an orb web (Lubin, 1973). Indeed, this is more

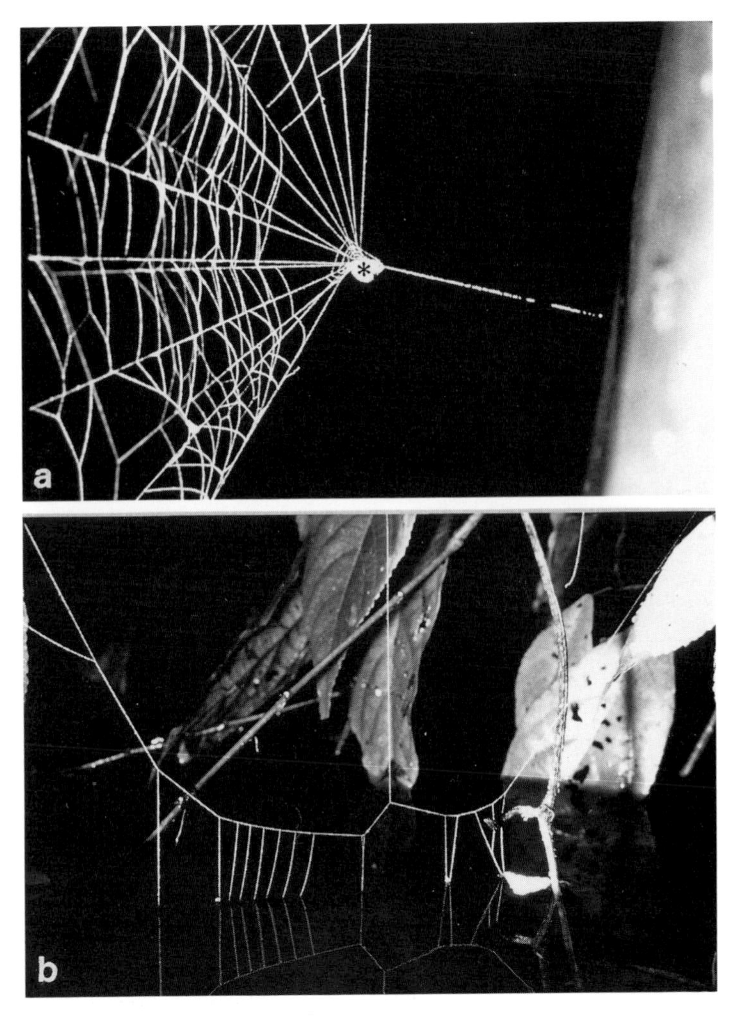

FIG. 113. (*a*) In the modified orb of *Theridiosoma* the hub of the web is pulled out by a tension thread and the spider (*) sits as a living link in between. (*b*) The theridiosomatid *Wendilgarda* builds a most unusual modification of a orb web, in which the sticky threads are attached to the water surface. (Photos: Coddington.)

likely, since *Cyrtophora* is morphologically (as determined by its genital structures and its spinning apparatus) a specialized araneid (H. M. Peters, 1993b).

Finally, an extreme variation of orb-web symmetry occurs in the ladder-web spider *Tylorida* from New Guinea (Robinson and Robinson, 1972). These webs are about 1 m high but only 15 cm in width. The hub is situated close to the top so that the spiral threads run almost horizontally—hence the name *ladder web*. A very similar web, yet turned 180° has been discovered in South America and Florida for *Scoloderus* (Eberhard, 1975; Stowe, 1986). Such elongated webs

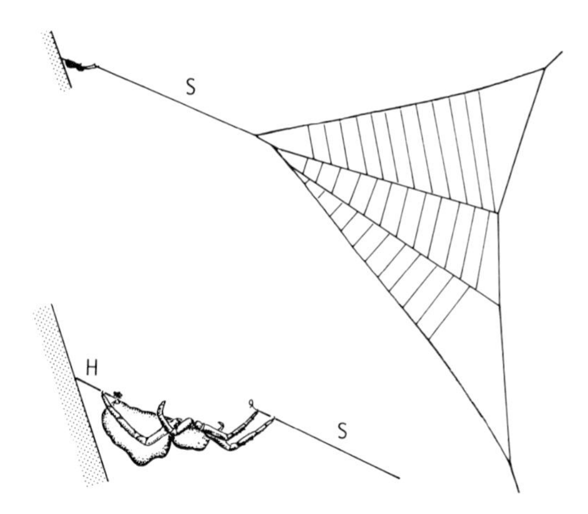

FIG. 114. The spring trap of *Hyptiotes*. The web consists of only three sectors of an orb. The spider itself forms a living bridge between a tension thread (S) and an attachment thread (H). (After Wiehle, 1927.)

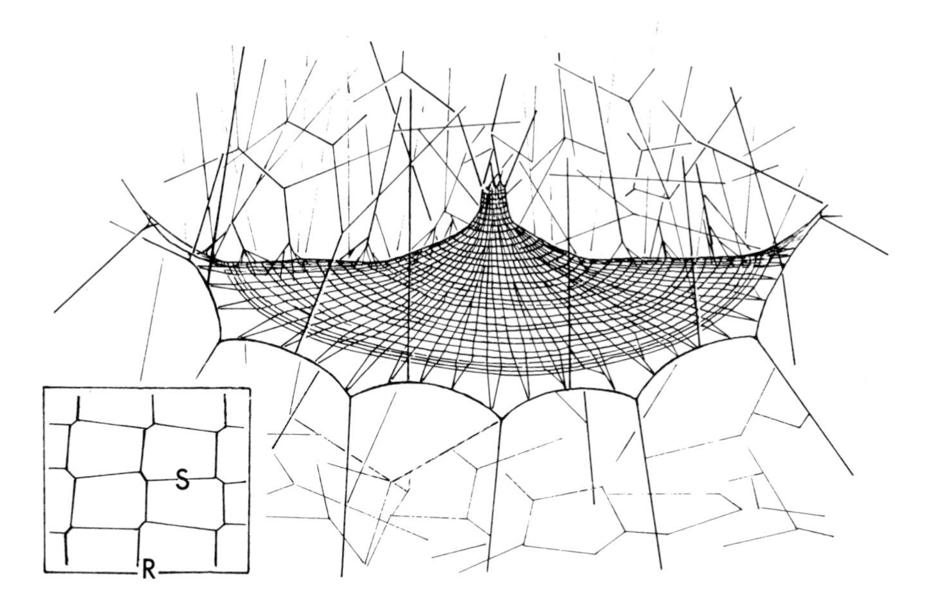

FIG. 115. Web of *Cyrtophora*. The horizontal sheet resembles an orb web; it is surrounded by an irregular mesh of threads above and below the sheet. The spider sits underneath the tentlike center of the sheet. (After Wiehle, 1927.) *Inset*: Fine structure of the horizontal "canopy" showing short fusion zones where radial (R) and spiral threads (S) cross each other. (After Kullmann and Stern, 1975.)

are most likely an adaptation for catching moths. Whereas moths can rather quickly escape from regular orb webs (see chapter 6), they slide down a ladder web; not only are they exposed longer to the attacking spider, but they also lose scales and thus become stuck more easily on the sticky threads.

Construction of an Orb Web

The building of an orb web follows three distinct phases: (1) the construction of the frame and radii; (2) the laying down of the auxiliary spiral; and (3) the establishment of the sticky (or catching) spiral. The following description of orb-web construction is based on the behavior of *Araneus* (H. M. Peters, 1939a,b).

Frame and Radial Threads The spider starts with a horizontal thread by simply exuding a silk thread into the air while sitting on an elevated point. This thread is carried away by wind or even the slightest draft until it gets caught on some object, a twig, perhaps. Actually, this horizontal thread is only a provisional structure; the spider bites through it (Fig. 116a, at X), grasps the cut thread with its front legs, and moves from point A toward point B. At the same time a new thread is reeled off behind the spider while the original thread AB is curled up by the front legs. About halfway (at M), the old and the new thread are connected. Since more silk is used for the new thread AM, the entire thread AMB sags a bit. At M the spider descends vertically until it reaches the surface to which it fastens the thread (Fig. 116b, at C). The resulting Y-structure is essentially the basic framework of the future orb, with M being the center.

The next step establishes additional radial threads. This is a rather complicated procedure, especially because the spider is constantly producing a new thread wherever it goes. The first radial threads are always made together with a frame thread, that is, in one move. The spider walks on the existing radius from M to B and fastens the new thread at B (Fig. 116c). There it turns around and uses the new thread MB as a guide line. Obviously, while walking back on MB, it is spinning a new thread again. After a short distance the spider attaches this thread to the new MB at the point D. The spider now keeps running via M toward the starting point A, where it attaches the thread (Fig. 116d). From A the spider returns to D, again doubling the thread AD (Fig. 116e). This produces the final frame thread AB and, at the same time, the radius DM, which is, however, replaced immediately, because the spider bites through at X (as it did with the first horizontal thread) while running back to the hub (M). The new thread XM becomes transformed into the final radius DM. The rather sharp buckle at D is somewhat compensated for by letting out more thread so that the new radius DM is longer than the previous one.

Following the same pattern, the next frame thread plus radial thread is executed (Fig. 116f). The basic scaffold of radial threads is then "filled up" with further radii. Again, these are produced twice; that is, an existing radius serves as a guideline while a new (temporary) radius is pulled behind as the spider walks from the center to the periphery. This radius is fastened only briefly to

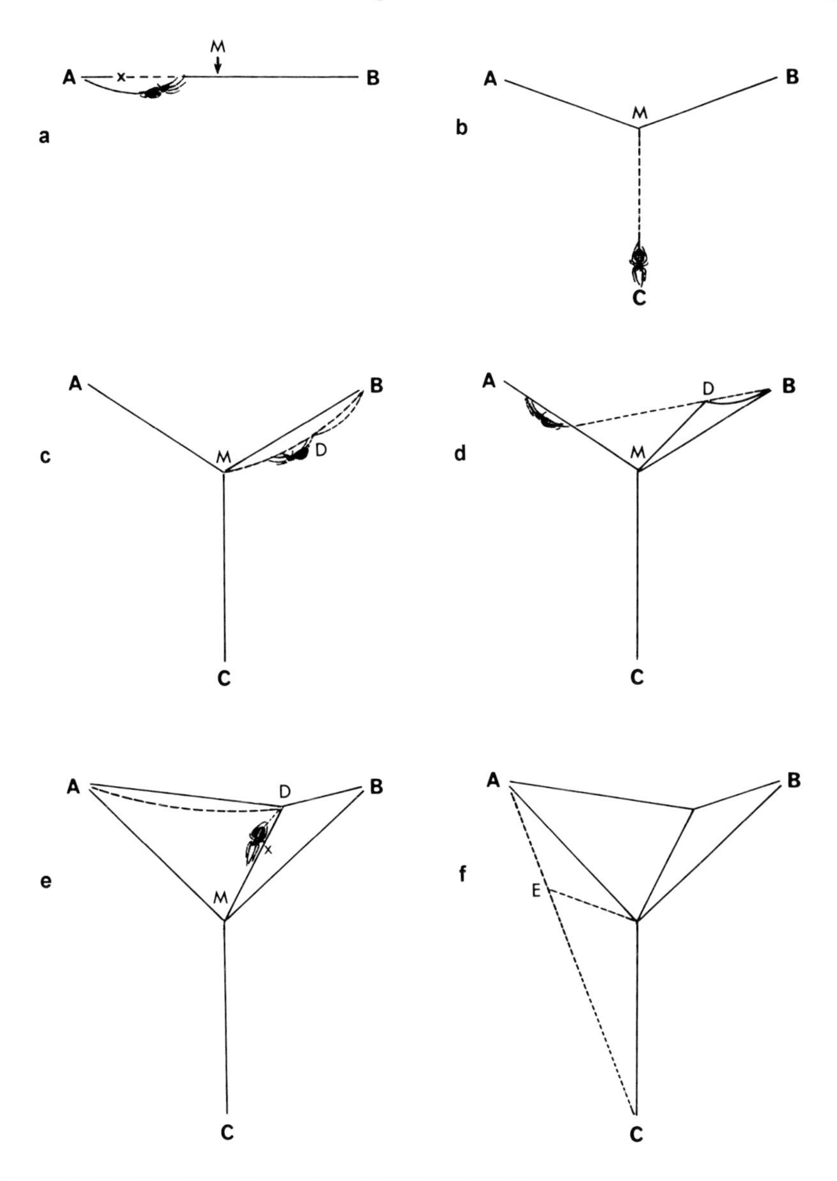

FIG. 116. Construction of an orb web. (After H. M. Peters, 1939a.)

the frame, then bitten through at the periphery and subsequently replaced while the spider returns on it to the hub.

The sequence of the placement of radial threads is by no means haphazard but follows two basic rules: (1) the direction of placement of a new radius alternates; and (2) the first radial threads, which fill the gaps between the original radii, are always constructed directly below an already existing (original) radius. This procedure of consecutive placement of radii becomes clearer in Fig. 117.

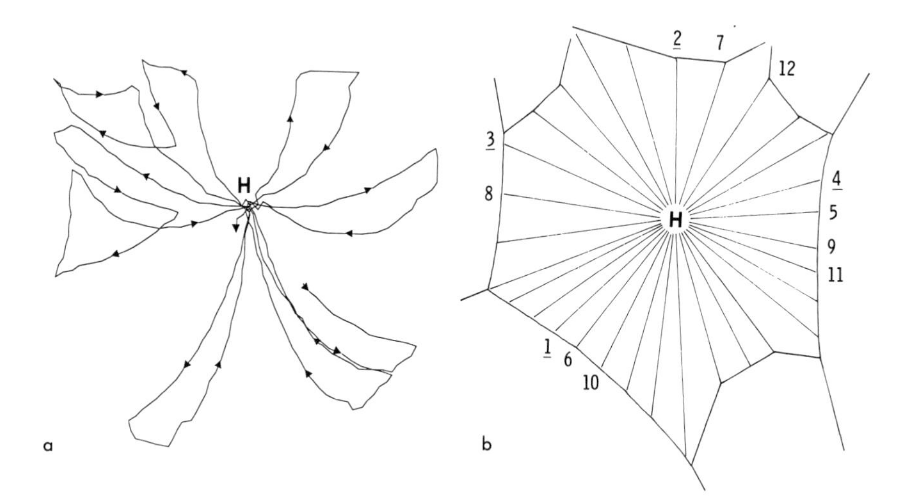

FIG. 117. (*a*) Laser-optical recording of the movements of a garden spider (*Araneus diadematus*) while building the first radial threads of an orb web. (After Zschokke, 1993). (*b*) Completion of frame and radii construction. The numbers give the sequence of placement of the first radial threads; note the opposite directions of radii 1–4. H = hub. (After H. M. Peters, 1939a.)

The spider always decides at the hub in which sector the next radius is to be placed. As soon as it has completed a radial thread and has arrived back at the hub, it briefly tugs on all the finished radii with its front legs. The observer gets the impression that the spider "measures" the angle between two neighboring radii. That an oversized angle would determine the placement of the next radial thread seems plausible, and various experiments, such as burning out radii while they were being laid down (Hingston, 1920), provided support for this hypothesis. However, other factors, like web tension, also seem to be involved.

Although the precise mechanism triggering the placement of radial threads is not yet known, it is clear that the front legs do have a measuring function (Reed et al., 1965). If the front legs are removed, the regularity of the angles between adjacent radial threads is impaired. Normally, the angles between the 20–30 radii of an orb web (for *Araneus diadematus*) are astonishingly constant, about 15°. It is true, though, that the upper half of the orb web contains fewer radii (and thus larger angles between neighboring threads) than the lower half.

Auxiliary Spiral While the spider is still busy building the radii, it is also interconnecting these threads by a few narrow circles in the center of the web. This hub region is then extended into the so-called strengthening zone by further peripheral turns. After construction of the radial threads has been completed, the strengthening zone is continued into a wide spiral toward the periphery, forming the *auxiliary spiral* (Fig. 118a). Although the auxiliary spiral has only

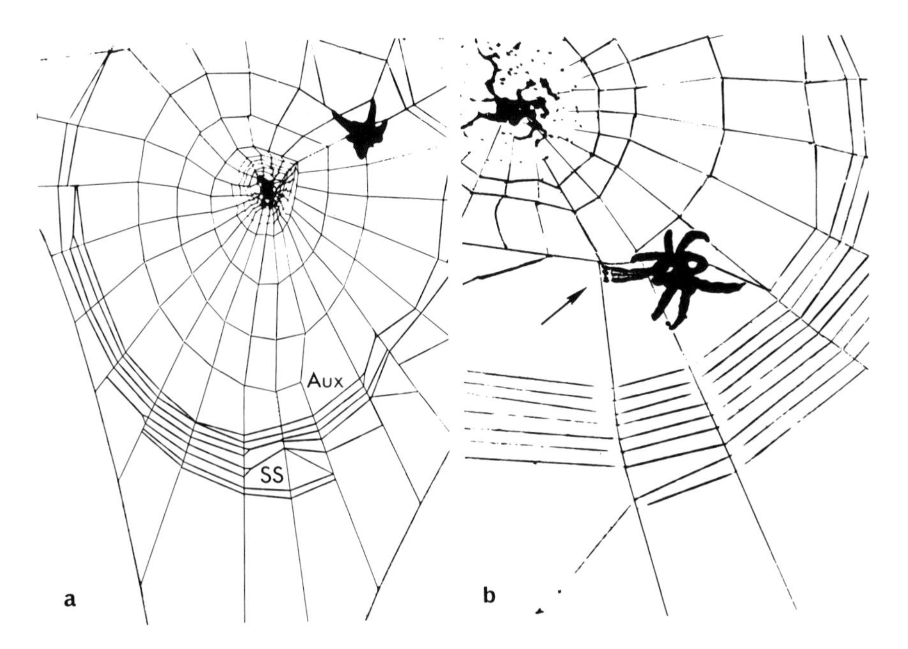

FIG. 118. (*a*) An orb web after completion of the auxiliary spiral (Aux). The first turns of the sticky spiral (SS) can be seen in the lower part of the web. (*b*) *Araneus diadematus* in the process of laying down the sticky spiral. Note that one front leg quivers briefly while touching the next radius (arrow). (Single frame from a movie by Witt.)

a few turns, it ties the radii together and helps stabilize the half-finished orb web. The auxiliary spiral thread also serves as a handy guide for the spider when it comes to laying down the catching spiral. Without an auxiliary spiral the spider can cross the radii only awkwardly, since each radial thread would sag greatly under the spider's weight. After the auxiliary spiral is finished, the spider pauses for a moment before it tackles the last step of orb-web construction.

Catching Spiral The auxiliary spiral begins at the hub and ends well before reaching the peripheral framework (Fig. 118a). When laying down the sticky spiral, the spider starts at the periphery and uses the auxiliary spiral as a guide line. While crossing the radial threads in a spiral fashion, it constantly fastens the sticky thread to each radius. In this process both the front legs and the hind legs play an important role. One front leg reaches for the next radius and touches it with a quick, brushing movement, apparently to ascertain its position. At the same time one of the fourth legs pulls the thread out of the spinnerets and dabs it against the radius (Fig. 118b). While doing so, the spider keeps the distance between successive turns about equal (Vollrath and Mohren, 1985).

As the construction of the sticky spiral proceeds, the auxiliary spiral is simultaneously taken down. In the finished web only the remnants of the crossing points can be recognized. The catching spiral is not continued into the

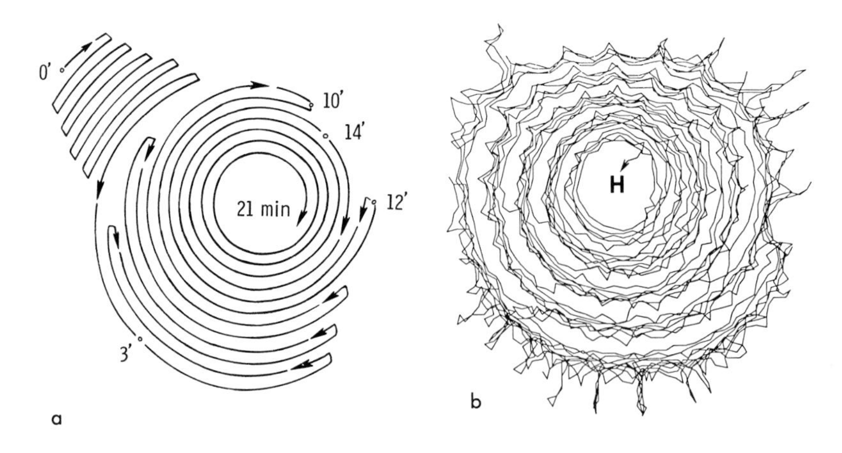

FIG. 119. (*a*) Placing of the sticky spiral. Spiral construction begins at the periphery (0'); the numbers indicate the time elapsed in minutes. (After Wiehle, 1927.)
(*b*) Laser-optical recording of the movements of a garden spider while placing the sticky spiral. The apparent bundling of spiral turns results from the spider following the five turns of the auxiliary spiral. (After Zschokke and Vollrath, 1995.)

center of the web but terminates shortly before it. This results in a rather open space between the catching area and the hub, termed the free zone. The sticky spiral is by no means a continuous spiral. It often reverses its course from clockwise to counterclockwise, and vice versa. Such turning points can easily be seen at many points in the web (Fig. 119). Since the position of the hub is slightly off center, that is, shifted toward the upper part of the web, most of the sticky spiral is confined to the lower half of the web.

The laying down of the sticky spiral is the most time-consuming stage of the orb web's construction. Whereas building the framework and the radii takes only about 5 minutes, the sticky spiral requires at least 20 minutes. That the entire orb web in all its complexity is created in less than half an hour is astounding. The total length of the threads is about 20 m. Since most threads are only 1–2 μm thick, the weight of an orb web is extremely low, between 0.1 and 0.5 mg. These values become even more impressive if one considers that a female *Araneus diadematus* easily weighs 500 mg.

It is also remarkable that the construction of the orb web is controlled only by the sense of touch, without any visual feedback. Even when all eight eyes are covered with opaque paint, the spider still builds a perfectly normal web. This achievement is quite routine for the spider, for it usually builds its web at night anyway. More surprisingly, perhaps, is the fact that gravity is also dispensable for the orb web's construction. This was most convincingly demonstrated by the two garden spiders that were sent into orbit with the space capsule Skylab in 1973. Despite complete weightlessness they still built regular orbs (Witt et al., 1977). However, if the orientation of the web is changed while the spider is building its web, then certain deviations in the web's geometry will occur (Vollrath, 1986a, 1988).

Incidentally, male orb weavers can build the same webs as females, at least until they reach sexual maturity. After that point they rarely make webs at all, and those are then rather small and rudimentary. Most notably, a sticky catching spiral is always lacking (Sacher, 1991; Müller and Westheide, 1993). Although this means that they can no longer catch prey, this matters little since their adult life lasts only about one week.

Relation of Structure and Function in the Orb Web

The ultimate purpose of any spider web is to capture prey—and the orb web is certainly no exception. Its highly geometrical construction suggests its special effectiveness and economy as a trap. As we have just seen, the orb web is built with a minimum of material (0.1–0.5 mg of silk) and time (20–30 minutes). The tightly strung radial threads converging in the web's center give the spider two advantages: (1) they transmit vibration signals from the periphery to the center, where the spider usually sits; and (2) they provide direct and quick access routes (Witt, 1975; Barth, 1982). The web can thus be regarded as an extension of the spider's sensorium. It has been shown that especially the longitudinal vibrations of the radii are important, since they follow a specific direction and are hardly attenuated. High-frequency energy sources—such as buzzing flies—emit the most effective signals toward the web's center (Masters, 1984a,b; Masters and Markl, 1981; Masters et al., 1986). The spider's first reaction to a fly that has hit the web is to jerk the radial threads several times with its front legs (Klärner and Barth, 1982). Only when the prey has been located in relation to the center does the spider rush out on a directed course—in most cases exactly on that radial thread leading straight to the prey.

Tensions in the orb web are not evenly distributed, but are highest in the periphery and lower toward the center. The forces acting on the mooring threads, frame threads and radii decrease in a ratio of 10:7:1 (Fig. 120; Wirth and Barth, 1992). This is also mirrored in the thickness of the respective threads: mooring and frame threads are about 6 μm in diameter, the radii only 2–3 μm. As a consequence, the actual stresses within a web are less, because higher tensions are balanced by stronger threads. It also seems that the amount of tension in a thread can be controlled by the spider and that adjustments can be made to environmental conditions (e.g., strong winds) as well as to its own body weight.

The radii serve as communication channels not only during prey capture but also during courtship, when the male approaches the female's web. Usually the small male sits at the edge of the web and rhythmically plucks the threads, apparently in order not to be mistaken for prey. The mechanical properties of the sticky spiral, especially its great elasticity, are also very important: it can be stretched several times its original length without breaking. This is of course advantageous when a vigorously struggling insect is trapped in the web. Aside from these mechanical stimuli, the web can also carry chemical signals, which stem from the spider's secretion of pheromones. It is known that a female's empty web may elicit courtship behavior of the male (Blanke, 1973a,b). The

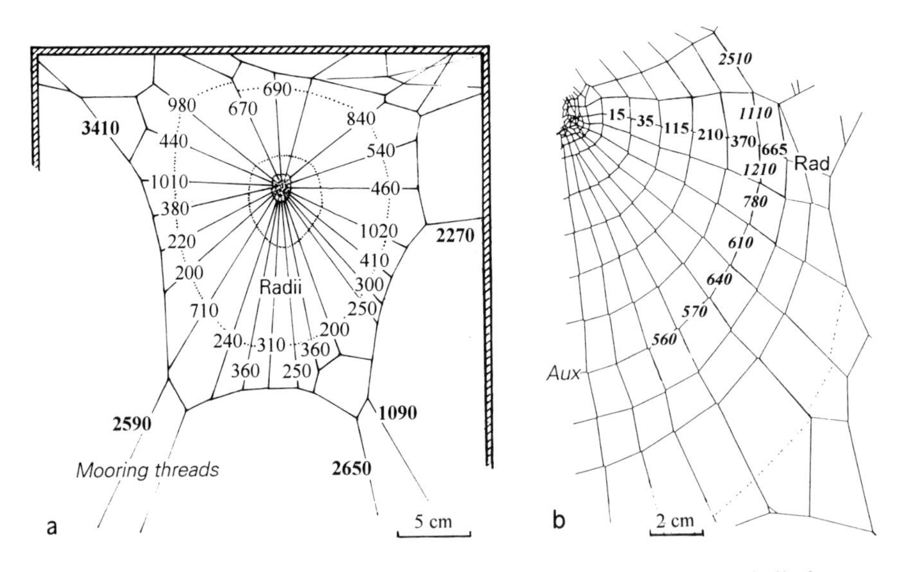

Fig. 120. (*a*) Forces in a completed orb web of *Araneus diadematus*. Pretensile forces (μN) are much higher in the peripheral mooring threads than in the radii. (*b*) Forces in a half-finished web. The high values of tension measured in the auxiliary spiral reflect its stabilizing effect on the whole web structure. Note that tension in a radial thread decreases markedly toward the hub. (After Wirth and Barth, 1992.)

carrying of chemical stimuli is not restricted to orb webs alone but applies also to the draglines of wandering spiders (Tietjen and Rovner, 1980).

The hub is usually slightly off-center, i.e., it lies above the geometrical center of the orb. This is probably not just accidental but serves a purpose: the spider sitting in the center can run a little faster downward than upward in the web. This means that prey in the (larger) lower half can be reached as quickly as in the (smaller) upper half of the web. Since there is a higher probability for insects to fly into the larger half, an asymmetric orb web will actually be a more efficient trap (Masters and Moffat, 1983).

Most threads in an orb web are quite thin (1–3 μm) and one would expect that insects could not really see them, because the resolution of their eyes is rather limited (Wehner, 1981). However, the threads do reflect light and are therefore noticed by many flying insects. They can be avoided, if they are seen early enough (Craig, 1986; Craig et al., 1985). However, the natural conditions are often in favor of the spider, e.g., under low levels of light or during windy weather, when the shaking web becomes practically invisible (Craig, 1988, 1990).

Most orb webs have a vertical orientation in space or, to be more precise, they are slightly tilted. The spider sits almost invariably on the "underside" of the hub, with its head facing down (two exceptions are *Verrucosa* and *Cyclosa*). In case of a disturbance or danger, the spider can simply drop to the ground on a safety thread—a response that is typical of many orb weavers.

Although orb webs have a species-specific structure, many Araneid spiders can move with ease in a foreign web. The pioneer French entomologist

J. H. Fabre successfully transferred an *Argiope bruennichi* to a web of *Argiope lobata* (cited by Wiehle, 1928). Similarly, *Araneus diadematus* has no difficulties catching prey in the foreign web of *Argiope aurantia* (Richardson, 1973). On the other hand, *Zygiella* does not feel at home in a web of *Araneus,* since it keeps climbing to the web's periphery and then dropping from the frame (Wiehle, 1927). Incidentally, it seems that some spiders can and do take over foreign webs under natural conditions (Enders, 1974; Eberhard et al., 1978).

In nature orb webs must be replaced often, because they quickly become damaged or destroyed by wind, rain, or early morning dew. As a rule, a garden spider (*Araneus diadematus*) builds a new web every day (Wiehle, 1927). Only the previous framework may be used again for the new construction. Taking down the old web is a rather systematic procedure. While walking out from the hub the spider destroys 3–5 radii together with the crossing sticky threads, thus producing an open sector in the orb (Carico, 1986). After 30–60 minutes the web is reduced to the frame and perhaps a few radii. At this point the spider starts building a new web. Some species of orb weavers practice a much faster technique (about 1 minute!) of taking their web down. Cutting the lateral frame threads makes the entire web collapse so that only the horizontal bridge thread will remain. The new web construction begins here by first establishing the basic Y-structure. Incidentally, the material of the old web is eaten by some orb weavers, other species simply throw it away, and some use it to wrap their egg cocoons (Carico, 1984).

If the web is only slightly damaged, as, for instance, after catching small prey, the spider may or may not mend the hole. In general, repair does not consist of reconstructing the original design but of patching the holes to make the web stable again. The cribellate orb weaver *Uloborus,* however, often performs elaborate repairs, sometimes replacing the entire half of a web in an orderly fashion (Fig. 121; Eberhard, 1972).

"Drug Webs"

In 1948 the German zoologist H. M. Peters, who had always been bothered by the fact that his garden spiders built their webs at an ungodly hour (between 2 a.m. and 5 a.m.), asked his colleague from pharmacology, P. N. Witt, for a stimulant that might shift the spider's web building to an earlier hour. Unfortunately, the drug (amphetamine) did not yield the desired effect. The spiders built at their usual hours, yet, surprisingly, the web structure was definitely altered: the radial threads as well as the catching spiral were placed irregularly. In the following years Witt tested a great variety of drugs (e.g., caffeine, mescaline, and strychnine) on orb-web spiders and investigated how these substances affected web construction (Witt, 1956, 1971). It was soon clear that certain drugs produced specific effects. For instance, a certain dosage of caffeine results in a typical "caffeine web" (Fig. 122), which is distinctly different from a mescaline web or an amphetamine web.

The drugs can affect the size and shape of the web, the number of radii and spiral turns, the regularity of placement of the radial and spiral threads, the

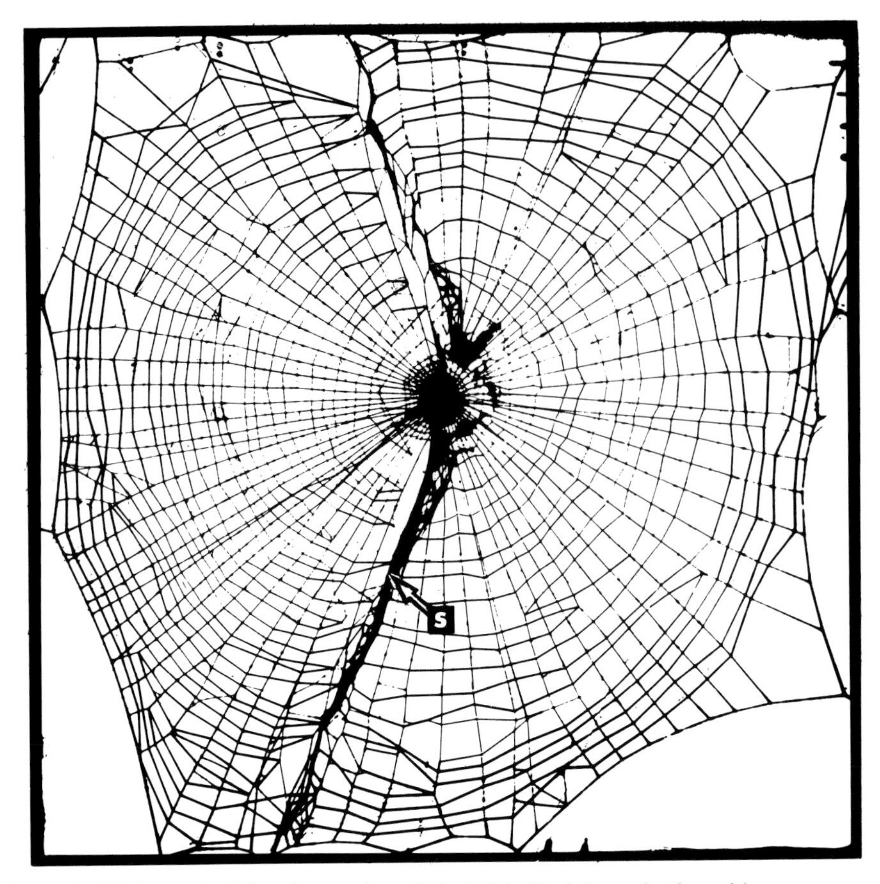

Fig. 121. Web repair. After destruction of the left half of the web, the spider (*Uloborus diversus*) first gathered all broken threads into a seam (S) and then started to rebuild the missing sector. Note the turning points of the spiral thread. The right half of the web was left unchanged, but the hub was built anew. (Photo: Eberhard, 1972.)

mesh width, and so forth. Relatively small doses of caffeine (10 μg/animal) result in smaller but wider webs than normal; the radii often form oversized angles, yet the regularity of the catching spiral is not affected. At higher doses (100 μg/animal), the alterations in the webs become much more noticeable and their regular design disappears. Most drugs tend to have a negative influence on web regularity. Only with very low doses of the hallucinogen LSD-25 could an increase in web regularity be observed. If the drug concentration was raised by one order (to 0.1–0.3 μg/animal) the regularity of the angles between radial threads decreased slightly.

Testing a drug on a spider is actually a rather simple procedure. The substances do not have to be injected but are simply dissolved in sugar water. A drop of such a solution is then touched to the mouth of the spider, and it is imbibed quickly. For quantitative studies, small drops of the drug of

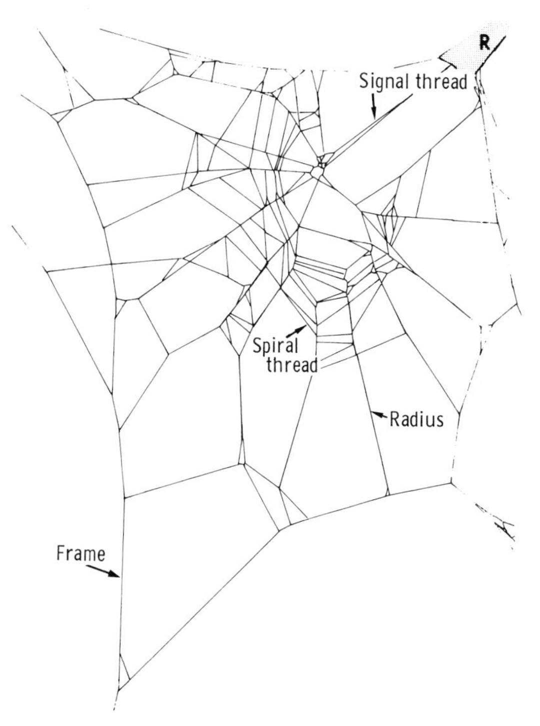

Fig. 122. Drug web of a *Zygiella x-notata* constructed after the administration of a relatively high dose of caffeine. Compare the normal web in Fig. 110a. R = retreat. (After Peters et al., 1950.)

known concentration and exactly determined volumes are delivered from a microsyringe to the spider's mouth. To study the effects of the drug, photographs of the spider's web are taken before and after the drug is administered. Structural alterations of the web can then be measured directly on the photographs. The use of statistical methods allows one to detect even very subtle changes that would not be apparent from a purely visual comparison of the control and "drug" webs (Witt and Reed, 1965).

The initial hopes that drug webs could be used in applied medicine (Groh et al., 1966), were not fulfilled, and this line of research was discontinued some 20 years ago. More recently, orb webs have also been used as bioassays for pesticide side effects (Samu and Vollrath, 1992). While some fungicides and insecticides had apparently no influence on spider webs, others suppressed web-building frequency, and severely affected web size and building accuracy. Again, it seems unlikely that these findings will have any impact on the actual use of insecticides in agriculture.

Is Web Building Genetically Determined or Learned?

The elaborate and highly regular structure of the orb web provokes the question as to whether the first webs of a spider exhibit the same complexity of those built by adults or whether some learning process leads by degrees to the "complete" orb web. The question is answered if we inspect the very first web built by a young garden spider: it is already a perfect, miniature orb (Fig. 110b). Detailed measurements have even shown that the early webs are often more regular than the webs of adults (Witt et al., 1972). Learning from previous experience does not seem to play a role in web building (Reed et al., 1970). This finding was further demonstrated in experiments in which young spiderlings were confined to narrow glass tubes so that they could not build any webs. When they were released after several months, they could weave orbs as regular as those of their normally reared siblings, which served as a control.

However, one cannot conclude that web building is an entirely stereotyped process. A certain plasticity of this behavior has been shown in many experiments. For instance, if one burns out a radial thread with a hot wire just after it has been constructed, the spider will replace it immediately. One may repeat this game more than 20 times before the spider gives up replacing the radial thread and proceeds to build the auxiliary spiral (Hingston, 1920). This shows clearly that already existing structures influence the further construction of the web. Apparently the spider can integrate the feedback from its sensory input and adapt its behavior to a given situation. On the other hand, the spider does not replace the burnt out radii indefinitely, but at some point simply proceeds to the next phase of web construction, even though the previous phase has not been completed.

Similar conclusions can be reached if one transfers a spider engaged in building a web to another web, which is at an earlier stage of construction (P. J. Peters, 1970). The spider then usually repeats phases (for example, laying down the auxiliary spiral) that it had already completed in its own web. In some cases, however, it was observed that important steps in web construction were not repeated, resulting in a rather irregular web. In the first instance the spider apparently reacted to its sensory feedback in completing the web; in the latter case the spider seems to have followed the original "program" coded in its CNS. This explanation is certainly an oversimplification, since other factors, such as sensory feedback from the silk glands, may also determine the extent to which stages of web building are repeated.

The statement that even very young spiderlings can already build perfect orb webs needs to be scrutinized. When spiderlings leave their cocoon, they do not build an orb web right away, although they are capable of silk production. Rather, they live communally in a mesh of irregular threads, and only after 2 weeks do they separate and start building their first orb webs (Burch, 1979). If the young spiderlings are taken out of the cocoon prematurely and kept isolated from one another, they can trail draglines and sometimes build a small retreat (H. M. Peters, 1969). Some days later, irregular orb webs appear. Successive webs are improved to perfection within a few days. Most likely, however, this improvement is not the result of a learning process but is rather a reflection of the fact that the CNS is still developing. At the time the spiderling hatches from the egg, its CNS contains only neuroblasts. The precise correlation between differentiation of the CNS and the type of web built at specific times in early development still needs to be established.

The juvenile webs of the cribellate orb weaver *Uloborus* are especially interesting (Szlep, 1961; Eberhard, 1977a). The spiderlings build an orb web immediately after leaving the cocoon, but this first web differs from the adult web in three major respects: (1) the catching threads are lacking; (2) the auxiliary spiral is not removed; and (3) instead of a catching spiral many "extra" radial threads are filled in between the original 15–20 radii. The main reason for this deviation from the normal orb web is that before the second molt *Uloborus* does not have a cribellum or a calamistrum; thus it cannot produce cribellate silk for the catching thread. After the second molt a cribellum and a calamistrum are present, and the spider weaves a regular orb web for the first time. Again it seems likely that the stage of maturation of the CNS plays a role in determining which kind of web will be built during the ontogeny of *Uloborus*.

In conclusion, then, juvenile spiders often build different webs than adults do. However, these early webs cannot be regarded as merely incomplete versions of the adult's web. Learning does not seem to be involved in the "adult" orb web. In the more specialized orb weavers (*Zygiella, Nephila, Eustala*) the juveniles show a stronger tendency to build the original orb web pattern than the adults (Eberhard, 1985).

Orb Webs and Evolution

The orb web is often considered the evolutionary summit of web-building spiders. It is rather hard to imagine how such a complex structure could have come about at all, yet it is even more difficult to explain how an orb web could develop in two different groups of spiders, the cribellate Uloboridae and the ecribellate Araneoidea (Araneidae, Tetragnathidae). Was the orb web "invented" in both groups independently or in only one group, which gave rise to the other (Kullmann, 1972a)? There are in fact good reasons to assume convergence in the evolution of the orb web. The two groups use quite different catching threads (cribellate silk versus ecribellate glue-coated threads), and it is hardly plausible that this difference developed only after the structure of the orb had been achieved (H. M. Peters, 1984, 1987; Kovoor and Peters, 1988). One hypothesis then is that the orb web developed independently twice, starting from simpler precursors in both groups (Fig. 123).

According to this hypothesis, the ancestral forms of web were merely a retreat with threads radiating from its opening. Cribellate spiders (like *Filistata* and *Amaurobius*) have a "hackle band" along the radial threads, whereas the corresponding plain tripping threads of ecribellate spiders (like those of *Segestria* or *Ariadna*) have only a signaling function (stage I). At the next level of evolution (stage II), the area of the catching plane is enlarged and extends somewhat into open space. In ecribellate spiders this led to the formation of three-dimensional webs (like those of linyphiids and theridiids). The next step

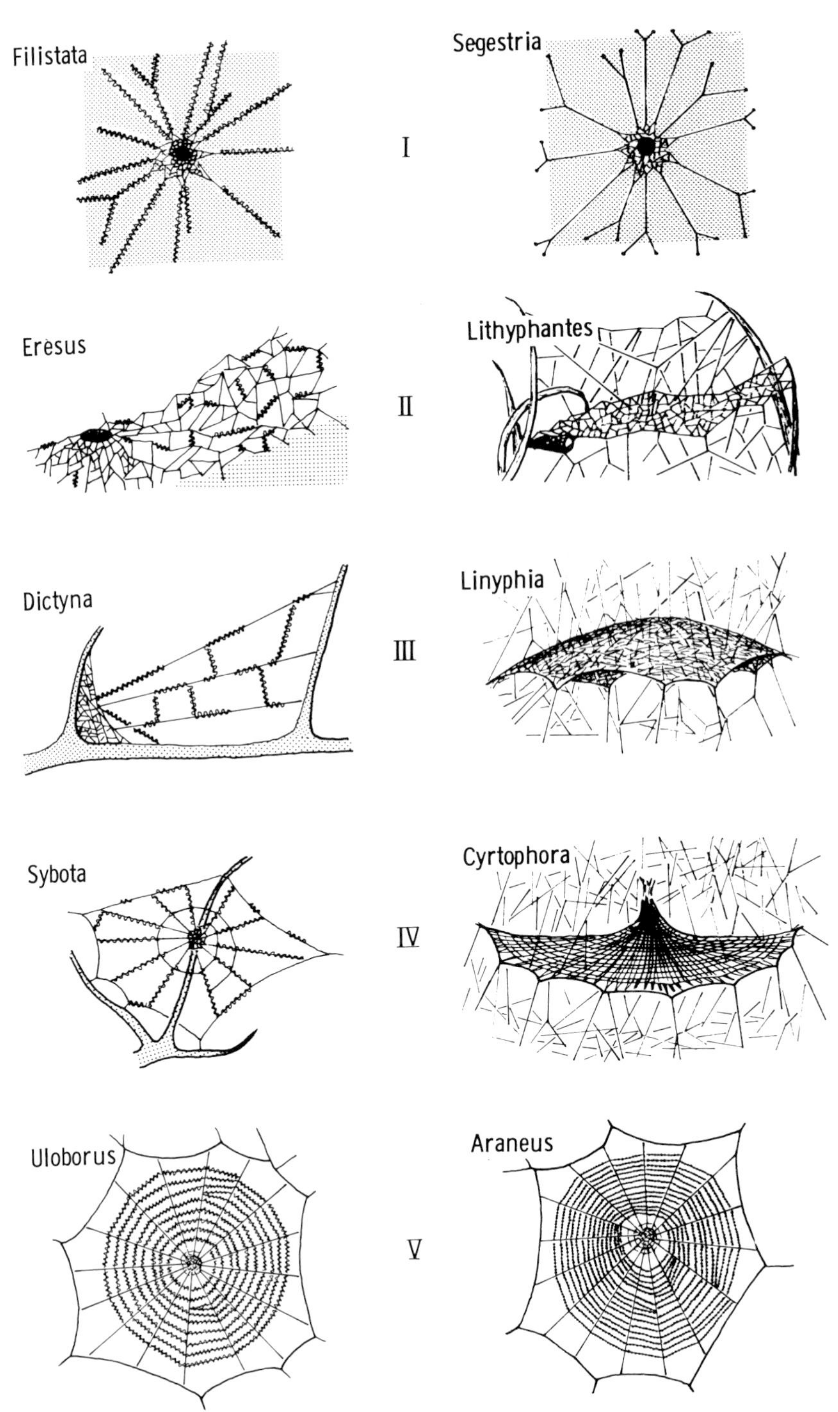

FIG. 123. Evolution of the orb web in cribellate (left) and ecribellate spiders (right). (After Kullmann, 1972a and Wiehle, 1931.)

postulates the omission of the primary retreat (stage III); secondary retreats may develop at the web's periphery or in its immediate surroundings. In the sheet webs of linyphiids and the cobwebs of theridiids we begin to find glue droplets on some threads to facilitate the ensnarement of flying insects. Still later (stage IV) the web becomes centric and the spider now sits in the hub. The last step would then lead to the typical orb web (stage V), which has a catching spiral with cribellate silk in cribellate spiders, while it is equipped with glue droplets in ecribellate orb weavers. A recent consideration of orb-web phylogeny suggests that the orb was already present in webs of common ancestors of uloborids and araneids, but subsequently evolved in different directions, one with cribellate silk, the other with sticky silk (Levi, 1980).

The hypothetical evolution of the orb web as pictured in Fig. 123 was well accepted 20 years ago but is viewed rather critically by most arachnologists today. Although the sequence of events seems logical on the cribellate side, it is rather questionable on the ecribellate side. For instance, the web of *Cyrtophora* is no longer considered as a precursor of the orb web but is most likely derived from it; similar views have been expressed for the horizontal dome web of the linyphiids (Levi, 1980). In other words, the orb web could be the ancient form from which the others have evolved, i.e., just the opposite sequence than the one shown in Fig. 123. In fact, the idea that the orb is the "primitive" state and evolved only once (*monophyletic*) has been proposed for several years (Levi, 1980; Coddington, 1986). Admittedly, this interpretation does not shed light on the evolution of the orb web itself. Yet it seems most likely that the precursors occurred among the cribellate spiders and that the cribellum became reduced while the typical glands for gluey capture threads developed. This hypothesis is supported by a cladistic analysis using 60 different characters in 18 families of "orb-web spiders" (Orbiculariae; Coddington, 1990). The resulting cladogram speaks more for a monophyletic evolution of the orb-web than for convergence.

Although nothing final can be said about the evolution of the orb weavers, we know now from newly discovered fossils of the Jurassic and Cretaceous time period (Eskov, 1984; Selden, 1990) that the orb web must be a rather "old" invention—probably in the order of 100 million years.

The orb web in fact is probably not the terminal point of the evolution of spider webs. Many orb webs of both cribellate and ecribellate spiders show various degrees of reduction and modification. The araneid *Zygiella* leaves the center of the web and uses one isolated radius as a signal thread leading to its retreat (Fig. 110). Since this signal thread is not crossed by the sticky spiral, two sectors of the web remain open. The uloborid spider *Hyptiotes* goes much further in reducing the original orb: only three sectors remain (Fig. 114). *Miagrammopes*, likewise a cribellate spider, spins only a single horizontal thread, which is dressed with catching silk in its middle (Lubin et al., 1978).

Corresponding reductions are also found on the ecribellate side, that is, among the Araneidae. The bolas spider (*Mastophora* in America, *Dicrostichus* and *Cladomelea* in Australia) uses only a short thread with a large drop of glue on its end (Fig. 144). This bolas is thrown after insects which fly close by. In

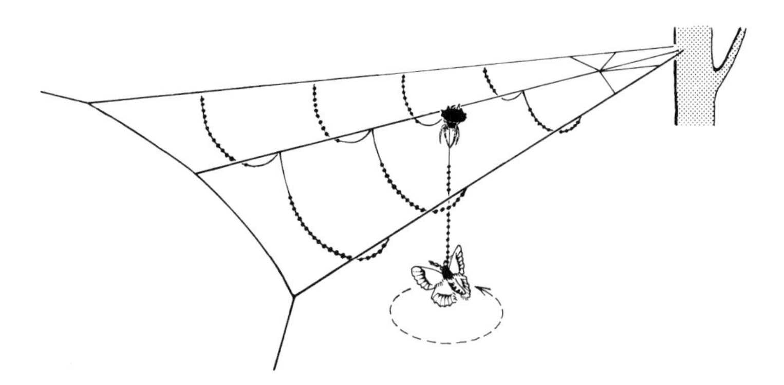

FIG. 124. *Pasilobus* from New Guinea constructs an extremely modified "orb" web. The sagging threads of the sticky "spiral" rupture when an insect brushes against them. (After Robinson and Robinson, 1975.)

Archemorus the orb web is also abandoned entirely. The spider sits patiently on the upper surface of a leaf, mimicking detritus or bird droppings. When an insect passes nearby, it is detected visually. *Archemorus* suddenly leaps at the victim and seizes it with its well-armored front legs (Robinson, 1980). Another curious "bird dropping" spider is *Celaenia* from Australia and New Zealand. While simply dangling on a short thread, it grabs moths that fly by (Forster and Forster, 1973). Since the moths caught are invariably males, it is believed that *Celaenia* exudes an attractive scent that mimics the sex pheromone of the female moth.

Other quite fantastic modifications of the orb web are known from the tropical *Poecilopachys* and *Pasilobus* (Robinson and Robinson, 1975). *Pasilobus,* for example, builds a triangular horizontal web consisting of three radii and 4–11 drooping, sticky threads. When a moth brushes against such a strand, the sticky thread ruptures at a predetermined point near one of the marginal radial threads. The insect becomes stuck to the broken sticky thread, and finds itself suspended and tethered in flight (Fig. 124). The alerted spider rushes out along the middle radial thread and hauls in its line with the attached prey. Orb webs without any sticky spiral are typical for the American araneid *Wixia* (Stowe, 1978, 1986). There are only a few radii and they are directly attached to twigs; they function as tripping lines for passing insects.

These examples show that evolution has not stopped at, but has gone beyond the orb web. In some instances the modifications and reductions have been so radical that there is no hint at all of the original structure—as in the case of the bolas spiders, for instance. Among the tetragnathids, only the juvenile *Pachygnatha* build orb webs, but the adults forsake web building altogether and return to a free, hunting way of life (Balogh, 1934; D. Martin, 1978). The specialized Hawaiian tetragnathid *Doryonychus raptor* also has no web, but impales small insects with the highly elongated tarsal claws of its front legs, while dangling from a single thread (Gillespie, 1992).

A very basic issue not dealt with previously is how spider webs per se could have evolved. It is speculated that the spinning threads were originally used only for wrapping the eggs in an egg sac (Kaston, 1964; Shultz, 1987b). Perhaps a retreat was also present at an early stage, and tripping threads emanating from it could alert the spider to insects passing by. At this stage, however, the silk threads were only a means to detect and not to trap prey. A definite improvement was achieved when special catching threads could hamper or even detain prey, making it much easier for the spider to overpower its victim. In conclusion, we can safely assume that snares developed from enlarged living quarters or retreats. Perhaps the silk was originally used to line retreats in the tidal zone, as a protection from flooding (Decae, 1984). It is known that submerged spiders can survive for days if they are enclosed by an air-filled silk tube. The reason is that the silk sheet functions as a "physical gill," allowing the constant diffusion of oxygen from the surrounding water into the air space (Rovner, 1986, 1987). The "modern" hunting spiders such as lycosids, salticids, oxyopids, and pisaurids, have probably secondarily reverted from snare building to a nomadic way of life.

6

Locomotion and Prey Capture

In keeping with their varied life styles and different habitats, spiders have developed diverse modes of moving about and capturing prey. Both are complicated by the fact that they can move in all three dimensions. Vertical locomotion is accomplished sometimes by jumping, but mostly, however, by climbing on vegetation or on their own silk thread. Many spiders are able to "fly," that is, they emit a line of silk that is caught by the wind, which, in turn, carries the spider aloft (Eberhard, 1987). Almost everyone has witnessed this phenomenon in autumn when watching gossamer strands drifting through the air. Usually, however, only the shiny silk threads are noticed, the tiny passengers are easily overlooked.

Walking

Any regular kind of walking is characterized by a specific *rhythm* or *stepping pattern*. If we look at all eight legs of a spider at the same time, it becomes obvious that two sets of legs always move alternately. For instance, legs 1 and 3 on the left side (L_1, L_3) are active together with legs 2 and 4 on the right side (R_2, R_4). In other words, while legs L_1–R_2–L_3–R_4 are moving, legs R_1–L_2–R_3–L_4 are at rest (Fig. 125a). Such a diagonal rhythm is quite commonly observed among arthropods, regardless of the number of legs they use. In insects, for example, the typical walking pattern is L_1–R_2–L_3 and R_1–L_2–R_3.

The movements of the different legs are not absolutely synchronized but show slight temporal delays. The onset of activation is quite obviously metachronous in slowly moving spiders, when a wavelike motion passes over the legs on each side of the body. As a consequence, we can also define the walking pattern by watching the stepping sequence on the right or left side. The most common stepping sequence in spiders is described by the formula 4–2–3–1 or 2–3–1–4, 3–1–4–2, or 1–4–2–3—since any leg can take the first step (D. M. Wilson, 1967; Seyfarth and Bohnenberger, 1980). If a spider is moving extremely slowly, the sequence 4–3–2–1 occurs, and the wavelike motion of the legs becomes especially apparent. In contrast, during running one set of legs moves practically synchronously and alternates more strictly with the other set of legs. This gives the impression of a very precise and "mechanical" run.

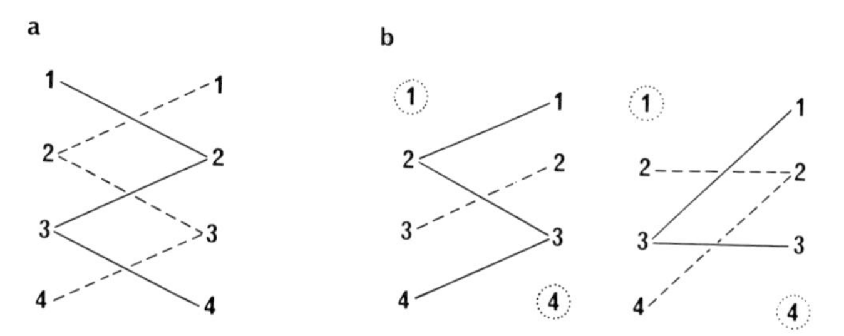

FIG. 125. (*a*) Regular walking pattern: diagonal rhythm. Synchronously moving legs are connected by solid or dashed lines. (*b*) Two possible walking patterns for a spider lacking the first left leg and the fourth right leg. The spider uses the stepping pattern on the right. (After D. M. Wilson, 1967.)

If a spider loses a leg, it quickly adjusts its gait to the new condition. First of all, the position of the remaining legs is corrected in such a way that the "gap" is more or less compensated for. The spider also modifies its walking pattern, and the length of the steps may increase. This behavioral plasticity is probably caused by an altered proprioceptive feedback: certain sensory inputs are now lacking altogether, and because of the new positions of the legs, the remaining joint receptors provide the spider with slightly different information than before.

Even if several legs are missing, a spider can still change its walking pattern to achieve smooth locomotion. An interesting experiment with a large "tarantula," from which two legs from opposite sides (L_1 and R_4) were amputated, should be mentioned here (D. M. Wilson, 1967). Theoretically two possibilities to compensate for the loss of these two legs exist (Fig. 125b). The spider could either retain a diagonal rhythm and move four legs (R_1–L_2–R_3–L_4) synchronously while balancing its body on the two legs R_2 and L_3, or it could move three legs at a time (R_1–L_3–R_3 or L_2–R_2–L_4), although this would completely offset the diagonal rhythm. The experiment showed that the spider adopts the second alternative, which is indeed the mechanically more stable. The fact that certain pairs of legs (R_2–L_2 and R_3–L_3) now move together proves that the diagonal rhythm is not rigidly "programmed." Apparently the basic pattern of the walking program is determined by a "central oscillator" located in the CNS, but can be modified by feedback from sensory organs in the legs (Seyfarth, 1985).

Most spiders run in a jerky fashion. At intervals of several centimeters the animal halts abruptly, and after a brief pause it starts again, often changing its direction slightly. When a wolf spider stops, it often freezes the position of its legs momentarily, making the diagonal walking pattern particularly noticeable. In contrast, when a jumping spider stops, it typically arranges its legs symmetrically around its body (Fig. 126); this stance allows for a sudden leap. When web spiders stop moving, many of them also draw their legs close to the body in a kind of resting or protective stance (Ehlers, 1939).

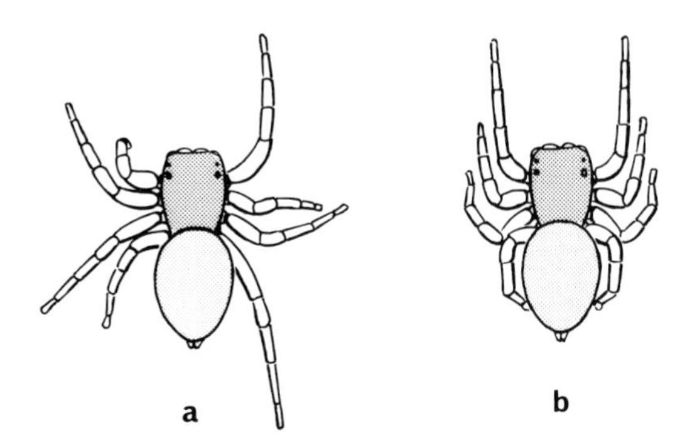

FIG. 126. (*a*) Walking and (*b*) jumping positions of the salticid *Philaeus*. (After Ehlers, 1939.)

Mechanics of Walking

Since all the legs of a spider are anatomically equal (except in length), they also move basically in the same manner. Yet quantitative differences may play a role, especially with respect to torsion of the leg axis. As a rule, legs 1 and 2 are directed forward and pull, whereas legs 3 and 4 point backward and push (Fig. 127).

During the motion of a single leg, one can differentiate a pulling or pushing phase (*remotion*) and a lifting phase (*promotion*, a forward motion made without contacting the ground). To move the animal ahead, the joints of the first two

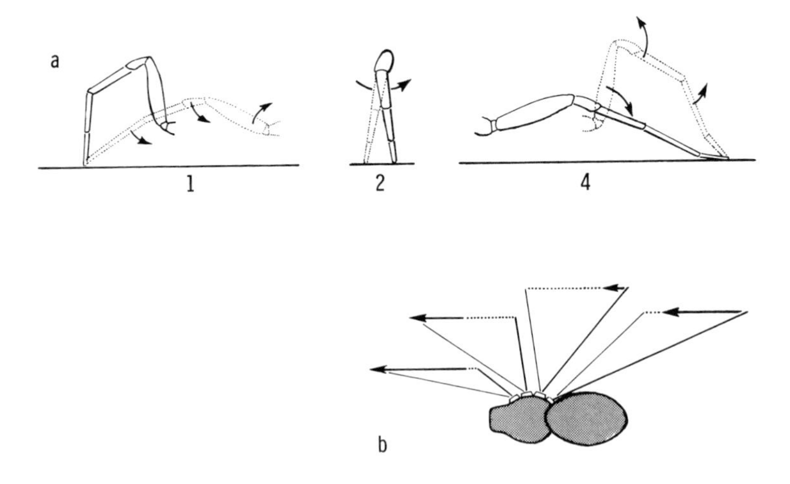

FIG. 127. (*a*) Contribution of the different legs to the propulsion of the spider. The initial leg position is shown with dotted lines. The first and fourth legs operate mainly by bending and stretching; legs 2 and 3 show torsion. Here only the torsion of the patellar joint is shown for leg 2. (*b*) Working range of the legs of a wolf spider during slow walking (5 cm/s). Dotted line = propulsion by torsion; vectors = relative contributions to propulsion by bending and stretching. (After Ehlers, 1939.)

pairs of legs must bend, while the joints of the hind legs (legs 4) must stretch. The third legs are laterally directed, so bending or stretching of their joints hardly contributes to forward movement; however, a torsion of their coxal joints does (Figs 127, 128).

The legs must move in a coordinated fashion, not only in time but also in space. For instance, if the spider makes a left turn, the legs on the right side must increase their stride (Ehlers, 1939). Walking is not necessarily limited to a forward direction. Many spiders—especially crab spiders and jumping spiders—can run adroitly sideways, or even backwards. In the latter case the walking pattern still follows the diagonal rhythm, but in reverse. This explains why jumping spiders can walk backwards just as fast as forwards. If the spider turns on the spot, all the legs on one side of the body move forward as usual while the legs on the other side move in the opposite direction (Land, 1972a).

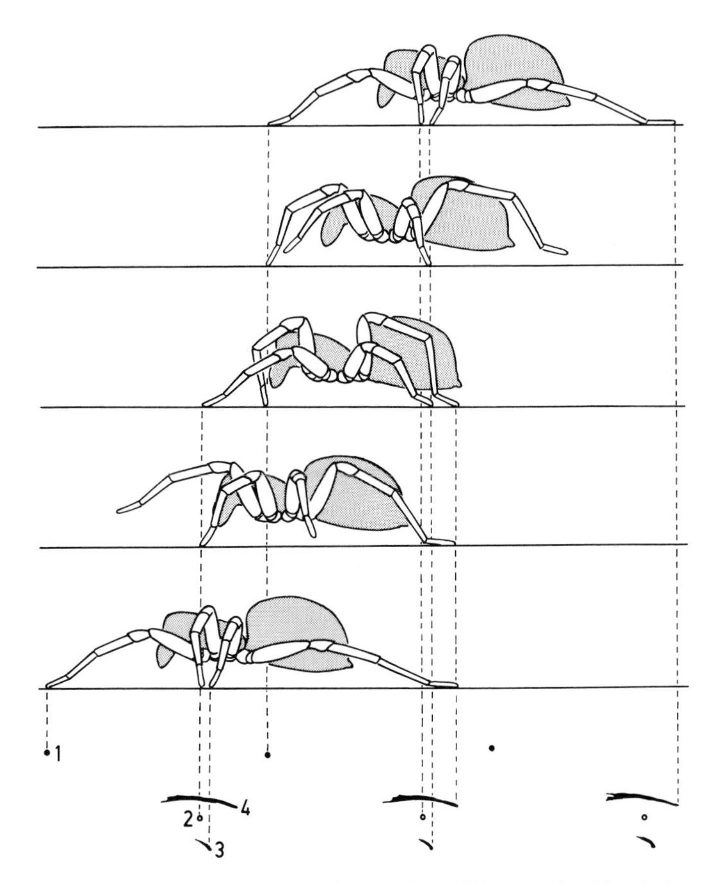

FIG. 128. Sequence of leg movements of a slowly walking wolf spider (*Alopecosa*). Only the left legs are pictured. Note the simultaneous movements of the first and third legs, and of the second and fourth legs, respectively. *Bottom*: Footprints as recorded on smoke paper. Whereas legs 1 and 2 touch the substrate only with their claws (dots), legs 3 and 4 leave long scratch marks. (After Ehlers, 1939.)

The walking speed of a spider is typically only a few centimeters per second, but large spiders such as *Lycosa* or *Tegenaria* can easily reach a speed of 40–50 cm/s. The length of the steps increases with the spider's speed—at 6–7 cm/s a *Tegenaria* takes steps of only 1 cm, but doubles that at 30 cm/s. As mentioned earlier, spiders can perform impressive sprints, although they become exhausted after a few seconds. The reason for this exhaustion seems to be connected to the fast breakdown (10–20 s) of aerobic, energy-rich phosphagen (Linzen and Gallowitz, 1975; Prestwich and Ing, 1982). Spiders can also gain energy from anaerobic glycolysis or from stored ATP (Prestwich, 1983a,b) but these sources are only used as a last resort. The complete exhaustion that occurs after about 2 minutes of high activity is probably caused by an accumulation of lactate. Since lactate is only slowly metabolized, recovery takes rather long, i.e., 10–20 minutes (Prestwich, 1988a,b). However, under normal circumstances there is no necessity for spiders to run at high speed for several minutes. A short "sprint" will usually suffice to reach a safe place.

In general spiders with legs that are equal in length are better runners than those with legs that vary in size. This can be easily confirmed if one compares the running of wolf and garden spiders. Garden spiders have a short third pair of legs and cannot run as quickly on the ground as wolf spiders can. However, while running on a thread the web spider can use its short third legs quite efficiently (Ehlers, 1939; Jacobi-Kleemann, 1953). Apparently the araneid's legs are well adapted for hanging from threads and less well for supporting the weight of the body above the legs. They also lack the traction that adhesive hairs (scopulae; see chapter 2) provide for efficient locomotion in wandering spiders.

Locomotion on a Thread

For a spider the simplest form of locomotion is to let itself drop by its own silk thread. This can be done without any help from the legs, but usually one of the fourth legs pulls the thread sideways when the spider wants to stop (Fig. 129a). Orb web spiders often use their fourth legs to reel out the thread

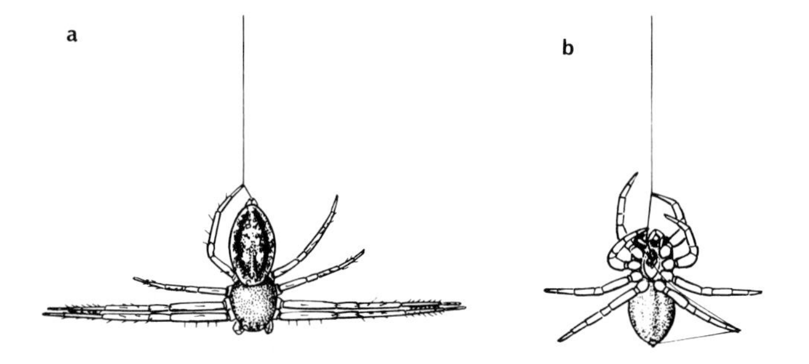

FIG. 129. (*a*) Crab spider *Diaea* dropping on its own thread. (After Bristowe, 1958). (*b*) Jumping spider *Philaeus* climbing back on its own thread. Only the two pairs of front legs grasp the thread, while the palps reel it in. (After Ehlers, 1939.)

from the spinnerets, thus regulating the speed of the downward motion (Ehlers, 1939). While descending by a thread, the garden spider stretches out its first pair of legs in anticipation of contact with the ground; the second pair of legs is spread horizontally to keep the spider from spinning around (Jacobi-Kleemann, 1953).

Climbing back up a vertical thread is accomplished mostly with the aid of the two pairs of front legs (Fig. 129b), while the third pair of legs or the palps gather up the loose thread. During this process the connection with the spinnerets is always maintained so that the spider can quickly drop down again. When moving upside down on a horizontal thread the garden spider uses all of its legs. In this case, the hind legs are moved more rapidly than the front legs, and the third legs often move twice as fast as the others (Jacobi-Kleemann, 1953). This is probably due to the fact that the third legs are rather short. Similarly, crab spiders also move their short hind legs at a faster rate than their long front legs (Ferdinand, 1981).

The dragline can also be used as a brake during a fast run. For instance, when the "raft spider" *Dolomedes* is hunting on the water surface and needs to stop fast, it grasps the trailing thread with the tarsal claws of its hindlegs. Since the thread is anchored at the river bank, the inertial movement is quickly reduced and the spider comes to a halt next to the prey (Gorb and Barth, 1994).

The mode of locomotion during web building does not differ in principle from the spider's normal walking in the completed web. It simply appears more complex because, aside from a merely locomotor purpose, some leg movements also function in placing the threads. For instance, a third and a fourth leg together always pull a radial thread close to the spinnerets so that the spiral thread can be fastened to the radius (Jacobi-Kleemann, 1953).

Running on a Sheet Web

It has always been puzzling how funnel-web spiders (Agelenidae) can move swiftly over their silken sheet (up to 45 cm/s; Fröhlich, 1978), whereas their insect prey can only walk clumsily on it. Bristowe (1941) explained that "the insect's position can be compared to a man trying to run through soft snow up to his knees pursued by an enemy on skis." Although this image seems convincing, a closer look at the running technique of *Agelena* reveals a different picture. As high-speed cinematography clearly shows, the spider does not run flat footedly, but puts the tarsi at a rather steep angle onto the sheet web (Fig. 130; Dierkes, 1988). Feathery hairs on the tarsi push against the densely woven fibers of the sheet web and prevent the feet from sinking in. In other words, *Agelena* is running on tip toe, thus minimizing the contact area between the legs and the sheet web. In contrast, most insects put their legs in a flat manner on the sheet web, which causes considerable adhesion between the leg cuticle and the silk threads. Consequently, their movements are severely hampered.

FIG. 130. High-speed cinematography of a running funnel-web spider (*Tegenaria*). Pictures (*a*) and (*b*) were taken at an interval of 1/60 of a second; running speed was 33 cm/s. Note how the tips of the legs (arrow heads) are put vertically onto the web. Only the hind tarsi form a rather flat angle, because they are in the pushing phase. (*b*) Tarsus 3 in an *Agelena*, set vertically on the web and actually indenting it (arrow)—yet without sinking in. (Photos: Dierkes.)

Jumping

Jumping is most obvious in the salticids, although many other hunting spiders (lycosids, clubionids, and oxyopids) are also capable of at least short leaps. We shall take a closer look at the mechanism of a salticid's jump.

It was believed formerly that the strong front legs are responsible for propelling the spider in a leap. Actually, however, the front legs are often lifted above the ground just before the spider takes off (Figs 7, 131). The front legs play a role only in landing and, of course, in seizing prey. The thrust for the takeoff comes from the hind legs, either from the fourth or third pair, and sometimes from both. Which technique is used depends on the species. *Evarcha* jumps off with the third pair of legs, *Salticus* with the third and fourth, and *Sitticus* with only the fourth. Just before taking off, the spider fastens a safety thread to the ground. Then the legs are pulled close to and symmetrically around the body. A very quick extension of the fourth pair of legs, a torsion and extension of the third pair of legs (or both), initiates the leap. The stretching of the legs is largely caused by an increase in hemolymph pressure and probably by a concomitant sudden relaxation of the flexor muscles.

Besides capturing prey, salticids utilize their jumping ability for escape. Normally they will jump only a few centimeters, but in some cases they cover up to 25 times their own body length (Ehlers, 1939). Compared with the jumps of some insects (grasshoppers or fleas, for instance) this may seem like a modest performance, yet after all, salticids do not have any oversized jumping legs.

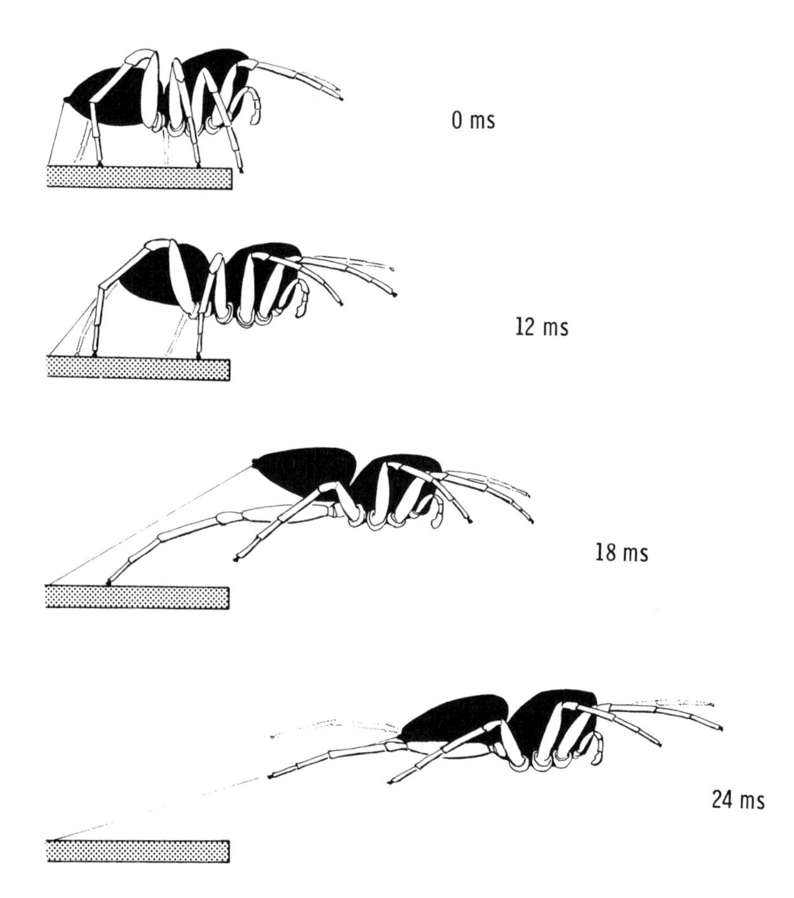

Fig. 131. Takeoff of the jumping spider *Sitticus*. The jump results from a rapid stretching of the fourth pair of legs. (Tracing of slow-motion pictures by Parry and Brown, 1959b.)

Short leaps are often interspersed in the normal walking pattern whenever the spider has to bridge a small gap. This leaping is also seen in other wandering spiders that live predominantly on vegetation. Again, the takeoff is caused by the action of the third and fourth pairs of legs (Ehlers, 1939).

Wheeling Locomotion

Humans are usually credited with the invention of the wheel. However, a few animals came up with the same trick long ago. Some dune spiders (e.g., *Carparachne*, Heteropodidae) can actively roll down sand dunes. They do so by flipping their body sideways and then cartwheeling over their bent legs (Fig. 132). The rotation is so fast—20 revolutions per second—that the spider appears only as a blurred ball, zooming downhill at a speed of more than 1 m/s

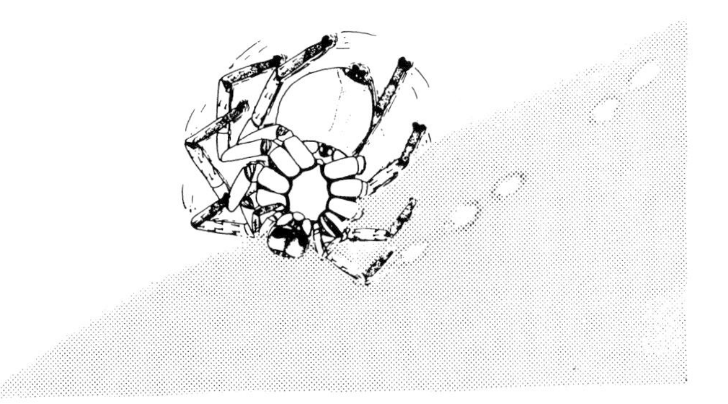

Fig. 132. Wheeling locomotion of the dune spider *Carparachne aureoflava.* The legs contact the ground only at the flexed tibio–metatarsal joints, thus leaving impact marks in the sand. (After Henschel, 1990a.)

(Henschel, 1990a). This behavior is successfully used to escape predatory wasps (Pompilidae), which try to dig into a spider's burrow. Wheeling behavior is not unique to *Carparachne* but was also observed in a salticid dune spider—which actually uses strong winds to propel itself over level sand. Incidentally, a similar somersaulting locomotion was observed in some beach crustaceans (Stomatopods), when they were washed ashore and tried to rush back to the sea (Caldwell, 1979; Full et al., 1993). Wheeling is certainly very exceptional and seems to occur only in animals that live on smooth, sloping surfaces, such as dunes.

Locomotion On and Under Water

Some spiders are not capable of walking on water; among them are the salticids, thomisids, linyphiids, and theridiids. Many pisaurids and lycosids (such as *Dolomedes, Pardosa,* and *Pirata*), however, can walk on water as well as they do on land. Propulsion is mainly by means of the second and third pairs of legs. The first legs are used as feelers and are stretched forward, while the fourth pair of legs is dragged behind. The smaller, lightweight species can raise their bodies high above the water (Fig. 133). The larger and heavier species walk rather flatfootedly on the water's surface so that the entire tarsus (and sometimes the metatarsus) is in contact with the water. In the large "raft spider" *Dolomedes*, the body rests on the surface of the water as well. Incidentally, *Dolomedes* does not make use of the diagonal walking pattern when moving on water. Instead, the third pair of legs and, after a short delay, the second pair of legs are pulled back in synchrony. *Dolomedes* thus moves in a manner similar to that of the water striders (Gerridae) among the insects. The legs are moved only at their insertion sites in the cephalothorax or at the coxal joints, the other leg segments, being neither bent nor stretched (Ehlers, 1939). Thus the spider rows across the water (Fig. 134; Barnes and Barth, 1991). During prey capture

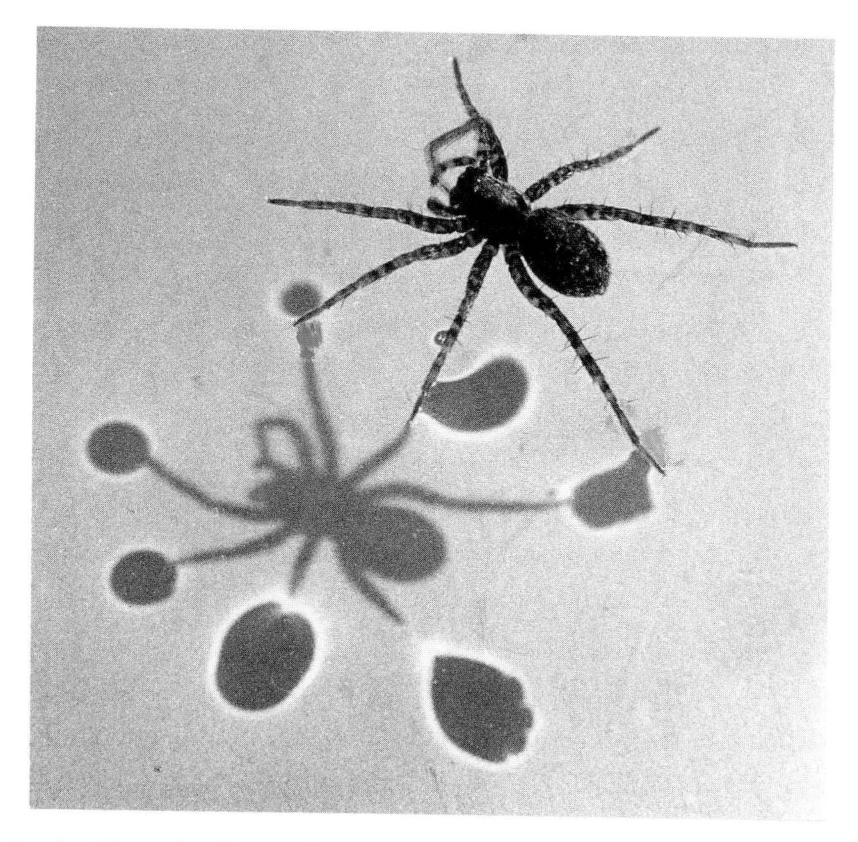

Fig. 133. The wolf spider *Pardosa amentata* afloat on the surface of water, cleaning its first right leg with its mouth parts. The tips of the other legs indent the water surface slightly. The sun rays project these indentations as large shadowy patches onto the bottom of the shallow pond.

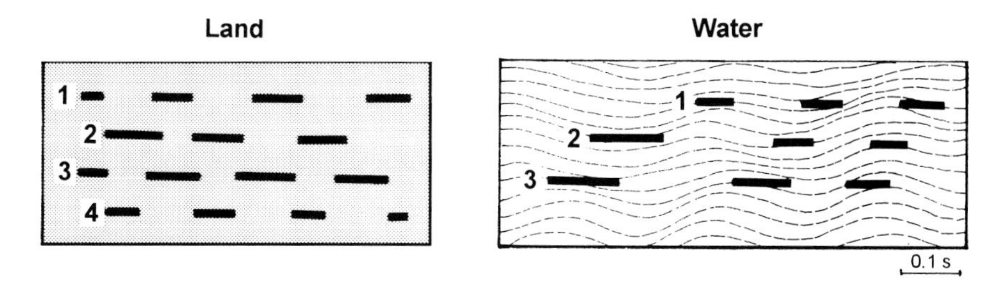

Fig. 134. Stepping pattern of *Dolomedes* while walking on land and rowing on water. Black bars indicate power strokes of legs 1–4 on one side of the body. Interspaces correspond to the return strokes. Note the alternating gait during walking as compared to the metachronal movement of legs 1–3 seen during rowing. (After Barnes and Barth, 1991.)

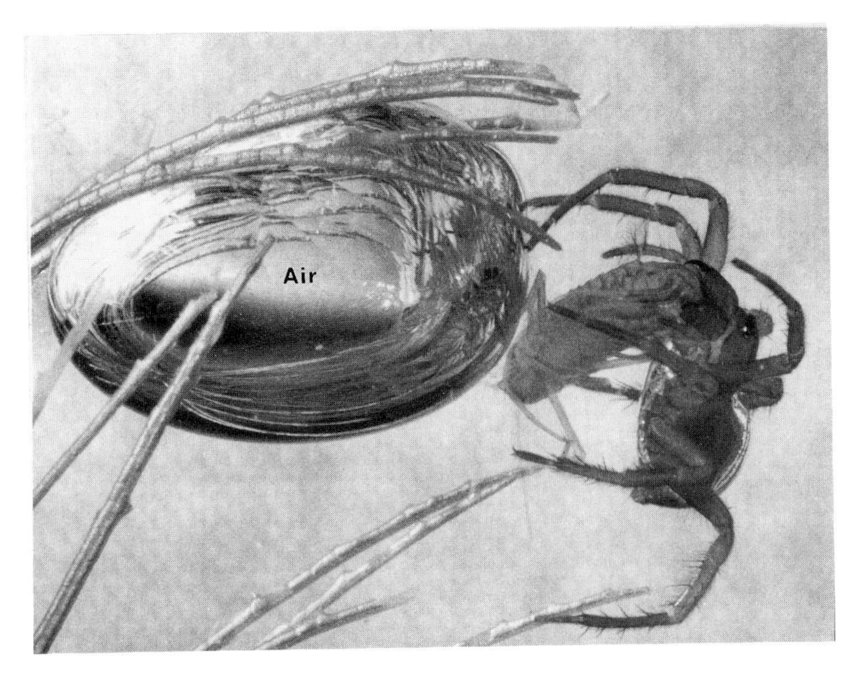

Fig. 135. The water spider *Argyroneta aquatica* carrying a mayfly larva to its diving bell (Air). (Photo: Vollrath.)

Dolomedes uses a more complex galloping gait that is five times quicker than ordinary rowing (75 cm/s; Bleckmann and Barth, 1984; Gorb and Barth, 1994).

The orb-web spider *Tetragnatha*, which is usually found near lakes or ponds, can also walk nimbly on the water's surface. Here only the first and second pair of legs move in the diagonal rhythm, whereas legs 3 and 4 are pulled behind passively. Despite this unusual technique *Tetragnatha* can attain the remarkable speed of 15–20 cm/s, faster than this spider walks on land (Ehlers, 1939).

The only spider that can walk and swim under water is the water spider *Argyroneta aquatica* (Fig. 135). Because of the air bubble that surrounds its abdomen, this spider is always in a labile, ventral-side-up position. The legs are in direct contact with the water and are not covered by air; this would increase buoyancy to such an extent that the spider could no longer dive. The legs are used not only for walking under water (along plants or silk threads), but also for free swimming. Although they move according to the diagonal rhythm, they also beat up and down, that is, the joints are bent and stretched, yet none of the customary lateral torsion of the coxal or body joints occurs (Ehlers, 1939).

Locomotory Activity

Most spiders are active at night. Notable exceptions are the jumping spiders, which move about during daytime but withdraw into their silken nests at dusk (Jackson, 1979). They are diurnal, of course, because of their dependence on vision. Similarly, lynx spiders (Oxyopidae), crab spiders (Thomisidae), and many wolf spiders (Lycosidae) are also active mostly during the day.

In contrast, many wandering spiders are hardly ever seen in the daytime; they venture from their retreats only at dusk. Before *Cupiennius* leaves its hideout, it goes through a fixed sequence of preparatory steps, each of which is correlated with a particular light level (Seyfarth, 1980). When the light intensity drops to about 15 lux, the animal turns around and stands above its retreat; there it waits until the illumination has decreased to 0.1 lux. Only then does the spider start to wander around. In females maximum activity is reached within two hours after dark, in males only around midnight (Fig. 136). Overall, males are more mobile, presumably because they are mainly out searching for females (Marc, 1990; Schmitt et al., 1990). In general, the night activity is advantageous for several reasons: (1) many prey animals (e.g., crickets, cockroaches, moths) are also nocturnal; (2) diurnal predators such as birds or reptiles are not encountered; and (3) high temperatures of the surface soil (over 70°C in the desert) and possible desiccation are avoided (Cloudsley-Thompson, 1961).

The duration of both the activity and the resting times is controlled endogenously. This can be concluded from the observation that *Cupiennius* shows almost the same activity rhythm, namely, of 24.9 hours, when kept continuously in the dark that it does under natural conditions. Such a circadian rhythm of locomotory activity is probably present in many other spiders (Cloudsley-Thompson, 1978).

The large desert "tarantulas" have also been known as strictly nocturnal animals (Cloudsley-Thompson, 1968, 1981). Although this is true as far as their hunting and free-wandering activity is concerned, it does not mean that they are completely inactive during day. Weaving activities, such as silking over the openings of their burrows take place during daylight (Minch, 1978).

The activity rhythm of web-building spiders has been studied mainly among the araneids (Ramousse and Davis, 1976; Ramousse and LeGuelte, 1979;

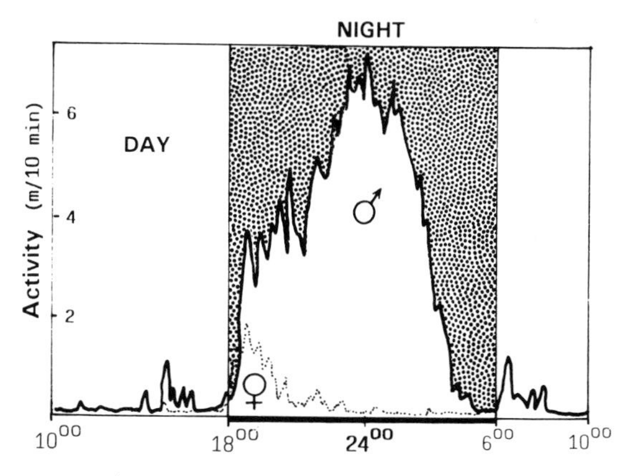

FIG. 136. Locomotory activity of the wandering spider *Cupiennius salei*. Males are active most of the night, while females roam only for 2–3 hours right after sunset. (Mean values of ten animals each; after Schmitt et al., 1990.)

Ramousse, 1980). Most species of *Araneus* build their webs preferentially at the beginning or at the end of the nocturnal period. Surprisingly, the first nymphs of *Araneus* are active mostly during light phases, but this behavior reverses itself after the spiderlings molt (Le Berre, 1979).

Many araneids are strictly nocturnal and may even destroy their web before daybreak (Eberhard, 1976; Stowe, 1978). Quite a few species have therefore been overlooked previously, since most spider watchers tend to be diurnal.

Prey Capture

The many ways in which prey is captured in a web have already been dealt with in chapter 5. The wandering spiders, however, do not rely on a snare, but must locate and overpower their prey directly. First we will take a close look at the predation strategy of wandering spiders and then compare the effectiveness of different methods of prey catching in some web spiders.

Wandering Spiders

The ctenid spider *Cupiennius* shall serve as a typical example for the events that take place during prey capture (Melchers, 1967). *Cupiennius* usually lies in ambush rather than hunts its prey actively. An attack is launched only when a prey animal comes very close. Most impressive is the speed of capture—the entire sequence, from attacking to subduing of the prey, can occur in less than 0.2 seconds (Fig. 137). One would expect that the spider's extremely fast grasping movements would follow a rigid motor pattern. However, high-speed

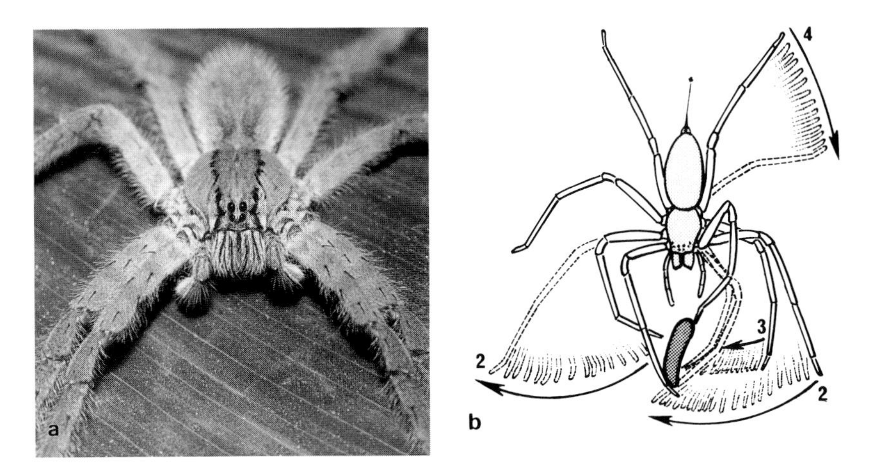

Fig. 137. Prey capture in *Cupiennius salei.* (*a*) The spider in its attentive posture on a banana leaf. (Photo: Seyfarth.) (*b*) Only the tarsi of the front legs handle the prey. The rapid change in the position of the legs (in only 25 ms!) is indicated by the dashed lines and the arrows. (After Melchers, 1967.)

cinematography (at 1,000 frames/s) has shown that these movements are by no means stereotyped but are well adapted to each specific situation.

First of all, the prey must be noticed and located. Visual cues play hardly any role for *Cupiennius*, for an experimentally blinded animal can catch its prey just as efficiently as an unblinded one. Instead, vibrations of the substrate or of the air (by wing beats), or the immediate contact with a victim elicit directed catching movements (Barth, 1982): the spider turns very rapidly toward the prey and grasps it with the front legs. The first gentle touch with the forelegs is quickly changed into a power grip. The legs are able to improve their secure hold by means of the adhesive hairs (scopulae) on the tarsi (Rovner, 1978). The prey is quickly pulled toward the spider's body and the chelicerae and their fangs move apart. Then follows the bite into the nearest part of the victim's

FIG. 138. Prey capture in a wolf spider (*a*, *b*, *Lycosa*) and a nursery-web spider (*c*, *Dolomedes*). (*a*) The wolf spider has seized a cricket and is catapulted through the air. (*b*) Often the spider lands on its back clamping the prey tightly above with all its legs. (Photos *a*, *b*: Rovner.) (*c*) After the bite, prey is held only with the chelicerae.

body. Immediately after the bite the tips of the legs release their grip so that the prey is held in the air only with the chelicerae, thereby minimizing danger to the spider from the prey's weapons. This too is advantageous because the victim has no contact with the substrate and therefore cannot apply any directed force to free itself (Fig. 138). Only after the prey has become immobilized by the poison does actual feeding (the chewing and the exuding of digestive juice) begin. As an extra measure, large prey items are fastened to the ground by some silk threads before the spider feeds.

In summary, the entire prey capture passes through the following stages:

1. Locating the prey.
2. Turning toward the victim and grasping it with the tips of the front legs.
3. Pulling the prey to the chelicerae and biting it (venom injection).
4. Releasing the grasp with the legs and holding the prey only with the chelicerae.
5. Fastening of some silk threads over the immobilized prey.
6. Feeding.

For most wandering spiders, mechanical vibration is the main releaser for prey capture. This is true even for vibrations transmitted on the water surface (Williams, 1979; Bleckmann, 1982). A *Dolomedes* resting at the edge of a pond can clearly distinguish between the ripples caused by wind and the surface vibrations generated by an insect (Fig. 139). A buzzing fly emits vibrations of high frequency (about 100 Hz) and small amplitude, whereas the natural "background noise" produced by gusts of wind is of low frequency (1–10 Hz) but high amplitude (Bleckmann and Rovner, 1984; Barth et al., 1988).

Some wolf spiders (for instance, some species of *Pirata*) use visual clues to orient themselves toward prey (Gettmann, 1976); most likely such spiders react to the prey's movements (Homann, 1931). Classic examples for visually guided prey capture are the jumping spiders (Fig. 140). Their highly developed main eyes can analyze shapes and therefore also recognize motionless prey (Fig. 72; Jackson and Tarsitano, 1993; for details see chapter 4).

As noted for Cupiennius, some hunting spiders place silken threads over the paralyzed victim and attach it to the substrate. The advantage of this behavior is that spiders that roam in the higher foliage strata can secure their

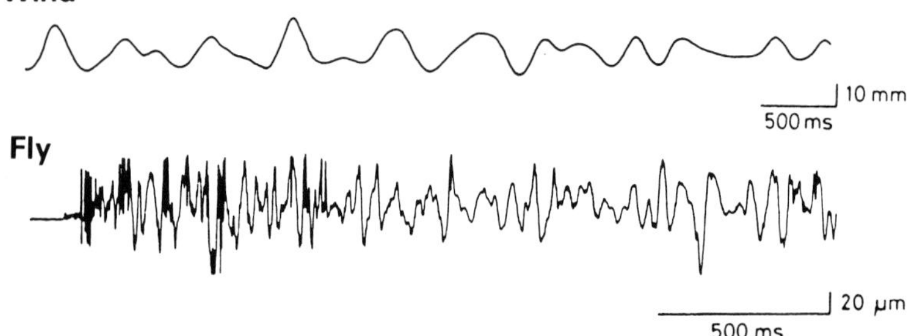

FIG. 139. Recording of surface waves caused by natural wind and by a buzzing blow fly that had fallen onto the water. Only the high frequencies of the prey (wing beat, about 100 Hz) trigger the capture behavior of the spider (*Dolomedes*). (After Bleckmann and Rovner, 1984.)

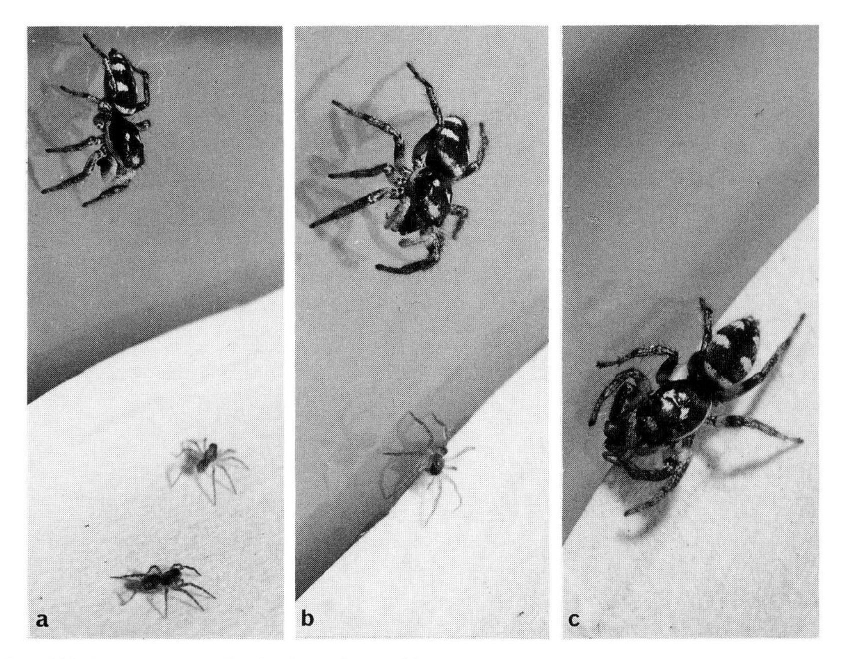

FIG. 140. Prey capture in the jumping spider *Salticus scenicus*. (*a*) The spider sits vertically on a glass wall and fixates its victim, a young wolf spider. (*b*) *Salticus* stalks the prey and is now within jumping distance. (*c*) The victim is seized with the front legs and the bite is applied.

prey near the catching site. If the spider is disturbed, the prey can then be left and still be relocated quickly (Rovner and Knost, 1974; Greenquist and Rovner, 1976).

The method of prey capture described for *Cupiennius* gives a general idea of how spiders hunt and capture their prey but does not apply equally to all wandering spiders. Even among the "ancient" Orthognatha we find many different levels of complexity with respect to predation (Buchli, 1969). On a rather simple level are some "tarantulas" (Theraphosidae) which merely move about until they come across potential prey. Other "tarantulas" (Avicularidae) climb in trees and jump at passing prey animals. The sedentary trapdoor spiders show a somewhat more complex behavior: they hide in a silk-lined, earthen tube, which is covered by a hinged lid, the "trapdoor." At night the spider lifts the trapdoor slightly and stretches out its front legs. The more "primitive" representatives of trapdoor spiders jump out and chase passing prey, whereas the more advanced species leave the tube only if prey is within easy reach. A further improvement was achieved when radial threads (signal or tripping threads) were added to the entrance of the burrow. This method is also used by the "ancient" Mesothelae (Klingel, 1966; Bristowe, 1975). Some Australian trapdoor spiders (e.g., *Aganippe, Anidiops*) attach a fan of "twig-lines" to the burrow's entrance (Fig. 141), which enlarge the foraging area and also aid in the transmission of vibrations (Main, 1976). Finally, in the orthognath Dipluridae we find sheet webs similar to those of the Agelenidae. As soon as

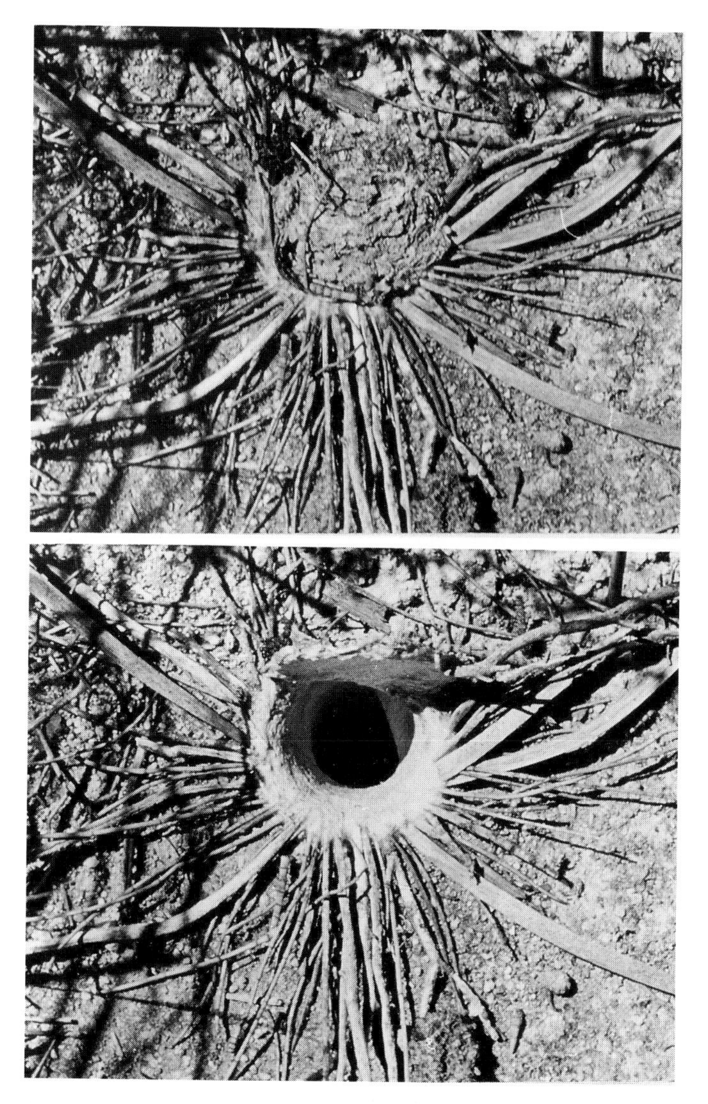

FIG. 141. The Australian trapdoor spider *Anidiops villosus* (Ctenizidae) digs deep burrows into the soil to protect itself from desiccation. When the door lid is closed, the entrance is hardly visible (above). Note the plant stems (Acacia) radiating from the entrance; they enlarge the capture area and also transmit prey vibrations to the burrow. (Photos: Main.)

a prey item falls on the web and sets up vibrations, the spider rushes out of its hiding place to attack it.

Web Spiders

Prey capture by web spiders does not differ in principle from that seen in wandering spiders. But there are many variations in the specifics of predation among the web spiders. For example, web spiders do not usually feed at the capture site but carry their prey to a safer place (into a retreat or to the hub of an orb web). The orb weavers (Araneidae) and the comb-footed spiders (Theridiidae) show another pecularity: they often subdue their prey by first wrapping it with silk before biting it.

The funnel-web spider *Coelotes* (Agelenidae) provides a good example of how sheet-web spiders capture their prey. *Coelotes* preys mainly on beetles, which blunder onto its sheet web (Tretzel, 1961a). The slightest vibrations in the web cause the spider to dart out from its subterranean retreat. In several quick sprints it takes a straight course toward the victim. When the spider has reached the beetle, it turns the insect on its back and tries immediately to puncture the segmental membranes of the abdomen with its chelicerae. The poison takes 8–10 minutes to produce any paralysis, and during this time the spider holds on firmly with its chelicerae. Only after the movements of the victim begin to weaken is the grasp somewhat loosened and digestive juice poured over the wound. When the beetle is severely paralyzed, its wings often open up, and the spider now has easy access to the softer parts of the body. The actual feeding begins only after the prey has been carried into the spider's retreat. There the immobilized prey may still be fastened to the ground with a few silken threads (Tretzel, 1961b), as was described above for some wandering spiders.

Prey capture has been most thoroughly studied in the orb weavers. Their methods of catching prey are even more variable than in other kinds of spiders. The common garden spider *Araneus diadematus* for example, often follows this sequence: (1) locating the prey in the web; (2) rapidly moving toward the prey; (3) immobilizing the prey; and (4) transporting the prey to the hub. Quite different strategies may be utilized, however, for the last two steps (H. M. Peters, 1931, 1933).

A fly buzzing close to a web does not elicit prey capture behavior directly; to produce this effect it has to touch the web itself (Klärner and Barth, 1982). When a fly becomes entangled in the sticky spiral thread of the orb web, it produces specific vibrations, which immediately excite the spider. Even if the fly then remains quiet, the spider will pluck several radial threads, apparently to probe the load on each radius. In other words, it tries to find the exact position of the prey. Even minute loads (0.1 mg) can be localized within the web (Liesenfeld, 1956). Especially if the fly moves its wings again, the spider will rush out of the hub using the exact radial thread that leads to the prey. The victim is briefly touched with the front legs and palps, then the hind legs wrap silk around it. Only thereafter follows a *brief* bite. Using its legs and chelicerae, the spider then cuts the neatly wrapped "package" from the web and carries it to the hub. There it is attached by a short thread before it is eaten. The feeding process always takes place in the hub, never at the actual capture site.

This rather general description must be made more specific. The main point is that the garden spider usually wraps its victim *before* biting it. This is in contrast to most wandering spiders. However, if the prey is very small (such as a fruit fly), it is simply grasped with the chelicerae and carried to the hub. Large insects, which set up strong vibrations in the web, are also bitten immediately, but then the bite lasts many seconds or even minutes. Such a *long* bite probably prevents a possible escape of strong prey animals. On the other hand, aggressive prey, such as wasps, are always wrapped first, and then bitten. Apparently in this situation it is safer for the spider to keep the dangerous prey at a distance. These few examples demonstrate that the biting behavior is not rigid but depends on the type of prey.

The same can be said for the spider's methods for transporting the subdued prey: if the victim is relatively light (about 10 mg), it is always carried with the chelicerae; heavier prey (about 80 mg) is always attached to a short thread and hauled behind with the fourth leg (H. M. Peters, 1933).

Other orb weavers (for example, *Argiope* species) behave like *Araneus*; that is, the prey is first wrapped and then bitten (Fig. 142). As a regular exception *Argiope* uses the bite first when dealing with moths (Table 3; Robinson and Robinson, 1974).

Many moths are stuck in the web only very briefly. They merely lose some wing scales on the sticky threads but escape quickly (Eisner et al., 1964). To capture a moth, then, the spider must be at the point of impact immediately. Even flies remain in an orb web for an average of only 5 seconds before they

Fig. 142. The tropical orb weaver *Argiope argentata* offensively wrapping a prey. The broad silk band emerges from the median and posterior spinnerets. (Photo: Robinson.)

TABLE 3. Strategy of Attack of *Argiope argentata* Toward Different Prey. (After Robinson and Robinson, 1974.)

	Strategy		
Prey	*Wrap→Bite*	*Bite→Wrap*	*Bite→Pull*
Grasshoppers	100%	—	—
Flies	74%	2%	24%
Moths	17%	75%	8%

can free themselves (Barrows, 1915). Usually, however, this is plenty of time for an alert orb-web spider to rush out for the capture. Once the spider has left the hub, most prey animals have hardly any chance for escape.

Unlike *Argiope*, the tropical orb weaver *Nephila* attacks *all* of its prey with a bite first and then performs the wrapping while sitting in the hub (Robinson and Mirick, 1971). This strategy can be inefficient and even risky if the "wrong" prey animals are encountered. This was shown in experiments with bombardier beetles (*Brachynus*, Carabidae), which were released into the webs of *Nephila* and *Argiope* (Eisner and Dean, 1976). Bombardier beetles possess a fantastic defensive weapon: they can aim explosive secretions toward an aggressor. When ejected from the beetle's abdomen, these secretions have a temperature of 100°C! Whereas a *Nephila* will be completely repelled by this weapon as it tries to bite the insect, an *Argiope* has already tightly wrapped the beetle, before the explosion can be triggered. Only when *Argiope* tries to bite through the swathing silk does the beetle try to defend itself. This has, however, little effect and leads only to further wrapping. Whereas *Argiope* can always overpower the bombardier beetle with the wrapping method, *Nephila* never succeeds with the biting strategy. Of course, it is true that bombardier beetles are not an essential item in the large prey-spectrum of araneids. On the other hand there are a number of insects that do use defensive chemical secretions. For instance, some stinkbugs (Pentatomidae) can defend themselves against a *Nephila* but not against an *Argiope* (Robinson, 1975).

In general, the offensive wrapping of prey yields two advantages for the spider: (1) there is less danger of being harmed by strong prey (a biting grasshopper or stinging wasp); and (2) it takes less time than applying the *long* bite. This second point becomes even more important when several prey items get caught in the web within a short period of time. The spider can fasten the first victim securely at the hub and can then rush toward the second. Such additional prey is also wrapped and bitten, but is left at the capture site as a "reserve."

Within the araneid family we find at least three related strategies of prey capture, which suggest successive steps in the evolution of their predatory behavior (Robinson, 1975; Fig. 143). *Strategy 1* is encountered sometimes in all araneids, especially if the prey is small. *Strategy 2* is typical of *Nephila*.

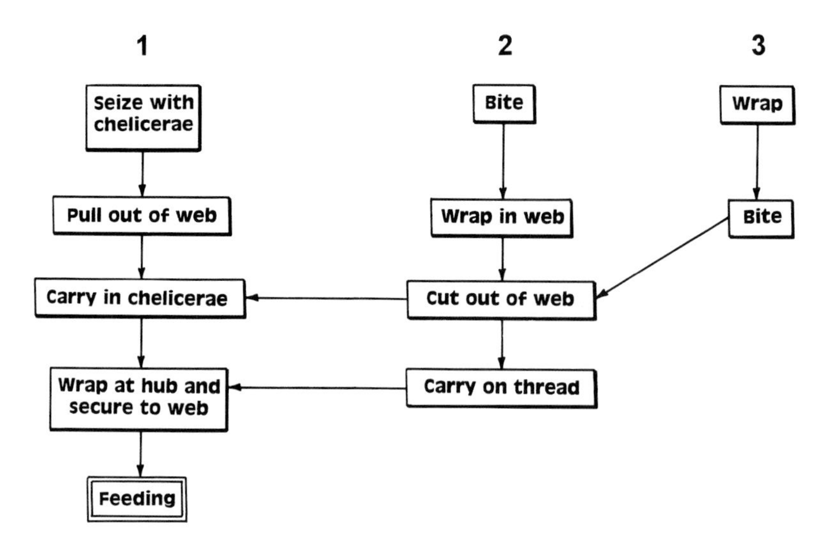

FIG. 143. Different strategies of predation in orb weavers (Araneidae). (After Robinson, 1975.)

Wrapping at the capture site makes it much easier to handle and carry bulky prey like butterflies or dragonflies. *Strategy 3* is preferred by *Araneus* and *Argiope*; the advantages have already been mentioned. The cribellate orb weavers (Uloboridae) have a fourth strategy: the prey is wrapped but not bitten at all. A bite would not be of much use anyway, as uloborids do not have poison glands. The offensive wrapping of prey is probably a more recent step in evolution, which might have derived from the defensive wrapping that takes place at the hub. In summary, about four different levels in the predation strategies of spiders can be distinguished (Eberhard, 1967):

Level	*Strategy*	*Type of web*	*Family*
1	Grasping and biting	No web	Theraphosids, clubionids, salticids
2	Grasping and biting	Ground web	Agelenids
3	Grasping, biting, and occasional wrapping	Aerial web	Linyphiids
4	Grasping, biting, wrapping or wrapping and biting or wrapping only	Aerial web	Theridiids, araneids, uloborids

Predation Specialists

So far I have tried to establish some general principles underlying the many different approaches spiders use to capture their prey. Some spiders, however, have developed highly specialized methods of predation that do not fit any general scheme. A few representatives of such spiders were mentioned earlier, for instance, the spitting spider *Scytodes* and the triangle spider *Hyptiotes*. The bolas spider *Mastophora* was briefly touched upon in chapter 5 but deserves a more detailed discussion here. One would think that the method of a bolas spider—to throw a sticky droplet after a flying insect (Fig. 144)—could hardly be successful. Such a strategy seems even more absurd if we consider that these spiders hunt at night. However, bolas spiders are quite adept at capturing certain moths (*Spodoptera*) or, to be more precise, only the male *Spodoptera*. The explanation of this baffling phenomenon is that bolas spiders apparently secrete a volatile substance that imitates the sex pheromone of the female *Spodoptera*. The male moths are attracted by this scent and fly in circles around the spider. Thus the spider has a much better target at which to throw its bolas. On the average, a bolas spider catches two or three moths a night. This is comparable indeed to the number of prey a regular orb weaver catches with its web (Eberhard, 1977, 1980).

Some spiders have become specialized for a certain prey type and have developed appropriate capture methods. Well-known examples are the pirate spiders (Mimetidae), which prey exclusively on other spiders, and the Zodariidae, which hunt only ants. The pirate spider *Ero* attacks other spiders by very briefly biting into one of their extremities. The injected poison acts very quickly—the victim is paralyzed within a few seconds. *Ero* then approaches again and starts sucking the prey from the bite hole. An even more specialized method is used by *Ero* when it is after the orb weaver *Meta segmentata*: *Ero* uses the courtship thread of the male and plucks it, supposedly imitating the courtship signals. The female *Meta* is apparently tricked, since she comes out of the hub—only to fall victim to *Ero* (Czajka, 1963). Other members of the mimetid family also use this method of "aggressive mimicry" by plucking the threads of web spiders.

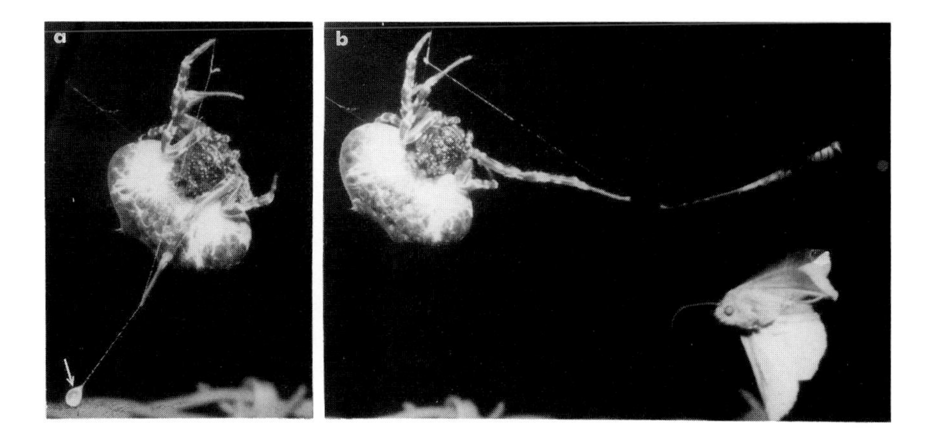

FIG. 144. Hunting method of the bolas spider *Mastophora*. (*a*) A front leg holds a short thread with a glue droplet (arrow) at its end. (*b*) When a moth passes close by, the bolas is hurled at it. (From Eberhard, 1977; Copyright © 1977 by the American Association for the Advancement of Science.)

The vibrations are presumably mistaken as coming from prey and, when the web owner rushes out, it is immediately bitten (Jackson and Whitehouse, 1986).

The ant hunter *Zodarium* (Zodariidae) attacks ants (*Cataglyphis*) at dusk, especially the guards at the nest entrance (Harkness, 1975, 1976; Harkness and Wehner, 1977). At first the ant tries to escape, but *Zodarium* will touch it quickly several times and then bite into one of its legs. After a minute the ant is paralyzed, and is carried away and eaten. Strangely enough, other ants do not come to the defense of the victim. It is believed that *Zodarium* camouflages itself chemically by producing a secretion that imitates the typical ant scent (Jocqué and Billen, 1987). Exclusive ant hunters are sometimes also found in other spider families, such as the Salticidae and Gnaphosidae (Edwards et al., 1974; Jackson and van Olphen, 1991). Many theridiids prey preferentially on ants (Carico, 1978) and one species is known to be specialized on termites (Eberhard, 1991).

One member of the gnaphosid family, *Callilepis nocturna*, has a highly specialized method of hunting ants (Heller, 1974, 1976). *Callilepis* runs in short bursts; the motion resembles that of certain ants (for example, *Formica*), but the body does not have an antlike appearance at all. Thus *Callilepis* cannot be considered as an ant mimic. The ant is always attacked head-on (Fig. 145a). The tarsi of the spider's front legs contact the ant's head and probe for the base of the antennae (Fig. 145b,c). Then follows a quick bite (0.2 seconds) at the base of one antenna, and the spider withdraws completely. A minute later *Callilepis* searches for the victim again and applies a longer bite. Initially the bitten ant is quite aggressive, but within a few seconds its injured antenna becomes limp and the ant starts walking in small circles (right-hand circles if the left antenna

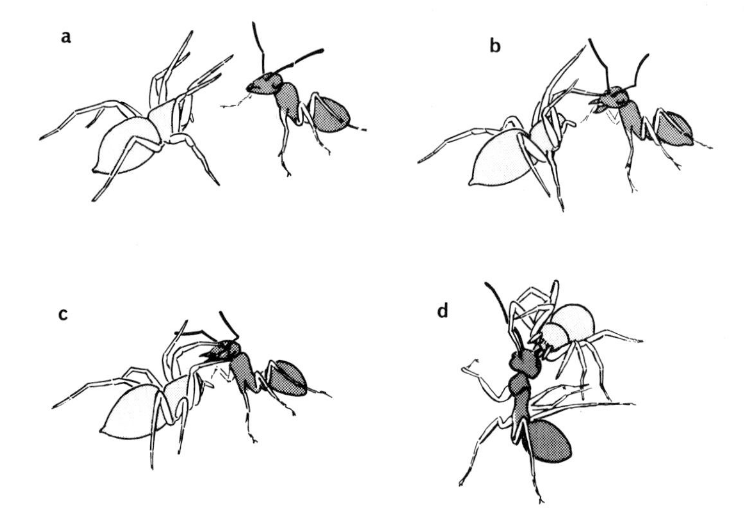

FIG. 145. Prey capture technique of the ant spider *Callilepis*. (*a*) The spider assumes a jumping position and raises its front legs. (*b*) One tarsus contacts the base of an antenna. (*c*) Both tarsi of the front legs locate the bases of the antennae. (*d*) Bite at the base of one antenna. (After Heller, 1974.)

was bitten and vice versa). Thus the ant hardly moves away from the spot where it was first attacked, and therefore it can easily be relocated afterwards. *Callilepis* tucks the paralyzed ant underneath herself and runs quickly for a hiding place. During this maneuver she is often attacked by other ants, but somehow she manages to dodge them. Once she has reached a safer place, she closes it off with a silken cover and starts feeding. The prey is never chewed, but is sucked from the "neck" and abdomen. After one or two hours the undamaged cuticular shell is all that remains of the ant.

The very rapid and precise capture method of *Callilepis* represents an adaptation to cope with strong and potentially dangerous prey. The bite is placed only at the vulnerable base of an antenna, and the legs are immediately removed from the reach of the ant's formidable mandibles. The predation strategy of *Callilepis* is rather rigidly "programmed:" if both antennae of the ant are cut off, the spider still attacks but never bites it. Apparently the bite is "blocked" when *Callilepis* cannot locate the antennal bases. On the other hand, if one glues two antennae on the ant's abdomen, the spider will bite there. The antennae must therefore be considered the key stimulus for the spider's bite (Heller, 1976).

Certain tropical jumping spiders seize tree ants (*Pseudomyrmex*) and then let themselves drop off a branch. While dangling on their safety thread they start feeding on the victim. Although the attacked ant releases an alarm pheromone that will recruit conspecifics, this is to no avail. The agitated helpers can only run back and forth on the branch above, but will not succeed in climbing down the safety thread towards the spider (Robinson and Valerio, 1977).

We want to conclude this chapter with a description of the unique prey capture of the ogre-faced spider *Dinopis*. This tropical spider hangs upside down in a vertical frame of threads, somewhat similar to an orb weaver (Robinson and Robinson, 1971; Coddington, 1986). The actual catching web, however, is held between the front legs (Fig. 146). Although the spider hunts mostly at night, it locates its prey visually. This is made possible by two huge posterior median eyes (1.4 mm in diameter), which are similar to the main eyes of jumping spiders (anterior median eyes; 0.4 mm diameter). Unlike the main eyes of jumping spiders, the enormous eyes of *Dinopis* do not have better resolution but are 2,000 times more sensitive to light (Blest and Land, 1977). For a nocturnal hunter this is, of course, a definite advantage.

When an insect comes within reach of *Dinopis*, its front legs spread the rectangular catching web and throw it like a sweepnet over the prey. The elastic web is jerked several times so that the victim becomes completely entangled in the hackle threads. Additionally *Dinopis* uses its hind legs to pull out silk threads from the spinnerets to wrap the prey. During this process the prey is bitten once, or several times, and thereafter actual feeding begins. It is curious that the spider often starts constructing a new catching web while it is still busy feeding. This is possible because the wrapped prey is handled only with the palps, and the legs are thus free for building a new capture web. Recent studies have shown that *Dinopis* can use two different hunting strategies depending

Fig. 146. Prey catching in the Australian ogre-faced spider *Dinopis*. The spider extends a rectangular web with cribellate catching threads between its front legs (*a*, *c*). (*b*) When an insect comes within reach, the web is swept over the prey. (From Kullmann and Stern, 1975.) (Inset: Land.)

on the type of prey. Flying insects, causing air vibrations, are attacked by flinging the capture net upwards; ground-living insects are perceived visually and are overwhelmed by throwing the net downwards (Coddington and Sobrevila, 1987).

Although spiders always catch live prey, this does not mean that they do not also feed on dead animals. In the laboratory they accept crushed insects and even animals that have been dead for several days (Knost and Rovner, 1975). Whether scavenging is common in nature has not been thoroughly investigated, yet there are some indications that it occurs regularly in some wandering spiders (Gettmann, 1978). In colonies of the social spider *Stegodyphus mimosarum* even cannibalistic scavenging has been observed (Lubin and Bar Shahal, 1995, pers. comm.). Given the variety of techniques that spiders have mastered to capture their prey, and the diversity of their habitats and diets, such a finding would not be surprising.

7

Reproduction

Spiders are always dioecious, that is, they invariably have separate sexes. Aside from a few exceptions (for example, *Argyroneta*), the females are larger than the males. This sexual dimorphism is especially obvious in many tropical orb weavers (such as *Nephila, Gasteracantha,* and *Micrathena*), where the males appear to be dwarfs (Fig. 147). There has been much speculation about the biological meaning of the small body size of male spiders, but so far no consensus has been reached (Gerhardt, 1924; Vollrath, 1980). In any case, the small spider males are very agile, and some can even "fly" on their own thread, just as young spiders do.

Because of their small body size, males need fewer molts to reach maturity than do females; consequently, males mature earlier. After their last molt the males have conspicuously thickened palpal tarsi and can thereby be distinguished easily from the females. The female palps represent simply a kind of shortened leg (without a metatarsus), but the male palps have tarsi that are specialized for the storage and transfer of sperm. This function of the male palps as copulatory organs is highly unusual, and nothing comparable exists in other arthropods.

In contrast to females, most male spiders change their habits after their last molt. They leave their retreats or webs, and become vagabonds; often they no longer even catch prey. As soon as they have charged their palps with sperm, they start wandering around, searching for a female. Usually they are rather cautious when approaching a female, because they always risk being dealt with as prey. Spiders have therefore developed a special courtship behavior that generally precedes mating. This courtship is species specific and ensures that hybridization is avoided. The common belief that spider males are eaten by the females during or after copulation is true only for a very few species. In most cases a male either walks away or beats a hasty retreat right after copulation. His palps are then refilled with sperm. This procedure, however, can be repeated only a few times, since most male spiders have rather short lives; many die soon after copulation. Females usually live much longer, since they must still lay eggs and build cocoons. In some species the females also exhibit brood care for the developing young.

FIG. 147. Size dimorphismus of the orb weaver *Argiope aemula* from New Guinea. The dwarfed male approaches the female sitting in the hub. (Photo: Robinson.)

Internal Sexual Organs

The internal sexual organs, the testes and ovaries, lie as paired structures inside the abdomen. The reproductive cells, the sperm and the eggs, are released to the outside in both sexes through a ventral opening (the epigastric furrow) which is situated between and slightly behind the book lungs (Fig. 148). The males exude their sperm through this opening onto a special sperm web and

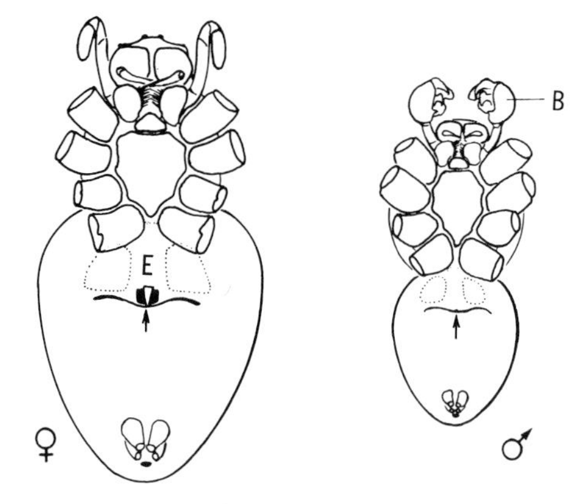

FIG. 148. Position of the external sexual organs of *Araneus diadematus*; ventral view. The genital opening lies inside the epigastric furrow (arrow). The epigynum (E) is situated in front of this furrow. Book lungs are outlined with dots. B = bulb of palp. (After Grasshoff, 1973.)

then transfer it to their palps. The internal sexual organs are more complex in females, which also have a special storage site for sperm, the seminal receptacles.

Testes and Sperm Cells

The two testes are highly convoluted tubes (Fig. 149) that merge into a common duct that leads medially into the epigastric furrow. The differentiation of sperm cells (*spermiogenesis*) takes place in the epithelium of the testicular tubes. The early stages of sperm cells (spermatids) are flagellated, although the flagella have a rather unusual arrangement of their axial tubules: spider sperm axonemes are assembled according to a 9 + 3 rather than the common 9 + 2 pattern (Fig. 150b; Osaki, 1969; Bacetti et al., 1970; Alberti and Weinmann, 1985). During spermiogenesis the flagellum looses its membrane and is apparently resorbed by rolling itself tightly around the nucleus of the sperm cell (Fig. 150; Reger, 1970). Concomitantly, the nucleus assumes a spiral shape (Fig. 149c,d) and the number of mitochondria becomes reduced (Lopez et al., 1983). The differentiated sperm cell is a disk, surrounded by a protein coat (Boissin, 1973).

Three kinds of coatings can be distinguished among spider sperm cells: (1) *cleistosperms*, in which each sperm cell has its own, separate coat; (2) *coenosperms*, where up to 100 sperm cells are enclosed by a common coat (Bertkau, 1877); and (3) *synsperms*, in which several spermatids fuse into one syncitium (Alberti, 1990). Mesothelae and Theraphosidae have coenosperms and are therefore considered as "primitive" (Alberti et al., 1986). Most Araneomorphs are cleistosperm and this is regarded as a derived character—as is synspermy, found in some haplogyne spiders.

The globular packaging of sperm cells can be interpreted as a kind of protection during storage or transport. When sperm is taken into the male

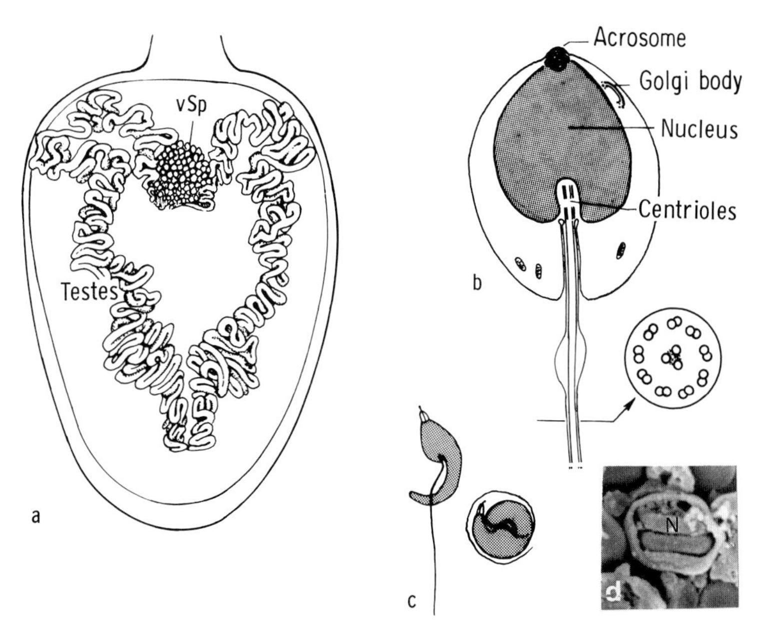

FIG. 149. (*a*) Testicular tubes in the abdomen of the tarantula *Grammostola*. The genital pore is covered by a ventral spinning field (vSp); these "epiandrous glands" contribute to the male's sperm web. (After Melchers, 1964.) (*b*) Spermatid of the lynx spider *Oxyopes*. The cross-section of the flagellum shows the unusual 9+3 arrangement of microtubules. 4,200 ×. (Drawn from electron micrographs by Osaki, 1972.) (*c*) Differentiation of the mature sperm cell by rolling the flagellum around the nucleus. (After Bösenberg, 1905.) (*d*) Sperm cell of *Pholcus* with partly removed surface coat, exposing the spiral-shaped nucleus (N). 5,100 ×. (Photo: Uhl.)

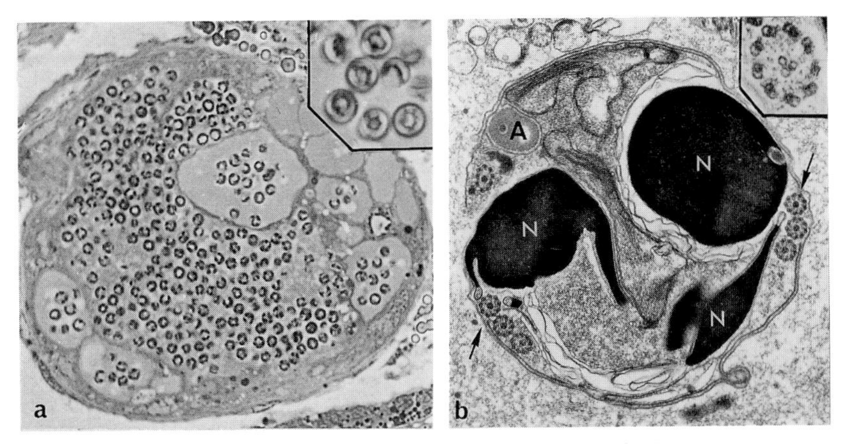

FIG. 150. (*a*) Cross-section of a testis of *Zygiella*. The lumen of the testicular tube is filled with ring-shaped sperm cells. 540 ×. *Inset*: 1,500 ×. (*b*) Differentiated sperm cell of *Pisaurina* as seen in the electron microscope. The flagellum (arrows) has been incorporated into the cytoplasm. A = acrosome, N = sections through the nucleus. 24,000 ×. *Inset*: Cross-section of flagellum showing the typical 9+3 arrangement of microtubules. 50,000 ×. [Photo (*b*): Reger.]

palps and also when it is transferred to the receptacula of the female, the sperm cells remain packed in little cysts and are therefore immobile. Only before fertilization does the protective protein coat dissolve, rendering the sperm cell motile (Brown, 1985). If water is added to freshly deposited spider sperm, vigorous movements can be observed (Bösenberg, 1905; Gerhardt and Kaestner, 1938).

In the course of spermatogenesis, cell divisions (mitosis and meiosis) can be studied particularly well. In those cells that are just in the process of dividing, the chromosomes can be made visible by special staining techniques. Under favorable conditions the entire set of chromosomes can be counted under the microscope. The spider species investigated to date (over 300) exhibit 20–30 chromosomes in the diploid stage. Crab spiders have about 22 chromosomes, wolf spiders 22–28, and jumping spiders 28–30 (Hackmann, 1948; Mittal, 1964, 1966; Painter, 1914).

Aside from the "regular" chromosomes (the autosomes, A), two special chromosomes represent the sex chromosomes (X chromosomes). Whereas female spiders possess a dual set of these X chromosomes ($2A+X_1X_1X_2X_2$), male spiders have but a single set ($2A+X_1X_2$). Consequently, after meiosis the haploid egg cells have $A+X_1X_2$ chromosomes, but the sperm cells are of two types: either they likewise possess $A+X_1X_2$ or they lack sex chromosomes altogether ($A+O$). Sex determination is thus dependent on the type of sperm cell that fertilizes the egg. Some spider species have only one X chromosome in the male, others even three. A Y chromosome, as is typical for many insects and mammals, is very rare among spiders (Maddison, 1982).

Many male spiders possess several glandular openings near the epigastric furrow. They are connected to small silk glands, which provide part of the sperm web, and also to some unicellular glands, which may produce pheromones. They have been variously referred to as *epiandrous glands* (Marples, 1967), or *ventral spinning field* (Melchers, 1964; Fig. 149a), or *epigastric apparatus* (Lopez and Emerit, 1988).

Ovaries and Egg Cells

The ovaries are paired, elongated structures situated in the ventral part of the abdomen (see Fig. 28). They are reminiscent of the ovaries of birds, because the egg cells (oocytes) bulge into the body cavity (Fig. 151a) and are connected to the ovarian epithelium by a short stalk (the *funiculus*). How the egg cells are released is still a matter of conjecture. The so-called oviducts represent the inside of the ovarian epithelium and are therefore not in an adequate position to take up the maturing egg cells to enter the lumen of the "oviduct" (Sadana, 1970).

In many spider families the egg cells contain a typical structure that has been called the vitelline body, or Dotterkern (von Wittich, 1845). It consists of several concentric lamellae (Fig. 151b), which have been identified as endoplasmic reticulum (Sotelo and Trujillo-Cenóz, 1957; André and Rouiller,

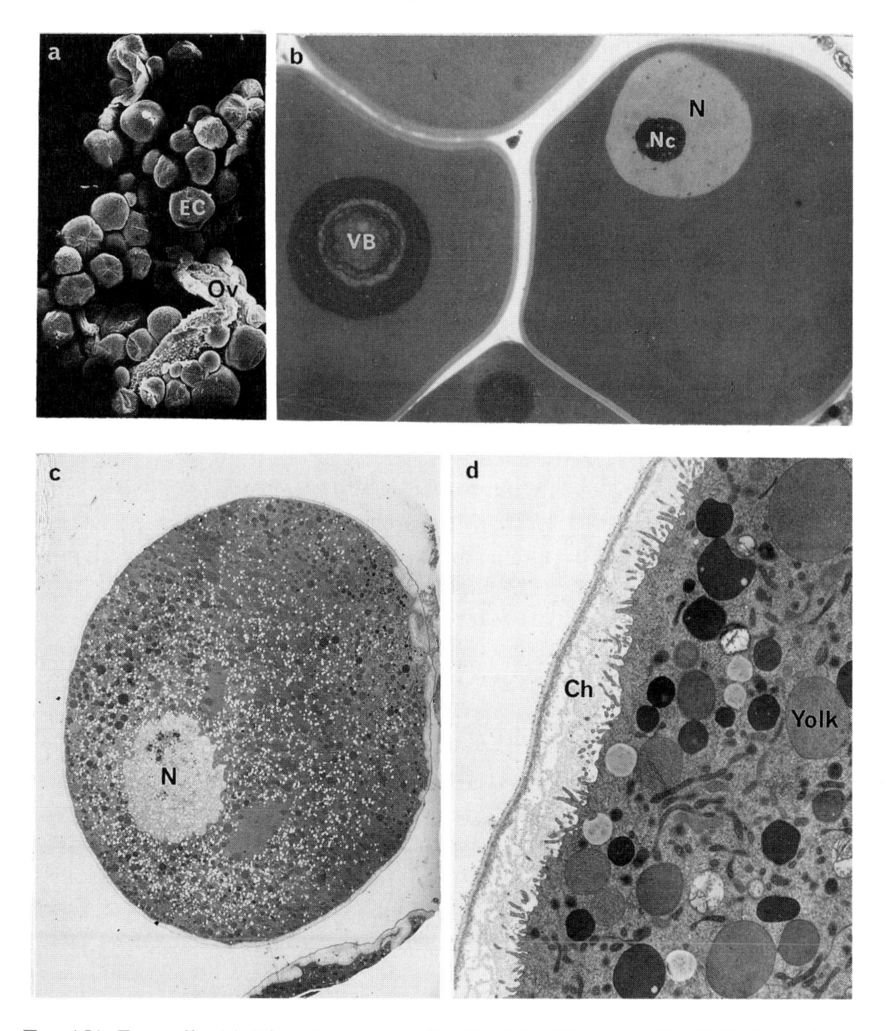

FIG. 151. Egg cells. (*a*) Dissected ovary of a *Zygiella*. The egg cells (EC) sit on the outside of the ovary (Ov). 60 ×. (*b*) Histological section of two young egg cells. A large pale nucleus (N) with a distinct nucleolus (Nc) is seen in the cell on the right; in the cell on the left a conspicuous vitelline body (VB) is surrounded by fine granular yolk. 380 ×. (*c*) The yolk granules are much larger in mature egg cells. 280 ×. (*d*) The periphery of a mature egg cell contains various yolk droplets and microvilli projecting into the chorion layer (Ch). 5,400 ×.

1957). Nothing definite is known about the function of the vitelline body, but ostensibly it represents an organizing center for yolk formation (Osaki, 1972). Very recent fine structural studies indicate that the vitelline body contains either mitochondria or bacteria (Hauser, 1994, pers. comm.). Most likely, these bacteria are symbiontic, but their precise role needs to be determined.

Yolk accumulation occurs in two steps (Seitz, 1971). At first, very fine grained yolk particles aggregate in the young egg cell until the cell reaches a diameter of approximately 100 μm. Only after copulation does a second accumulation of yolk begin, this time in the form of much larger granules. The second phase of yolk accumulation can take place only if enough food is available. After copulation, female *Cupiennius*, for example, voraciously attack their prey. Within 2 weeks the volume of an egg cell increases 10–12-fold. Accordingly, the female's abdomen swells visibly, and just prior to egg deposition four-fifths of her opisthosoma is taken up by the ovaries.

The total mass of the eggs that a female spider produces is impressive. *Cupiennius,* for instance, produces 1,500–2,000 eggs for her first cocoon, equivalent to a weight of 1.5–2.5 g (Melchers, 1963). The subsequent four or five cocoons are each made at intervals of 45 days; they contain successively fewer eggs.

It is noteworthy that the yolk of an egg cell must contain the entire energy that is needed for embryonic development, hatching, molting, as well as all the activities of a young spiderling before it actually catches its first prey. If the embryonic development happens to fall into the cold season, this will mean up to 200 days without any external energy input (Schaefer, 1976b). It is certainly advantageous that the metabolic rate during early development is much reduced, i.e., only 10–20% of the adult metabolism (Anderson, 1978).

External Sexual Organs

Male spiders lack primary copulatory organs. Instead, their pedipalps have been modified to transfer sperm. In many "primitive" spiders (Orthognatha, Haplogynae—but not Mesothelae) the males have only slightly modified pedipalps and the females correspondingly rather simple copulatory organs. On the other hand, in entelegyne spiders a good correlation between the highly differentiated palp of the male and a complex epigynum of the female is evident. The external sexual organs are often so specific to a species that systematicists use them as decisive characters for species identification (Fig. 216). It is tempting to postulate that the highly specialized "lock-and-key" mechanism of male and female copulatory organs serves to prevent cross-breeding between different species. However, we have no experimental proof for this assumption. More likely, hybridization is already inhibited at the behavior level: courtship is discontinued if one of the spiders does not react in a species-specific way. In fact, many male spiders do try to court females of the "wrong" species (Fig. 152); they are, however, usually unsuccessful, because the females are responsive only to males of their own species (Crane, 1949; Grasshoff, 1973).

A good example is seen in the courtship of several closely related wolf spiders, so-called *ethospecies*, which differ very little in their morphology but can be distinguished by their behavior (Hollander and Dijkstra, 1974; Kronestedt, 1990; Uetz and Denterlein, 1979; Uetz and Stratton, 1982). *Schizocosa ocreata* and *Schizocosa rovneri*, for instance, exhibit almost identical genitalia, yet do not cross-breed under natural conditions, because females accept only conspecific males (Stratton and Uetz, 1981). However, if the females

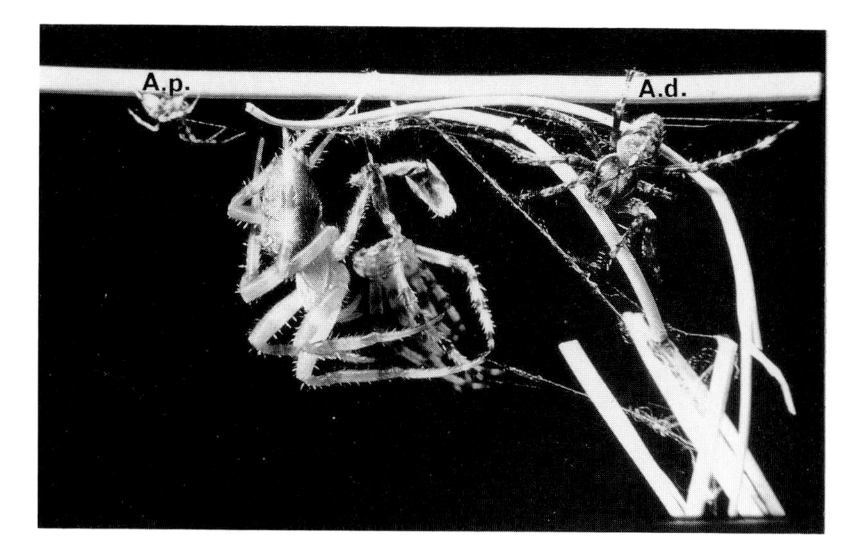

FIG. 152. A female garden spider (*Araneus pallidus*) immediately after her final molt. Note the two males waiting: on the left an *A. pallidus* (A.p.), on the right an *A. diadematus* (A.d.). (Photo: Grasshoff.)

were briefly anesthetized with carbon dioxide, males of the "wrong" species did mount them and copulated successfully. Such females laid fertile eggs from which hybrid spiderlings developed (Stratton and Uetz, 1983, 1986). Under natural conditions there are further barriers that prevent cross-breeding, e.g., a slightly different habitat of the two species or a seasonal separation of the courtship periods (Fig. 194). At any rate, behavioral differences are apparently more important for the isolation of species than genital morphology, as the above experiment clearly shows. It cannot be excluded, however, that interspecific hybridization may occur occasionally in the wild, e.g., in *Tegenaria* species (Oxford and Smith, 1987; Oxford and Plowman, 1991).

Male Copulatory Organs

The simplest form of a male palp is seen in the haplogyne and orthognath spiders. The tarsus of the palp (the *cymbium*) carries an extension in the form of a pear-shaped bulb, or palpal organ, which acts as a reservoir for the sperm (Fig. 153a). A blind duct spirals through the bulb and opens at its very tip; the narrow portion of the tip is called the *embolus*. The bulb functions like a pipette and can take up a droplet of sperm. The sperm cells are then stored inside the duct (*spermophor*) until mating. It is not quite clear how the sperm are expelled, since an elevated hemolymph pressure does not seem to be involved. Presumably the mass of sperm becomes displaced by a glandular secretion which is discharged into the spermophor (Harm, 1931).

The male palps are much more complex in entelegyne spiders (Fig. 153b), because the wall of the palpal organ consists of hard, sclerotized parts (the

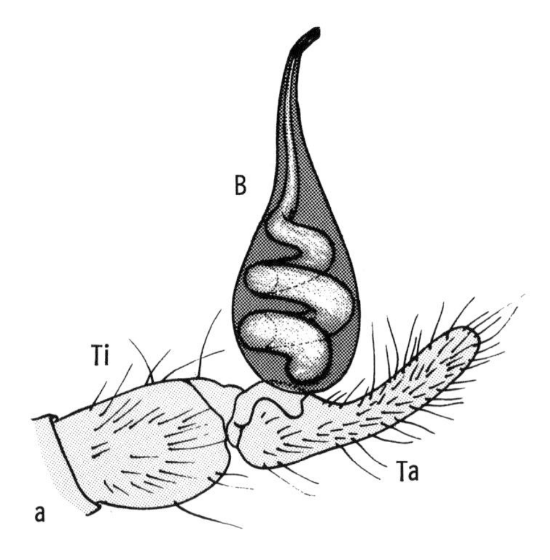

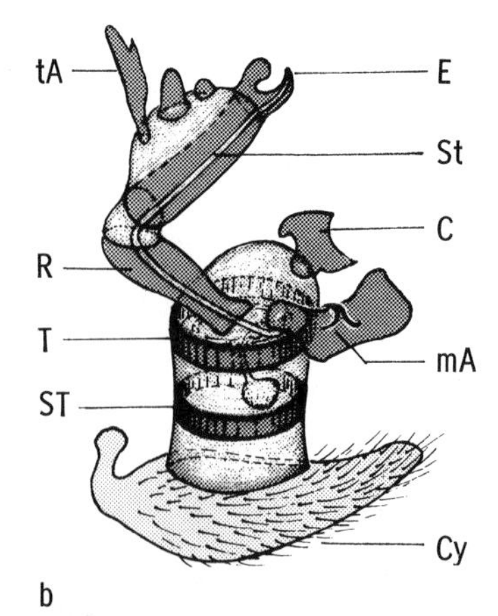

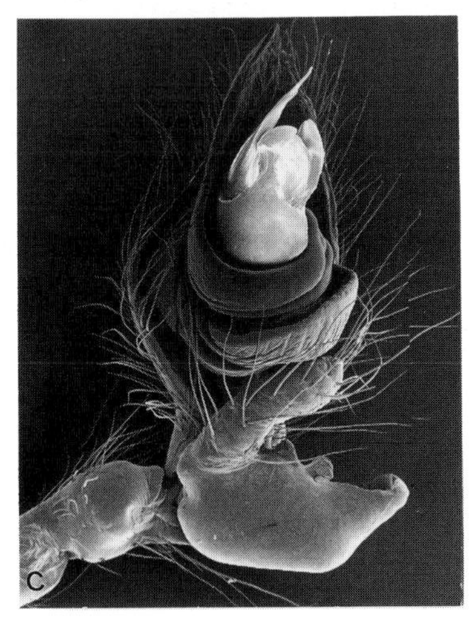

FIG. 153. (*a*) Simple type of a male palp in *Segestria florentina*. The spiraled spermophor can be seen through the bulb (B). Ta = tarsus, Ti = tibia. 78 ×. (After Harm, 1931.) (*b*) Complex type of a male palp in *Araneus diadematus*. The diagram shows the expanded state. Cy = cymbium. Sclerites: C = conductor, E = embolus, mA = median apophysis, R = radix, St = stipes, ST = subtegulum, T = tegulum, tA = terminal apophysis. The extensible hematodochal areas are lightly stippled. (After Grasshoff, 1968.) (*c*) Scanning electron micrograph of a palp of a male *Meta menardi*. 35 ×. (Photo: Opell.)

sclerites) and soft areas (the *hematodochae*); both can bear special protrusions (*apophyses*) which play an essential role during copulation. Such complex genitalia can only be understood by regarding them as functional units (Grasshoff, 1975; Weiss, 1982). Recently, a freeze-fixation of mating spiders with liquid nitrogen and subsequent serial sectioning of the locked copulatory organs have provided further insights into their function (Huber, 1993, 1994).

The soft hematodochae are inflatable and allow the palpal organ to expand hydraulically (Homann, 1935). Normally the entire bulb is collapsed so that the fragile embolus is protected from mechanical damage. When the hemolymph pressure is increased, the hematodocha expands, and the sclerites become erect and project from the rest of the palp (Fig. 154). Theses sclerites assume defined spatial positions, and this is a precondition for a secure coupling with the female's epigynum.

The various sclerotized parts of the palpal organ have their own names. The spoon-shaped base of the tarsus is referred to as the *cymbium*, the basal

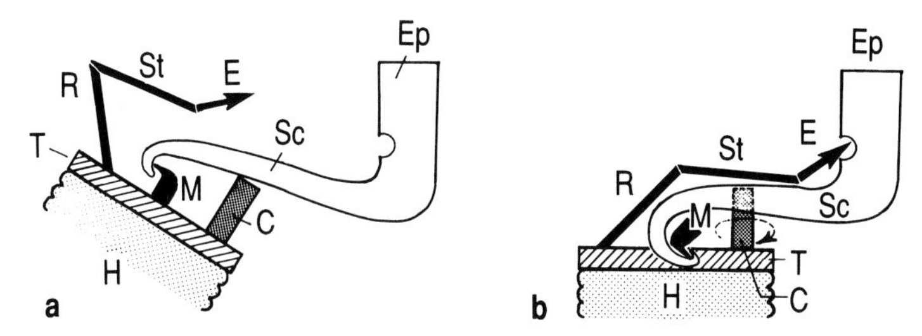

FIG. 154. A mechanical diagram depicting the coupling of the bulb and the epigynum in Araneus. (*a*) The spine of the median apophysis (M) hooks onto a process (Sc, the scapus) of the epigynum (Ep). (*b*) The median apophysis twists and becomes enveloped by the scapus. The sclerites R, St, and E shift their position so that E (the embolus) becomes situated in front of the epigynal opening. The inflation of the hematotocha (H) causes a rotation of the tegulum (T), which in turn presses the conductor (C) to one side of the scapus and the embolus into the epigynum. (After Grasshoff, 1973.)

appendage of the tarsus as the *paracymbium*. The arrangement of the different sclerites (*tegulum*, *subtegulum*, *median apophysis*, *conductor*, embolus) is shown in Fig. 153b. In some entelegyne species the embolus can be drawn out into a long, spiraled thread, sometimes several times the length of the male's body. These lengthy emboli often break off during copulation and remain stuck in the epigynum of the female. This happens regularly in the golden silk spider *Nephila clavipes* and in the black widow *Latrodectus* (Wiehle, 1961). Until quite recently it was believed that this would prevent further copulations, but it has been shown that females with emboli that have been broken off in their epigynes as well as males with mutilated emboli (stumps) can actually mate again (Breene and Sweet, 1985). It is believed in such multiple matings that sperm from the first male fertilizes most of the eggs (*first-male sperm priority*; Austadt, 1984; Watson, 1991), but there are also exceptions (Masumoto, 1993).

Female Copulatory Organs

The terminal part of the oviduct is referred to as the *uterus externus*. In the Orthognatha and the Haplogynae it ends in the primary genital opening (*gonopore*) and is situated between the anterior book lungs. The spermathecae, or seminal receptacles, connect directly with the uterus externus, which is also the site of fertilization. Thus, in the Haplogynae the copulatory duct is at the same time a fertilization duct (Fig. 155). In the Entelegynae the females possess a special copulatory organ located in front of the genital opening, the *epigynum*. This is a slightly raised sclerotized plate with several cuticular infoldings, which constitute the connecting ducts (sperm ducts) and the seminal receptacles (Fig. 156).

A closer look at the epigynum shows that it is quite complex. Aside from the primary genital opening (gonopore)—which lies hidden in the epigastric furrow—we find other orifices that play a role in reproduction. Those of the

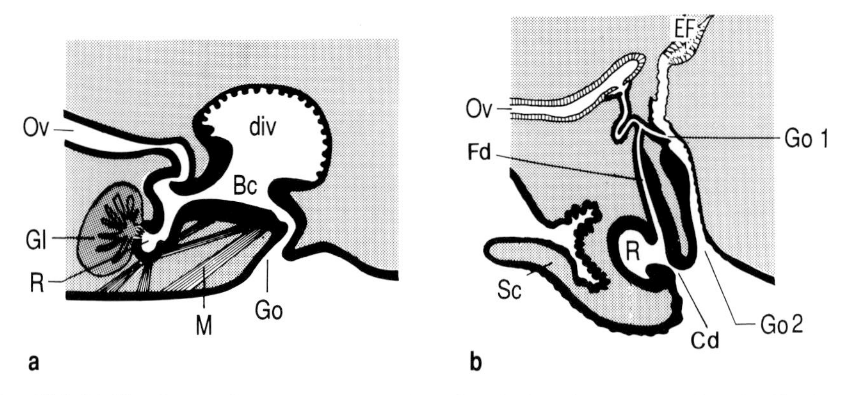

Fig. 155. (*a*) Simple copulatory apparatus in a female *Dysdera crocata*; longitudinal section. Bc = bursa copulatrix with a dorsal evagination (div); Gl = gland attached to the seminal receptacle (R); Go = genital opening; M = muscles; Ov = oviduct. (After Cooke, 1966.) (*b*) Complex copulatory apparatus in a female *Araneus diadematus*. Cd = copulatory duct; EF = epigastric furrow, Fd = fertilization duct; Go 1 = internal genital opening; Go 2 = external genital opening; Sc = scapus. (After Gerhardt and Kaestner, 1938.)

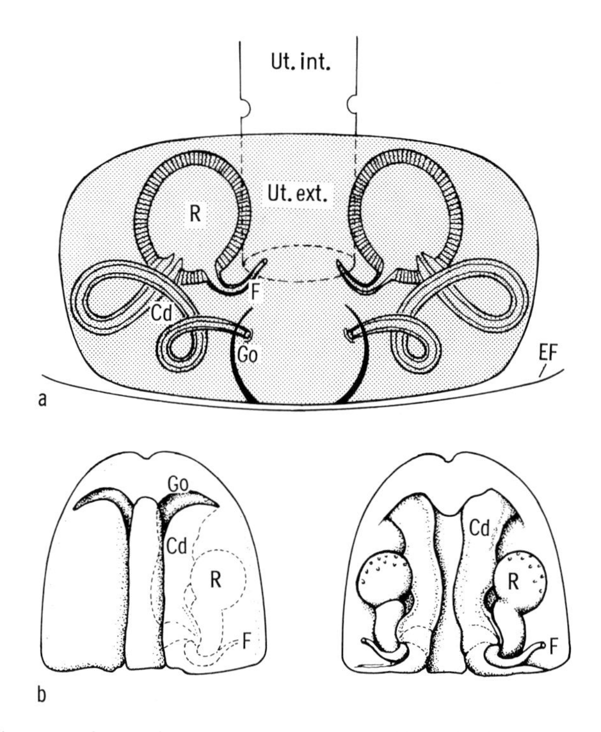

Fig. 156. (*a*) Diagram of an epigynum, ventral view. Cd = coiled ducts; EF = epigastric furrow; F = fertilization duct; Go = genital opening, R = seminal receptacle; Ut. int., Ut. ext. = uterus internus and externus. (After Wiehle, 1967.) (*b*) Epigynum of *Cupiennius salei*. External view of the sclerotized parts on the left, internal view on the right. Abbreviations as in (*a*). (After Melchers, 1963.)

coiled connecting ducts (or sperm ducts) lead to the seminal receptacles (Fig. 157), and from there, special fertilization ducts connect with the uterus externus. During mating the male's embolus is inserted through the external genital opening into the sperm duct. Both structures must match; that is, a long, convoluted sperm duct usually requires a lengthy, coiled embolus. The embolus may penetrate right up to the seminal receptacles, where the sperm cells are deposited. Only when the female starts laying her eggs do the sperm in the receptacles become active and migrate through the fertilization ducts to the egg cells. The expulsion of the sperm cells is most likely caused by special glands whose secretions displace the sperm mass (Cooke, 1966; Coyle et al., 1983). In haplogyne species without receptacula (e.g., *Pholcus*) the glandular secretion serves to store the sperm mass prior to fertilization (Uhl, 1993, 1994). Several batches of eggs (cocoons) can be fertilized, even if only a single copulation has taken place.

Fertilization takes place in the uterus externus. One hour after oviposition, the sperm nucleus is already moving toward the center of the egg. At that time the female pronucleus has just completed the second maturation division at the egg's periphery and will start its migration toward the center (Suzuki and Kondo, 1994a,b). The actual conjugation of the two nuclei takes place 1–2 hours later, being followed by the first cleavage division.

There is some evidence that *parthenogenesis* may occur in spiders. For instance, female *Dysdera hungarica*, which were kept isolated from males,

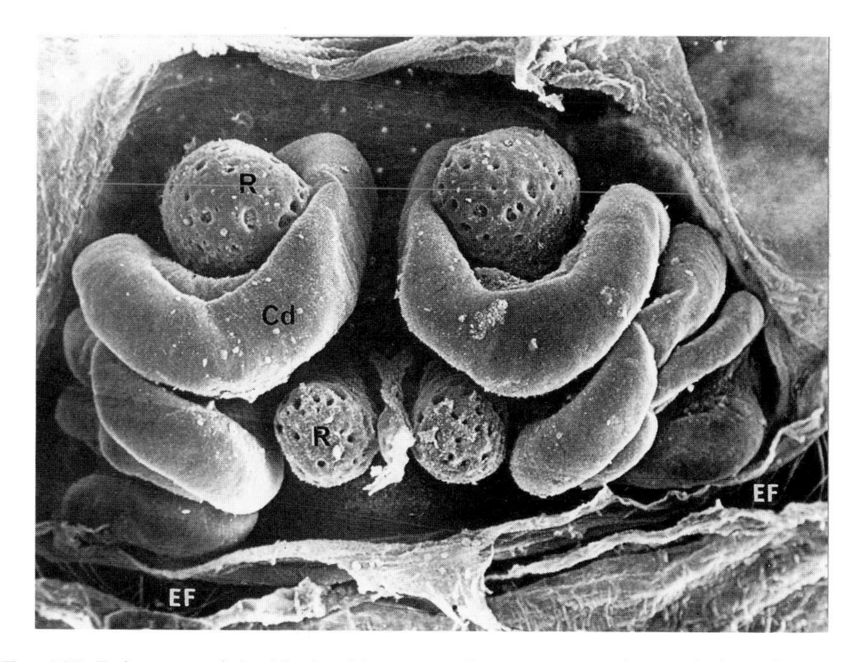

FIG. 157. Epigynum of the black widow *Latrodectus mactans;* internal view of sclerotized parts after dissolution of soft tissue with potassium hydroxide. Cd = coiled ducts; EF = epigastric furrow; R = seminal receptacles. 200 ×.

produced egg sacs and apparently a few of the (unfertilized) eggs developed into spiderlings and even adult spiders (Gruber, 1990). In the linyphiid *Pityohyphantes,* 1–2% of the embryos were found to be haploid ($n=14$), which would also indicate a development from unfertilized eggs (Gunnarsson and Andersson, 1992).

The sperm ducts and seminal receptacles are lined with cuticle. In some spiders, which continue to molt even in their adult stage (Mesothelae and Orthognatha), the lining of the seminal receptacles is also shed. This means that female "tarantulas," for example, which have mated previously, become "virginal" again after each molt; they need to mate again in order to produce fertile eggs. Male tarantulas do not molt once they have reached sexual maturity. Usually they die within a few years after their last molt.

Filling the Palps ("Sperm Induction")

Before a male spider goes in search of a female, he charges his palpal organs with sperm. First he spins a sperm web, which is usually a small triangular structure (2–4 mm^2) that is suspended horizontally. The male presses his abdomen against the rim of the triangular web, and moves his abdomen up and down until a drop of sperm emerges from his genital opening. Thereafter the male (of many species) moves to the underside of the sperm web and reaches with his palps around the margin of the web. The palps are dipped alternately into the sperm drop (Fig. 158). Most likely capillary forces play a role in the uptake of sperm, since the tips of the palps are often "soaked" between the mouth parts before the bulbs are filled (Blest and Pomeroy, 1978). This process is sometimes called "palpal lubrication."

Sperm can be taken up either directly—as just described—or indirectly, in which case the sperm cells must be sucked through the mesh of the sperm web (Gerhardt, 1929). Different spider families use different methods of transferring the sperm to the palps. In some species (*Scytodes*, for example), the sperm web

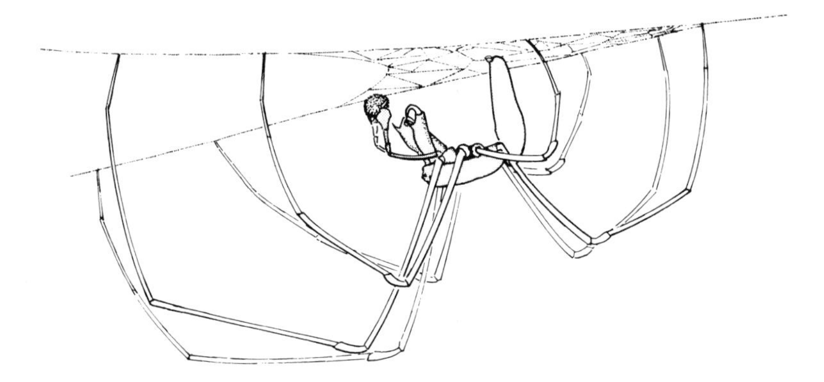

Fig. 158. A male *Tetragnatha* hanging suspended beneath his sperm web, charging the palps with a droplet of sperm. (After Gerhardt, 1928.)

is reduced to a single thread, which is held by the third pair of legs and drawn across the genital opening. The exuded sperm droplet adheres to this thread and is then drawn into the bulb (Gerhardt, 1930).

For a long time arachnologists thought that "sperm induction" was a precondition for the courting behavior of the male. In a series of elegant experiments this idea has now been refuted. Male spiders with their palps removed or with a covered genital opening display themselves quite normally toward a female (Rovner, 1966, 1967). It is likely that the initiation of courtship is triggered by the central nervous system, probably via hormones, some days after the spider's last molt. Perhaps this explains why newly molted males, which have not yet filled their palps with sperm, do not court in the first days following molting.

Courtship

Courtship can be defined as those ritualized behavioral patterns that are preparatory to mating. It is paramount for male spiders to avoid being mistaken for prey. Furthermore, females, which appear rather passive at first glance, need to be sufficiently stimulated before copulation can take place. Of course, female spiders are not completely passive during courtship. Some female wolf spiders (*Lycosa rabida*), indicate their willingness to mate by vigorously waving their legs (Rovner, 1968a). Even more active is the female *Alopecosa cuneata*: she grasps the male's front legs (tibiae) with her chelicerae and pulls him slowly towards her (Kronestedt, 1986); only after slackening of this ritual grip does copulation follow. It is believed that pheromone glands located in the male's thick tibiae trigger this peculiar behavior (Dahlem et al., 1987; Juberthie-Jupeau et al., 1990).

Since almost every spider species has developed its own courtship, it is hardly possible to make generally valid statements about this behavior, but to give at least an outline we shall group the courtship behaviors into three categories or levels. Each level is defined by the mechanism that triggers the male's courting (Platnick, 1971): level 1 requires direct contact between male and female; level 2 needs female pheromones to stimulate the male's courtship behavior; and level 3 postulates a visual recognition of the female by the male.

Level 1

Crab spiders (Thomisidae) and sack spiders (Clubionidae) have only a very modest courtship. Typically, the male simply crawls over the female, pulls the female's abdomen close, and inserts one palpal organ into the female's genital opening. Many Haplogynae (Haupt, 1977, 1979) and "tarantulas" also show little in the way of preliminaries (Baerg, 1928; Gerhardt, 1929). After direct body contact, the male and female exchange a brief interplay with their front legs and palps before they copulate. Some crab spiders (*Xysticus*), however, have developed a rather interesting courtship: the male "ties" the female with some

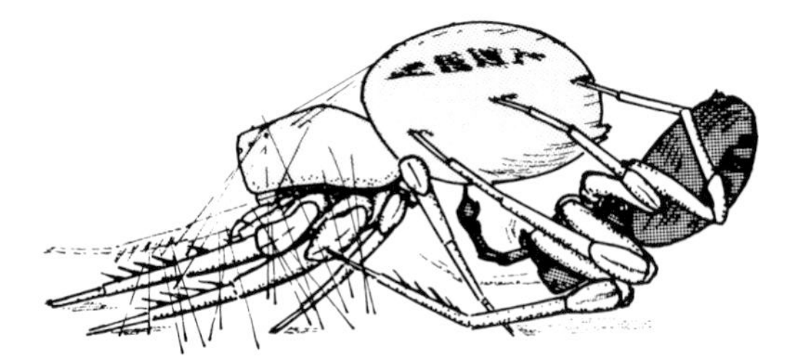

FIG. 159. Copulation of the crab spider *Xysticus*. The male "ties" the female to the ground and then crawls under her abdomen to start mating. (After Bristowe, 1958.)

strands of silk before mating (Fig. 159). These threads do not hold the female captive, though, for she has no trouble freeing herself after copulation. Similar "ties" are also used by *Nephila* (Araneidae) when the minute male places a few threads over the legs of the large female (Robinson and Robinson, 1973); again, these ties can have only a "symbolic" significance.

Level 2

Courtship at this level is somewhat more complex. Often male spiders (e.g., *Filistata, Segestria,* and *Amaurobius*) will lure females by pulling or drumming on their webs. Female pheromones are usually involved as well. The males use their contact chemoreceptors, and perhaps also olfaction, to recognize the silk of the females. Draglines of female wolf spiders are known to induce "following behavior" and courtship in the male (Tietjen, 1977, 1979a,b; Pollard et al., 1987). In *Amaurobius ferox* direct body contact is necessary to elicit courtship behavior, whereas in *A. similis* and *A. fenestralis* the presence of pheromones alone suffices to stimulate the male. The males of the different *Amaurobius* species use various vibratory signals, which they generate with their legs, palps, and abdomens. Each species has its own code (Fig. 160; Krafft, 1978), and a male of *A. similis,* for example, will court only at a web of its own species. The specific signals of the males are often answered by specific vibrations of the

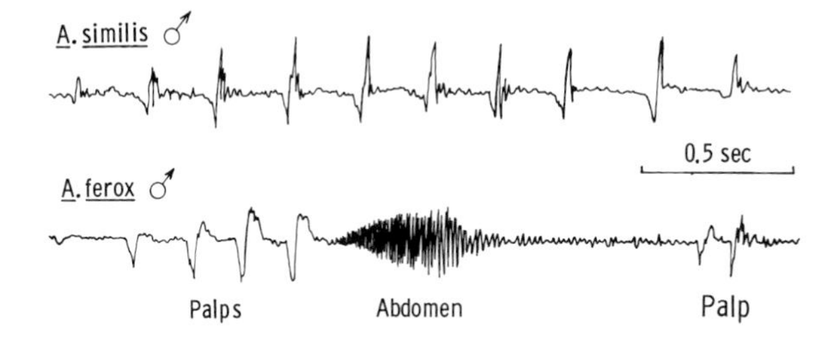

FIG. 160. Recording of male vibratory signals from two different Amaurobius species courting at the webs of the respective females. A. similis drums only with his palps, whereas A. ferox strums with one or two palps and additionally vibrates his abdomen. (After Krafft, 1978.)

female; for instance, in *Cupiennius* the male signals have a frequency of 75 Hz, those of the female of only 40 Hz (Schüch and Barth, 1985). Furthermore, the vibration pattern is very structured into "syllables" and "pauses" typical for each species. This ensures reproductive isolation, since females tend to respond only to signals of their own males (Barth, 1993).

In some spiders the males apparently recognize the nests of the immature females and spin their own nests right next to them (Bristowe, 1941; Jackson, 1986a; Toft, 1989). Immediately after the final molt of the female—that is, during a period of time when the female is essentially defenseless—copulation takes place. To refer to such behavior as "rape" may just reflect our anthropomorphic point of view. It may well be that copulation is possible only during that time, when the cuticle of the female is not yet fully sclerotized. Mating right after the female's final molt is also common among many orb weavers, such as *Argiope*.

When a male web spider (for instance of Araneidae, Agelenidae, and Linyphiidae) approaches a female in her web, he is particularly prone to being dealt with as prey. Another risk is that the female might disrupt the courtship because prey has entered the web or because another male has appeared on the scene. The male tackles these problems in different ways, mostly, however, by somewhat restraining the mobility of the female.

Araneid males initially remain at the periphery of the web and make their presence known by plucking the threads with their front legs. Their actual approach of the female can be a tedious and lengthy procedure. Usually the male attaches a special thread, the *mating thread*, to the female's web. By plucking and beating this thread rhythmically with his front legs, and especially with his third pair of legs, the male entices the female to move from her web onto the mating thread (Fig. 161). If she is receptive, she will advance toward the male. After a short exchange of foreleg dabbing, the female assumes a typical copulatory posture (Fig. 167).

Pheromones, which are emitted a few days after the last molt (Blanke, 1972), often play a role in helping the male find the opposite sex. Male araneids apparently can recognize the female pheromone chemotactually on the web and also seem to be guided by olfaction, at least for short distances of about 60–80 cm (Blanke, 1975b). It is noteworthy that female spiders can also detect pheromones. Black widows (*Latrodectus hesperus*), for instance, show courtship behavior in response to male silk (Ross and Smith, 1979). In the related species *L. revivensis,* females were observed to react aggressively toward the silk of conspecific females (Anava and Lubin, 1993).

Although the courtship behavior of araneids may appear rather uniform at first glance, we find some remarkable variations from the usual pattern. The males of *Nephila* and of some *Argiope* species do not use any mating threads but climb directly into the hub of the female's web, where they start a tactile courtship. In other *Argiope* species the male also crosses the female's web but then cuts a hole near the hub, attaches a short mating thread, and commences a vibratory courtship. From an evolutionary point of view a tactile courtship at the web's hub may represent the original type of courtship in araneids (Robinson and Robinson, 1978, 1980).

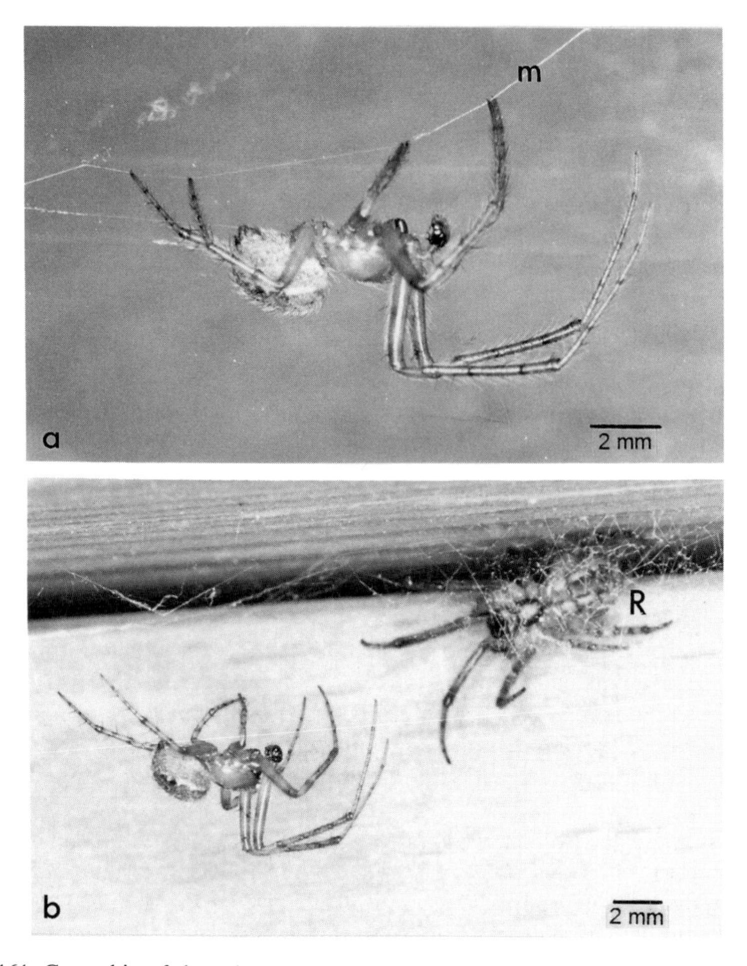

FIG. 161. Courtship of the orb weaver *Zygiella x-notata*. (*a*) The male's front legs beat with a high frequency (60 Hz) on the mating thread (m); (*b*) the male's plucking entices the female to leave her retreat (R) and to approach the male. (Photos: Blanke.)

Normally a rather lengthy courtship is necessary before the female spider allows the male free access to her web. In the case of *Linyphia triangularis,* however, the male is always dominant and conquers the female's web without encountering any resistance (Rovner, 1968b). The female never attacks him; on the contrary, if prey happens to fall on the dome of her web (Fig. 105), the male will chase the female away and feed alone. During courtship the male destroys most of the female's web; within a few minutes the dome may be reduced to a small silken band. The female remains entirely passive during this action and her passivity probably signals to the male that she is willing to accept him. At any rate, copulation occurs immediately after her web has been destroyed. The purpose of the web reduction is apparently to remove the female sex pheromone, so that no other males will be attracted (Watson, 1986). Webs of females (*Linyphia litigiosa*) that have already mated are not reduced by the male,

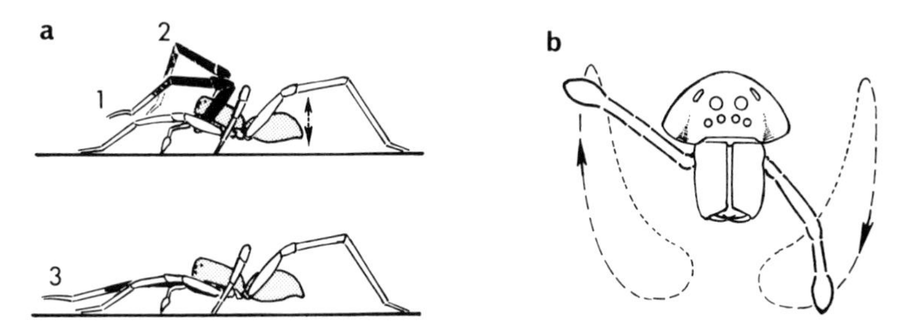

FIG. 162. (*a*) Courting attitudes of the male wolf spider *Lycosa rabida*. In the initial phase (1) the first pair of legs is raised and flexed; the tibiae are extended horizontally; the abdomen vibrates vertically. (2) Stretching of leg 1. (3) Fully extended leg 1. (After Rovner, 1968a.) (*b*) Palp rotation as an element in courting by male wolf spiders. (After Bristowe and Locket, 1926.)

presumably because the silk threads no longer contain female pheromones. However, webs of other female linyphiids (e.g., *Frontinella pyramitela*) elicit courtship in males even after 30 days, and even if they are unoccupied; in such cases it seems likely that a contact pheromone is involved (Suter and Hirscheimer, 1986).

Courtship is somewhat more complex for many wolf spiders (Lycosidae) and nursery-web spiders (Pisauridae). The behavior of the male is influenced not only by tactile and chemotactile stimuli, but also by vibratory and visual signals.

When a male *Lycosa rabida* notices a female passing by, he assumes a typical courtship posture (Fig. 162a): the front legs are raised, the palps touch the substrate, the body is lowered (Rovner, 1968a). The subsequent behavior of the male consists of characteristic courtship movements and intermittent periods of resting. First the palps are waved alternately in a circular fashion, called palpal rotation (Fig. 162b); then they contact the substrate again. One front leg is stretched forward, and is raised and lowered several times. At the same time the abdomen vibrates up and down. The palpal rotation is soon changed into palpal "drumming," which is audible to the human ear. It has been shown that the palpal drumming includes a stridulation that is produced in the tibio–tarsal joint of the palp (Rovner, 1975): a cuticular spur scratches over parallel ridges (the "file") of the tibia, thus producing sounds. After the stretched front leg is snapped back into a flexed position, the male remains inactive for about 15 seconds, or he may approach the female slightly. It is during those pauses that the female becomes active. About 5 seconds after the male's palpal drumming, the female responds by briefly waving her front legs. Usually she also takes a few steps toward the male. The male now repeats and intensifies his courtship movements—leg stretching, palpal rotation, abdominal quivering. When he is within reach of the female, he cautiously extends one front leg, but he does not actually touch the female. The first contact is always initiated by the female and apparently signals her readiness for copulation.

From this description it is evident that *Lycosa rabida* exhibits a reciprocal courtship; that is, both partners take turns exchanging signals. The visual signals are largely confined to perception of motion. This can be concluded from the observation that a male often initially directs his courtship toward other males (although he receives only a threat in response). Males without their palps or front legs can still court successfully and be accepted by the female. This means that the acoustical and vibratory signals are not necessary components of courtship. On the other hand, leg waving by the females has been elicited by playing back the acoustical signals of the male over a tiny loudspeaker (Rovner, 1967). It is plausible that acoustical and vibratory signals play a major role in courtship only in the dark. In fact, *Lycosa rabida* is active at night as well as during the day.

Chemotactile signals seem to be the most important factors in the male wolf spider's courtship behavior. When male spiders are put into empty boxes that have previously housed conspecific females, they start exhibiting their typical courtship behavior (Bristowe and Locket, 1926; Kaston, 1936; Rovner, 1968a; Dijkstra, 1976). They show no reactions, however, if males or immature females have been kept in these boxes. If a male wolf spider comes across the dragline of a mature female, he will stop immediately, and will commence probing the substrate with the tips of his legs and palps. His front legs start quivering and the entire courtship sequence may follow, just as if the female were present. He will take the female's dragline between his palps (Fig. 163) and then follow it toward the female (Tietjen and Rovner, 1980, 1982). A similar trail following can be seen in *Dolomedes* (Pisauridae), where the male tracks the female's dragline even on the surface of the water (Roland and Rovner, 1983). Draglines from males or other species are largely ignored, which indicates that a sex pheromone elicits this trail following behavior. However, it seems that these pheromones are not really species specific but rather trigger a first arousal in males.

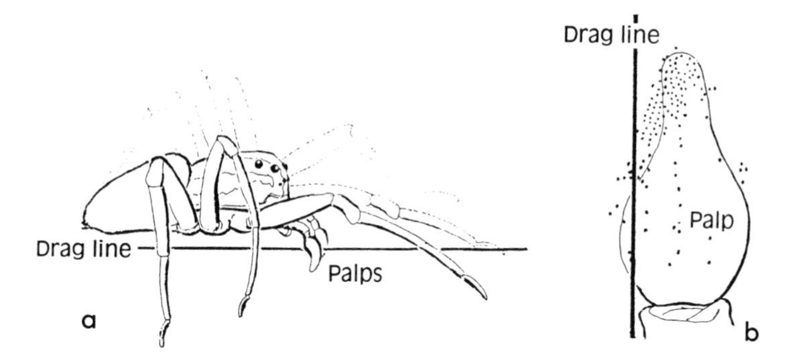

FIG. 163. Trail following in the wolf spider *Lycosa punctulata*. (*a*) The male takes the female's dragline between his palps and uses it as a guide line toward the female. (*b*) Distribution of chemosensitive hairs on the tarsus of the male palp (dorsal view). Most of the sensory hairs are located on the inner side of the palp, where they have maximal contact to the dragline. (After Tietjen and Rovner, 1980.)

Other species of wolf spiders show courtship movements that are similar to yet distinctly different from those of *Lycosa rabida*. Legs and palps can be waved together or alternately; the palps may be held in a certain position, or vibrate excitedly, or describe aerial loops. Each species has its own, typical repertoire. This has been studied in detail in five species of the wandering spider *Cupiennius* (Barth, 1993). In addition to an audible palp drumming, males of *Cupiennius salei* also exhibit a rapid leg quivering, which causes specific substrate vibrations. These vibrations are transmitted on plant leaves over distances of 1 m and more, and cause a receptive female to respond with leg vibrations of her own (Fig. 164). The male can then orient himself and gradually approach the female. Again, this example shows clearly the reciprocal communication that takes place between male and female spiders during courtship (Barth, 1985b).

The nursery-web spider *Pisaura* is famous for its extraordinary courtship. The male first catches a fly, wraps it in silk, and carries it in his chelicerae to the female. If the female is receptive, she seizes the "bridal gift" and starts feeding on it (Nitzsche, 1981). At the same time the male begins copulating with her. Despite such tricks males are not always safe from female attacks; *Dolomedes* (Pisauridae) males are often eaten, probably because they approach the female from the front and without the necessary caution (Gerhardt, 1926). Incidentally, a synchronization of the male's courtship with feeding in the female is known to occur in several other spider families, for example, in the araneid Meta segmentata (now Metellina segmentata) (Blanke, 1974a), and the uloborid *Uloborus geniculatus* (now *Zosis geniculatus*) (Gerhardt, 1927).

Level 3

In salticids and oxyopids, visual signals are the primary releaser for courtship, although tactile and chemical stimuli are also involved (Jackson, 1977, 1978). It is presumed that courtship of these spiders has developed from courtship levels 1 and 2.

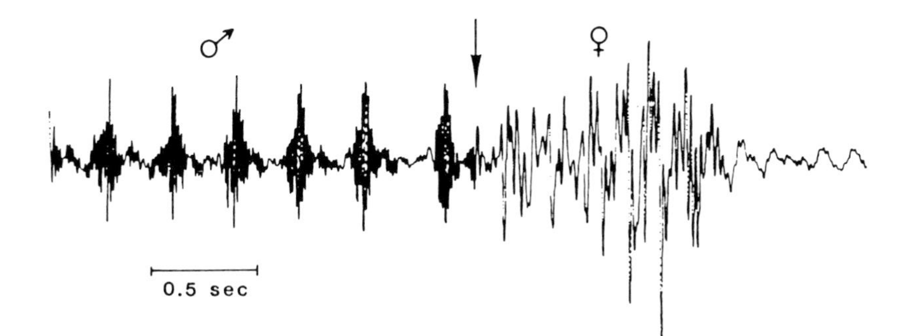

FIG. 164. Reciprocal communication in the courtship of the wandering spider *Cupiennius*. Pulses of leg vibrations in the male are answered by typical leg vibrations of the female (starting at arrow). (After Rovner and Barth, 1981.)

All male jumping spiders display their specific courtship movements in front of the females (Richman and Jackson, 1992). These movements range from the simple lifting of a leg to the complex, sequential movements of several extremities. Usually, those extremities, which are mostly used in the courtship display, are also conspicuously colored (Bristowe, 1929; Kaston, 1936; Crane, 1949). When a male notices a female, he approaches her in a zigzagging dance. At the same time, he displays species-specific courtship movements such as raising his front legs, vibrating his palps, and twitching his abdomen (Fig. 165). The female remains rather passive, yet she watches the male. In some species the female responds by vibrating her palps, and sometimes she may even weakly imitate the male's courtship routine. During this initial phase of courtship the primary objective of the male is to identify himself as a mate of the "right" species. If the female is receptive toward the performing male, she assumes a quiet, crouching position. The next phase, which leads directly to copulation, is very similar for all jumping spiders. The male extends his forelegs so that they are parallel and touches the female. After further fondling, he simply climbs on the female's back and begins copulation.

The strong dependence of a jumping spider's courtship on visual clues becomes evident when one covers a male's main eyes with black paint. He still shows signs of excitement if a female is present, but he no longer performs the typical courtship dances (Homann, 1928; Crane, 1949). Conversely, when the main eyes of the female are painted black, she will not accept a courting male. This demonstrates that the courtship in salticids is indeed a reciprocal visual affair. If a male addresses the female of another species—and this occasionally happens in nature—the female withdraws at some point in the first phase of courtship. Apparently, without the species-specific movements of the male, there will not even be a chance for him to copulate.

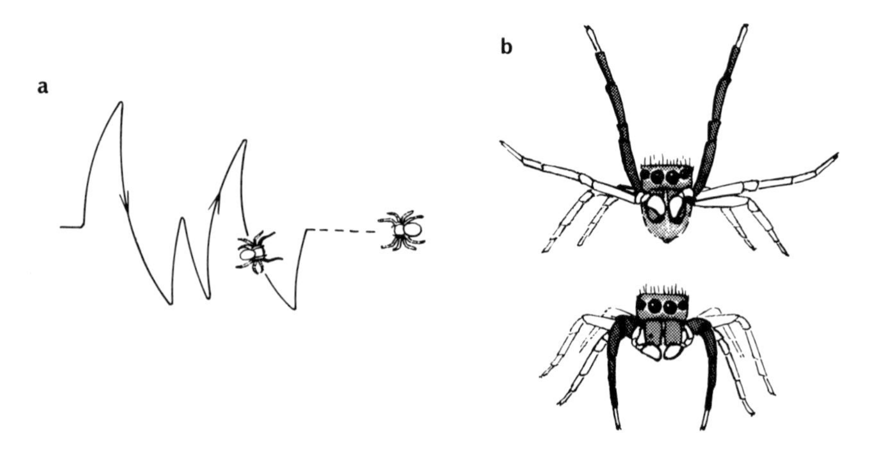

FIG. 165. (*a*) Zigzagging dance of a male jumping spider in front of the female. (After Homann, 1928.) (*b*) Two courting stances taken by the male jumping spider *Euophrys frontalis*. (After Bristowe, 1958.)

Aside from the important visual signals (movement, and size and shape of the mate), chemical stimuli also play a role in salticid courtship. Chemical cues alone do not suffice to trigger the courtship dances—as they do in lycosids—yet contact with the female's silk still seems to have a stimulating effect on the male. Furthermore, it seems that very young females (1–2 weeks after their final molt) are preferred by the courting males. Older females (6 weeks postmolt) are less attractive to the males, probably because their pheromone production has already decreased.

Within the large family of jumping spiders, we can find species with a rather "primitive" courtship, which relies on the spider's chemotactile or olfactory sensitivity, as well as gradually more "advanced" species that use visual cues almost exclusively when they court (Crane, 1949; Pollard et al., 1987). Recent investigations have shown that, even within the same species, different courting strategies are employed (Jackson, 1977)

For instance, if a male *Phiddipus johnsoni* encounters a female outside the nest, he will perform the usual visual type of courtship. If the same male comes across a nest in which a female is hidden, however, he shows a completely different courting behavior: the tarsi of his front legs are rubbed over the surface of the female's nest, and his whole body oscillates up and down. After a while, the male enters the nest and mating ensues. Since empty silken nests of the females can also trigger the male's courtship, it is assumed that some volatile pheromones adhere to the silk threads (Jackson, 1981, 1987). It seems interesting that only the first type of courtship is dependent on light, but not the second. When both sexes are put together under dark red light, no courtship behavior is elicited from either spider, even if they bump into each other. As soon as a white light is turned on, however, the male begins the visual type of courtship within a few minutes.

Lynx spiders (Oxyopidae) also can recognize each other from a distance of several centimeters. The males often possess darkly colored palps, which they wave in front of the females (Gerhardt, 1933; Bristowe, 1962). Each species has its own sequence of palp, leg, and abdominal movements; they also show an exchange of leg stroking before mating.

Copulation

The mechanics of the copulatory organs were already briefly explained earlier. During copulation, the palpal organ of the male is inserted into the female's genital opening, and the sperm is deposited in her seminal receptacles. Haplogyne spiders insert the entire bulb of the palp into the female's genital opening (Fig. 166); in entelegyne spiders only the bulb's tip, the embolus, enters the copulatory duct. Characteristic of the Entelegynae are their extensible hematodochae; such spiders can inflate the normally collapsed palpal organ and thereby push the embolus into the sperm duct of the female's epigynum. Only after the palpal organ has been coupled with the epigynum do the hematodochae swell to their maximum size. This hydraulic action also leads to

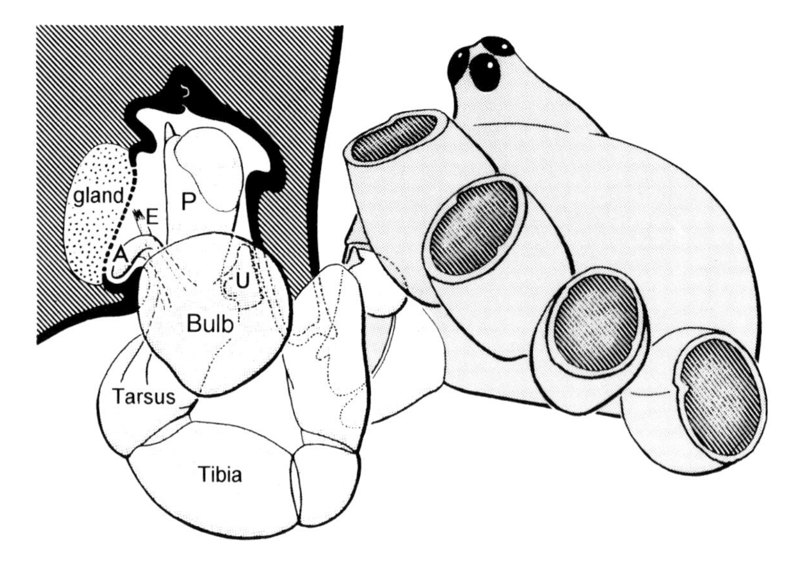

FIG. 166. Locking mechanism of male and female genitals in *Pholcus*. The male bulb becomes fixed in the female genital cavity by means of several sclerites (A = appendix, P = procursus, U = uncus). The embolus (E) releases the sperm mass directly into a secretion of the adjacent gland. Actual copulation occurs upside down (see Fig. 167b) but is shown here inverse for better comparison with Fig. 155a. (After Uhl et al., 1995.)

the erection of the leg spines. In some species the hematodochae swell only once and very briefly before the palps are withdrawn from the epigynum. This is the case in most species of araneids. In other species (as in most linyphiids) the hematodochae pulsate rhythmically for several hours. There are also differences as to whether the palps are inserted simultaneously or consecutively: most Haplogynae employ both palps at the same time, whereas the Entelegynae insert first one and then the other palp.

The mating behavior of spiders and its impressive diversity have been studied extensively from a comparative point of view (Gerhardt, 1911–1933; Bristowe, 1941). The many variations of the mating positions of spiders can be reduced to three or four basic types (Fig. 167).

Type 1 is characteristic of the "primitive" wandering spiders (Mesothelae, Orthognatha, and Haplogynae). The male approaches the female from the front (Fig. 167a), the female raises her prosoma, and the male inserts one or both palps into the female's genital opening. Afterwards both spiders separate cautiously. *Type 2* is widespread among web spiders. The spiders mate in essentially the same position as in type 1, but with both partners hanging upside down from the web (Fig. 167b). The male's palps are inserted into the female consecutively. *Type 3* is found among the "modern" wandering spiders (e.g., the Clubionidae, Lycosidae, Salticidae, and Thomisidae). The male climbs over the prosoma of the female, and then turns toward either the left or right side of her opisthosoma. The female twists her abdomen in such a way that one of the

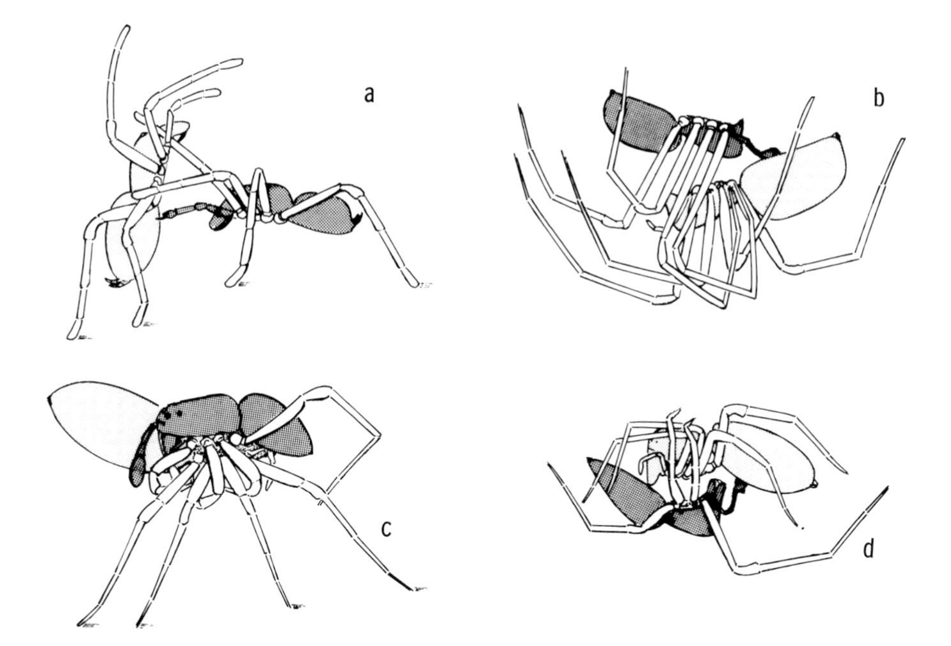

FIG. 167. Mating positions of various spider families: (*a*) "tarantulas," (*b*) linyphiids, (*c*) lycosids, (*d*) clubionid *Cheiracanthium*. (After von Helversen, 1976.)

male's palps can be inserted (Fig. 167c). Each time the other palp is used, the male changes to the other side of the female's abdomen.

Type 1 is most likely the original mating position for spiders, and types 2 and 3 are derived from it (von Helversen, 1976). *Cheiracanthium* (Clubionidae) uses a modified, fourth kind of position: the spiders face each other with their ventral sides in contact (Fig. 167d). It should also be mentioned that a certain kind of mating position is not necessarily stereotyped within one family. Most agelenids, for instance, copulate according to type 3, yet some *Tegenaria* species follow type 1.

The supposed aggressiveness of the female spider toward the male is largely a myth. When a female is ready for mating, there is little danger for the male. If a male courts at the wrong time, however, he may well be attacked and eaten by the female. In some clubionid species, the female allows only one copulation and reacts aggressively toward any further advances of males (Pollard and Jackson, 1982). In general, most species separate quite peacefully after copulation. Only in some exceptional cases does the male fall victim to the female, for example, in certain *Argiope* and *Cyrtophora* species. The infamous black widow (*Latrodectus mactans*) does have a bad reputation, but—quite contrary to popular belief—the male usually withdraws unharmed (Kaston, 1970; Ross and Smith. 1979). In some *Latrodectus* species the males live for several weeks in the female's web and may even feed on her prey (McCrone and Levi, 1964; Anava and Lubin, 1993).

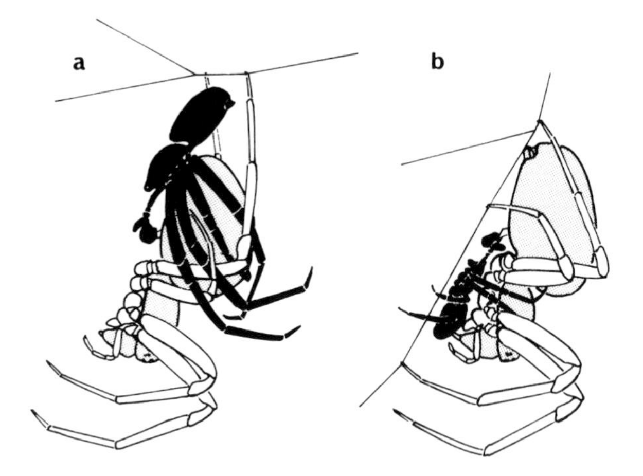

Fig. 168. Copulation in two *Araneus* species. (*a*) *A. diadematus*: the male embraces the female's abdomen and inserts one bulb. (After Gerhardt, 1911.) (*b*) *A. pallidus*: the small male turns a somersault of 180° and falls directly in front of the female's chelicerae. He is bitten immediately and thus fixed in the mating position. (After Grasshoff, 1964.)

The small male of *Araneus pallidus* is not so lucky, however. He initiates copulation by jumping toward the ventral side of the female's abdomen and fixing one palp to the epigynum. During this process he tumbles backward so that his abdomen rests right underneath the female's prosoma. The female immediately seizes his abdomen with her chelicerae and within a few minutes starts feeding on him (Fig. 168). Apparently, this is more of a technical necessity than pure cannibalism. If the female is prevented from biting the male, he constantly slips off her abdomen and is unable to insert his palp (Grasshoff, 1964). A very similar mating behavior has also been observed in *Cyrtophora cicatrosa* (Blanke, 1975b).

Agonistic Behavior

So far we have dealt only with those reproductive behaviors that take place between sexes. Of course different types of behavior are elicited when two males of the same species meet during courtship. All those activities that are related to a competitor (e.g., threatening, fighting, or fleeing) are referred to as agonistic behavior. Such behavior occurs among spiders in both sexes but is more elaborate in males.

When two male wolf spiders (*Pardosa*) encounter each other in the presence of a female, they assume specific threat postures and sometimes start to fight (Fig. 169). Usually the attacking male will be the winner. Once he has established his dominance, he is likely to maintain it during further bouts (Dijkstra, 1970). The loser tends to slink away, while the winner flaunts about jerkily. He will then soon encounter the female and circle her, walking sideways. Rising as if

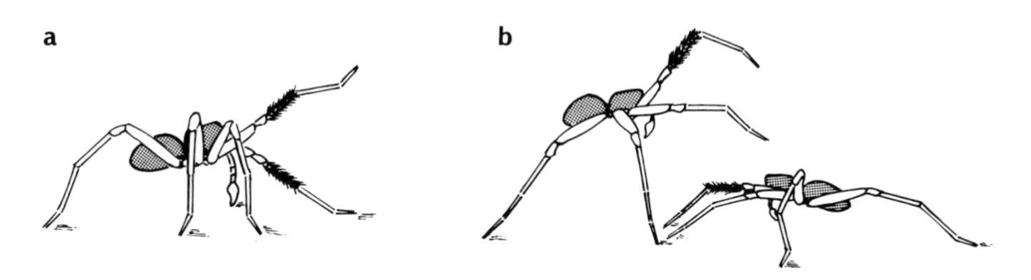

Fig. 169. Agonistic behavior of male wolf spiders (*Schizocosa crassipes*). (*a*) The primary reaction to another male is the cautious extension of one front leg. (*b*) The dominant male (left) threatens the submissive male which remains in a resting position. (After Aspey, 1975.)

on tiptoe, he waves one or both palps in the air and then starts his usual courtship. The agonistic behavior has a selective advantage for the species: the winner has a better chance of mating with the female (Aspey, 1975) and thus of passing on his genome.

When two male linyphiids (*Linyphia triangularis*) meet, they also exhibit specific threat and fighting behaviors, which they never show toward females (Rovner, 1968b). Both males spread their large chelicerae (Fig. 170a), extend their front legs perpendicularly from their bodies, and vibrate their abdomens rapidly. If neither of the two opponents is impressed, they will move closer and finally begin to fight. First they push each other with the tips of their legs, and later they interlock their chelicerae. The loser drops quickly from the sheet web on which they had been vying for the female; usually the larger male turns out to be the winner. The males may also start to fight without a female present. It seems that the first male to reach the female's web will defend "his" territory, the dome of her web. Females will defend their web only against other females, but not against courting males.

In most jumping spiders, threatening behavior between males is strongly ritualized (Fig. 170b); actual fights are rare (Crane, 1949; Jackson, 1982). By no means does the larger male always win the ritualized contest, but simply the more aggressive male. If two males (*Marpissa marina*) encounter each other in the vicinity of a silk nest, then the "home owner" is always the superior one (Jackson and Cooper, 1991). The threat signals are visual, as are normal courtship signals, and they consist mainly of raising and lowering the legs. In the more "primitive" jumping spiders the threat movements are identical with regular courtship movements, probably because the male cannot distinguish between a female and a male spider at greater distances (Bristowe, 1929). In the most "advanced" jumping spiders, in contrast, the threat movements clearly differ from the courtship movements (for instance, different legs are used). A key stimulus for a male's threatening behavior lies in certain contrasting patterns on the palps or on the carapace of a potential competitor. Such markings occur only in the males (Crane, 1949). When a female is not ready for mating, she may also threaten a male: she moves her legs up and down as she walks sideways in a zigzagging fashion (Forster and Forster, 1973).

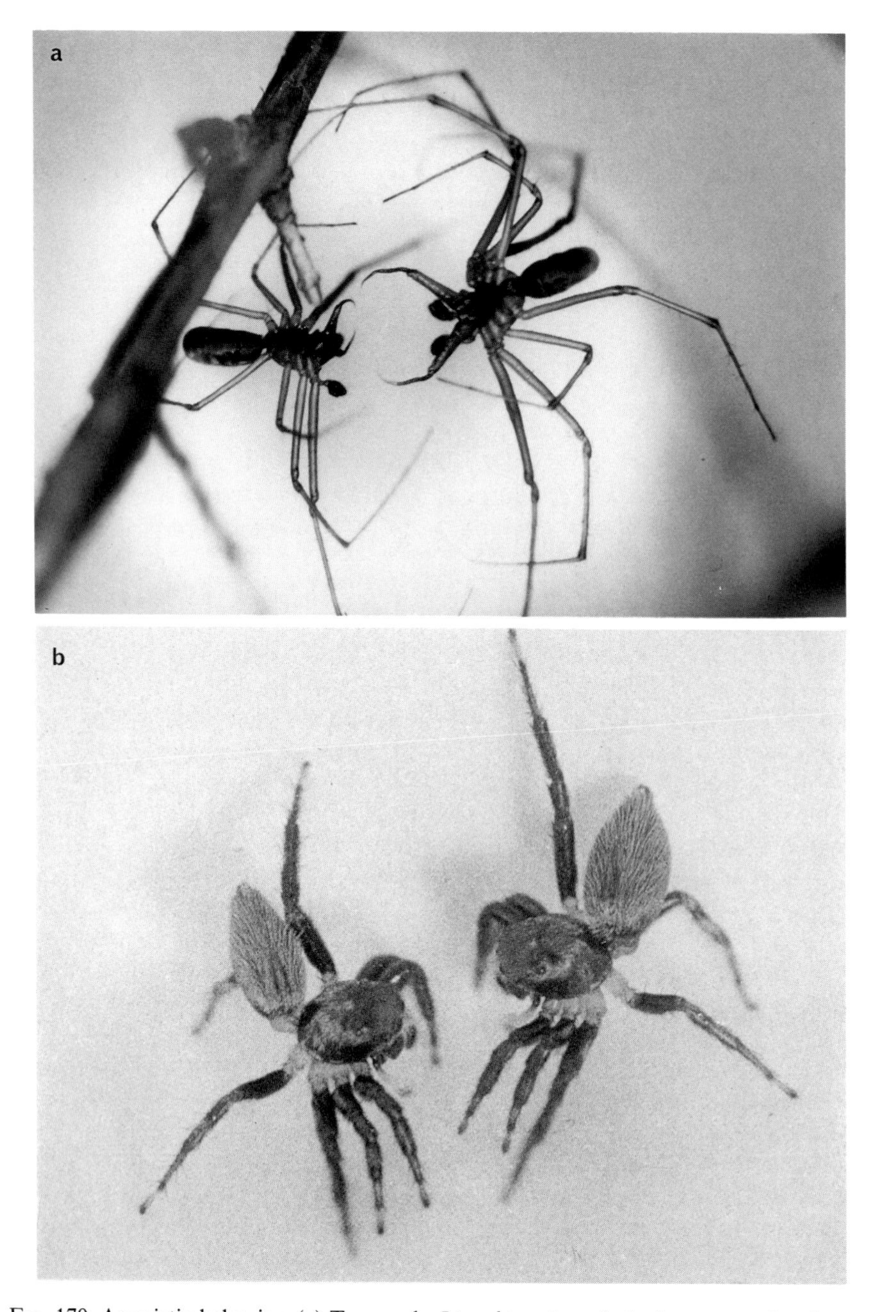

FIG. 170. Agonistic behavior. (*a*) Two male *Linyphia triangularis* threaten each other with wide open chelicerae. (Photo: Rovner.) (*b*) Two male jumping spiders (*Euophrys parvula*) face each other with their third legs raised and their abdomens swiveling sideways. (Photo: L. Forster.)

Agonistic behavior between female spiders has been studied in a few species. In the wandering spider *Cupiennius*, the female–female interactions are actually more aggressive than the strictly ritualized fights among the males (Schmitt et al., 1992). Female wolf spiders threaten each other by raising their front legs and opisthosomas, and by opening their cheliceral fangs, yet bite only rarely (Nossek and Rovner, 1984). The entire behavior is aimed to prevent any further approach or attack by the opponent, and to maintain a certain "territory" of their own. This is certainly true for web spiders, which often have to fend off invading competitors (Buskirk, 1975a; Riechert, 1978, 1982). They use vibrations of the web as warning signals. In linyphiids the web owner is usually at an advantage and manages to defend its web even against larger females.

Brood Care

Within a few weeks after copulation the female is ready to begin laying her eggs and produce an egg case (or cocoon). Fertilization takes place just before the eggs are deposited; the sperm cells that had been stored in the seminal receptacles are released as the eggs pass through the uterus externus. The fertilized eggs are then squeezed one by one through the genital opening. Generally, they are surrounded by a clear, viscous liquid, and after a short time this fluid dries up, cementing the eggs together. Normally egg laying takes only a few minutes, even when there are very many eggs. *Araneus* lays almost 1,000 eggs within 10 minutes (Pötzsch, 1963), *Cupiennius* as many as 1,500–2,500 in about 8 minutes (Melchers, 1963)—quite an impressive performance, especially if one considers that each egg measures at least 1 mm in diameter (Fig. 171a,b). Evidently, egg laying is a rather energy-consuming process. This is reflected in the increase of the heart rate, which in *Cupiennius* is elevated from a normal 60–70 beats/min to about 160 beats/min during egg laying. After all the eggs have been deposited into the egg sac, the female looks rather skinny because of her shriveled abdomen.

Egg-case Construction

The eggs are never directly exposed to the environment but are always protected by silk. In the simplest case only a few silk threads are wrapped around the eggs (Fig. 171a). A typical cocoon, however, consists of a basal plate and a cover plate, which enclose and protect the egg mass. One of the main advantages of an egg case is the favorable humidity level that is maintained inside; hence the eggs can develop without depending as much on climatic conditions. Most important in this respect is the outer layer, which has a texture similar to strong paper (Hieber, 1985). It may also serve as a mechanical barrier against egg parasites (flies and wasps), and as a thermal insulation against fluctuating temperatures.

The silk threads making up a cocoon are of varying diameter (1–6 μm; Opell,

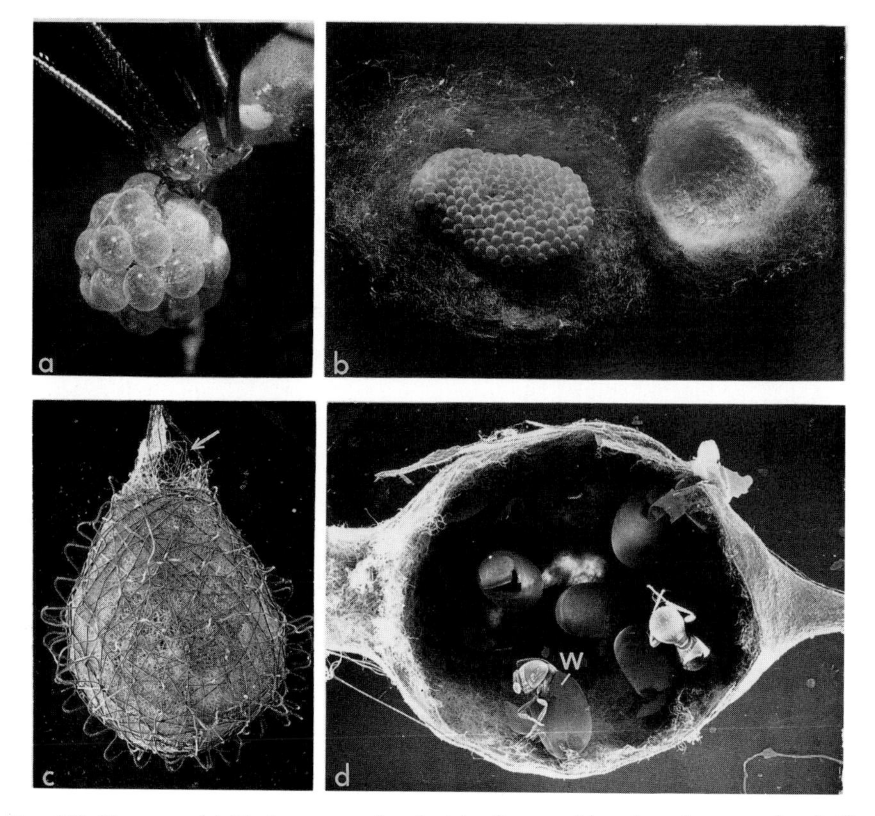

FIG. 171. Egg sacs. (*a*) *Pholcus* wraps her batch of eggs with only a few strands of silk and carries it in her chelicerae. (*b*) An opened cocoon of *Araneus diadematus*. The basal plate is seen on the right; the tightly packed eggs rest on the cushion of downy silk threads. (*c*) The cocoon of the pirate spider *Ero furcata*. The closely woven cocoon wall is covered with stronger, sinuous silk threads (see Fig. 172). The young spiderlings have emerged through a hole at the top of the egg sac (arrow). 17 ×. (*d*) Opened cocoon of the theridiid *Argyrodes*, parasitized by wasps (W). 28 ×. (Photo: Foelix and Vollrath.)

1984) and originate from different silk glands (mainly the Gl. tubuliformes and piriformes, but also in part from the Gl. aciniformes; Peters and Kovoor, 1989). The firmness of the cocoon wall is due to the tightly criss-crossing silk fibers; furthermore, single threads seem to fuse into stronger, composite fibers (Fig. 172).

In order to understand the various types of cocoons, we shall take a close look at their construction; the cocoon of the orb weaver *Araneus quadratus* will serve as an example (Crome, 1956a). Before the spider starts making her egg sac, she withdraws for several days into her retreat. She then spins a thin layer of single tightly woven threads. Using only abdominal movements she molds this first layer into a disk, the basal plate (Fig. 173a). This sheet is covered on both sides with flocculent, curly silk threads. The spider then crawls underneath the basal plate and keeps turning around in circles, spinning

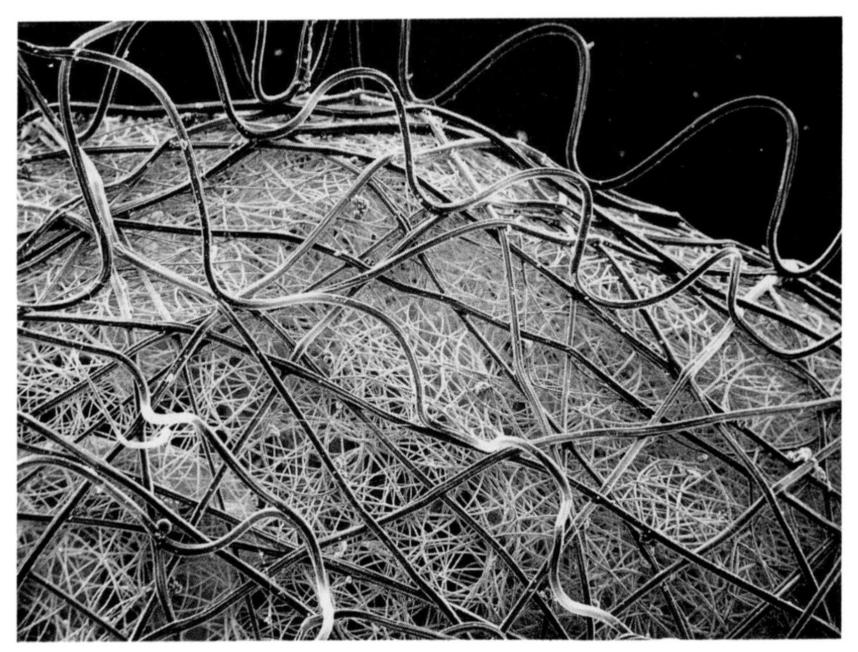

FIG. 172. Silk threads from the cocoon wall of the pirate spider *Ero furcata.* The thick sinuous threads, which rise above the surface of the egg sac, apparently consist of several fused single threads. 85 ×.

continuously. This motion leads to the construction of a cylindrical wall (Fig. 173b). The palps are constantly held in contact with one side of this wall while the side opposite is dabbed with the spinnerets (Fig. 175a). Thus the cylindrical wall gradually grows until it has reached a height of 5 mm after about 2 hours. The size of cocoon is directly correlated with the size of the spider but not necessarily with the number of eggs that will be deposited.

The spider pauses for a few minutes before she starts laying the eggs. When she has finished, she abruptly pulls away from the egg mass and begins to cover

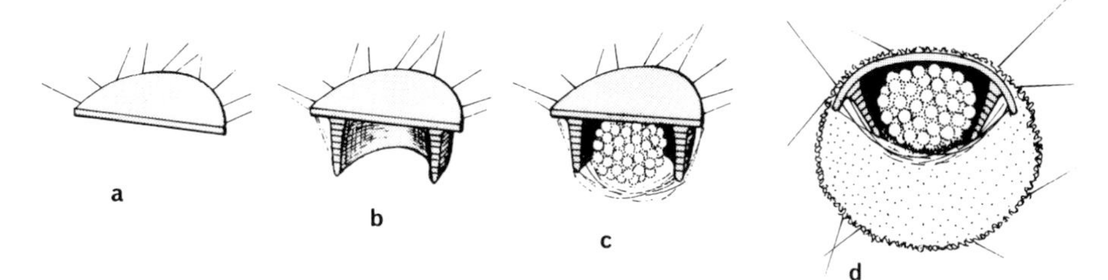

FIG. 173. Stages in the construction of an egg sac in *Araneus quadratus.* (*a*) Basal plate with attachment threads. (*b*) Addition of the cylindrical wall. (*c*) Egg chamber with a cover plate below. (*d*) Section through the finished cocoon. The lower part is filled with a loose mesh of silk threads. Attachment threads suspend the cocoon. (After Crome, 1956a.)

it with silk threads (Fig. 173c). This gradually becomes the cover plate, to which the spider keeps adding loops of thread. A loose mesh thus ultimately envelops the entire surface of the egg sac. Finally, the spider spins a few suspension threads to attach the egg sac to neighboring leaves or twigs (Fig. 173d). During the next days the spider will stay near the cocoon. If some suspension threads are severed, she will replace them and, if necessary, also tackle larger repairs. Most female orb weavers die only a few days after having built their egg sac. Outwardly the cocoon remains unchanged from October until spring, while inside the spiderlings develop. It is not until May that the offspring will emerge.

In summary then, the construction of an egg sac passes through the following stages:

1. Weaving a horizontal silk disk, the center of which is condensed into the basal plate.
2. Adding the cylindrical wall.
3. Egg laying.
4. Closing the egg chamber with a cover plate.
5. Surrounding the egg case with a loose mesh of threads.
6. In some species, spinning an additional tougher cocoon wall to which suspension threads are attached.

Wolf spiders follow basically the same sequence of steps as araneids in constructing their cocoons. However, instead of crawling beneath the basal plate the spider stays on top of it to form the cylindrical wall, and she sits above the egg chamber when laying her eggs. Originally the egg case is white, but it is then tightly wrapped with bluish green threads rather than encased in a downy meshwork. The finished cocoon is lentil-shaped and has a tough, paperlike surface. It is attached to the spinnerets and carried around until the young hatch (Fig. 176a). If one tries to take the egg case away from the spider, she will vehemently defend it and will often even pierce her chelicerae into the cocoon wall in an effort to hold onto the cocoon. If one does remove the cocoon, the spider will search for it for hours. Often she will accept artificial substitutes like small paper balls or snail shells, which she then attaches to her spinnerets.

The construction of a spider's egg sac is often cited as an example of rigid or "programmed" behavior. For instance, if a *Cupiennius* is stopped while building her cocoon and placed on another cocoon that is in a more advanced stage of construction, the spider will resume her building at the point at which she was interrupted (Melchers, 1963). This may lead to the construction of another cylindrical wall, even though that structure was already present. On the other hand, one can also observe some degree of plasticity in this behavior. For instance, when *Cupiennius* is disturbed while building her own cocoon, she never returns to her half-finished product but starts making a new one. This means, however, that she has now to repeat the initial phases of egg-sac construction. Since this is done in a somewhat abbreviated manner, the result is a functional yet rather odd-looking egg sac. If that second cocoon turns out to be too tattered, the spider will simply eat it. From this behavior we can conclude that the spider is quite capable of perceiving the condition of the cocoon's surface.

Although egg-sac construction appears to be rather "programmed" in *Cupiennius*, this is by no means true for all spiders. If in *Cyrtophora,* for instance, the egg sac is tilted by 90° while it is under construction, the spider must deposit her eggs from the side rather than from below. *Cyrtophora* adapts easily to this condition and finishes a normal cocoon. Even if the cocoon is turned upside down, the spider tries to compensate for its new position; thus there is no question that her behavior can be flexible enough to meet new situations (Kullmann, 1961).

Types of Cocoon

Many spiders lay their eggs inside their retreats, which can then be called egg nests. The jumping spider *Heliophanus* (Fig. 7b) spins only a few threads around her 20–40 eggs; she then guards them within her closed silken retreat (Fig. 174a). Another jumping spider, *Marpissa*, builds 3–5 flat cocoons, which she stacks inside her nest (Fig. 174b). Each cocoon contains 20–40 eggs, and is encased by a basal plate and a cover plate. Thus different types of cocoons may occur within the same family of spiders (here, in the Salticidae). The same is true of the Gnaphosidae, where some genera have only simple cocoons inside their nests, and others build more complicated egg sacs with impregnated and camouflaged cocoon walls. When the walls are camouflaged, the female's guarding of the eggs becomes superfluous (Holm, 1940).

One spider that makes a camouflaged cocoon is *Agroeca brunnea* (Clubionidae). Its egg sac is one of the most unusual types of cocoons and looks like a little Japanese lantern. The bell-shaped egg sac is attached by a short stalk to grass blades or to twigs. In its unmasked condition it is pure white, but usually the spider glues soil particles or plant material to the outside so that it blends well with its surroundings. The interior of the cocoon is divided by a horizontal "wall" (which is really the cover plate) into an egg chamber and a molting

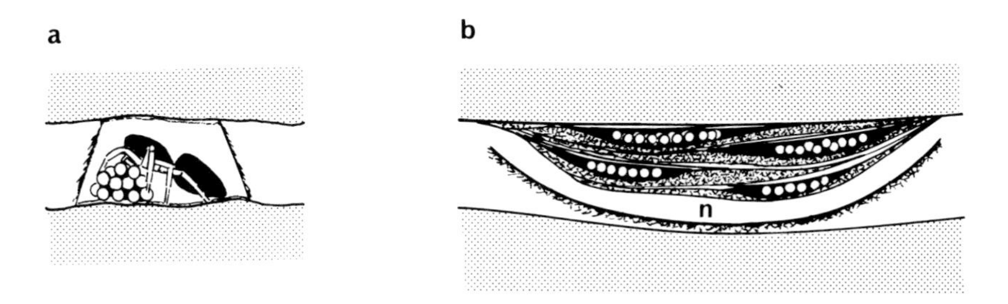

FIG. 174. (*a*) Section through an egg nest of the jumping spider Heliophanus cupreus. The female guards the 20–30 eggs inside the closed silk nest. (*b*) Egg nest of the jumping spider *Marpissa rumpfi*. Cocoons are arranged in several layers and enclosed in an open silk chamber (n). (After Holm, 1940.)

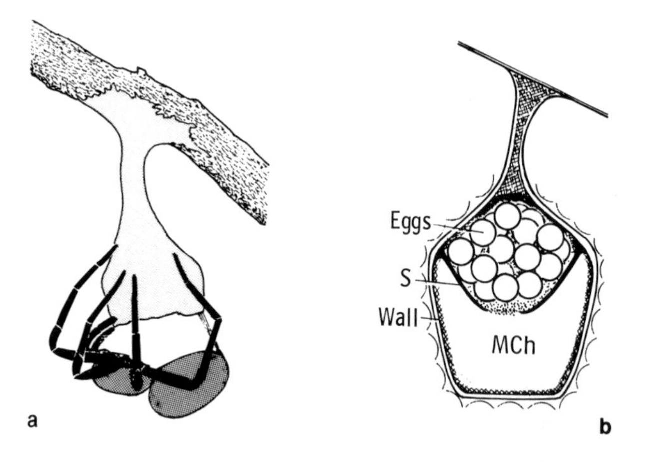

Fig. 175. Cocoon of the clubionid *Agroeca brunnea*. (*a*) Construction of the cylindrical wall. The diameter of the cocoon corresponds to the distance between palps and spinnerets. (Drawn from a photo by Lüters, 1966.) (*b*) Section through the finished cocoon. A septum (S) divides the cocoon into an egg chamber and a molting chamber (MCh). The cocoon wall is studded with camouflaging soil particles. (After Holm, 1940.)

chamber (Fig. 175). About one week after hatching, the 40–60 young spiderlings crawl from the egg chamber into the molting chamber, where they stay for 2–3 weeks (Lüters, 1966). Some days after molting they cut a circular hole into the cocoon wall, and the spiderlings emerge one after the other.

It is true that the various spider families do not always have one type of cocoon, yet we can differentiate two basic types of cocoons among the entelegyne spiders. The Trionycha usually have an egg sac that is surrounded by a loose mesh of threads, whereas the Dionycha possess a tough, thin-walled egg sac (Holm, 1940). If the egg cases are rather simple, as in *Heliophanus* or in *Pholcus* (Fig. 171a), it must always be considered that this does not necessarily represent a primitive stage, but that it could as well be the result of a secondary reduction.

Brood Care

Many spiders do more for their offspring than just construct a cocoon. As we have seen, some species camouflage, guard, and defend their egg sac; others even care actively for the newly hatched spiderlings, either by protecting them in special ways or by directly feeding them.

A wolf spider mother will assist her offspring during "hatching" by opening the rim of the cocoon with her chelicerae (Fig. 176). Without this help the young spiderlings are unable to leave the egg sac (Fujii, 1978). Normally the newly emerged spiderlings climb immediately onto their mother's back, where they hold fast to her abdominal hairs (Rovner et al., 1973). In all, there may be over

FIG. 176. Brood care in the wolf spider *Pardosa amentata.* (*a*) Carrying the cocoon (C) attached to the abdomen. (*b*) Opening the seam (S) of the cocoon wall with the chelicerae. (*c*) Spiderling, just hatched from the cocoon. (*d*) Two spiderlings climbing the legs of their mother. (*e*) About 80 young spiders sit closely packed on their mother's abdomen.

100 tiny spiderlings resting on top of each other in several layers (Higashi and Rovner, 1975). During the 7 or 8 days spent on the mother's back, the spiderlings live exclusively from their yolk supply. The mother likewise does not feed during the entire period of brood care. Whether the spiderlings gain any particular advantage by being carried around is still unclear. Certainly they enjoy some kind of protection from predators and to some extent from adverse microclimatic factors, yet experiments have shown that the young can survive easily when raised isolated from their mother and from each other (Engelhardt, 1964).

Several other experiments have demonstrated that young wolf spiders do not climb exclusively onto their own mother but will also try to climb onto male spiders—even those of a different species (Meyer, 1928; Engelhardt, 1964; Higashi and Rovner, 1975). In such experimental situations the males often simply eat the spiderlings. It is also possible to trick a female spider into accepting a foreign cocoon or young other than her own. Even spiderlings of another species are readily accepted, and they will soon settle down on the back of their "stepmother." From these experiments it is evident that a mutual recognition between the mother spider and her offspring does not exist (Engelhardt, 1964).

Nursery-web spiders (Pisauridae), which are related to wolf spiders, also care for their young. The female carries the globular cocoon in her chelicerae. Before the young hatch, the mother constructs a tentlike web and suspends the egg sac inside it. The emerging spiderlings stay for several days in this "nursery web." After molting they gradually disperse and start their regular free-hunting life.

In species where the mother spider provides food for the young spiderlings, one can differentiate several levels of organization: (1) leaving caught prey passively at the disposal of the spiderlings; (2) actively feeding them liquefied food from her midgut; and (3) as a supplementary food source the mother can lay an additional batch of small eggs, upon which the young ones feed (*oophagy* or egg *cannibalism*; Gundermann et al., 1991; Tahiri et al., 1989). Active brood care that involves feeding of the young is found in only a very few spider species (in about 20 of the 34,000 known species), namely, in certain theridiids, eresids, and in some agelenids, such as *Coelotes terrestris* (Tretzel, 1961b; Gundermann, 1986; Gundermann et al., 1988).

The young *Coelotes* remain for about one month in the funnel web of their mother. When the mother catches prey, the young "beg" for food by touching their mother's chelicerae with their palps or legs. Usually the mother simply drops the softened prey, and one of the young spiders quickly grabs it and carries it into a hideout to feed on it. As long as the prey is still struggling, the mother refuses to turn it over to a spiderling; her "message" comes across as she vigorously beats her hind legs against the sheet web. The mother can distinguish her young (and their vibrations) from prey or enemies quite well. Prey animals produce vibrations of higher amplitude and higher frequency than those of young spiders. Nevertheless, the mother often "checks" her young with one palp; only thereafter can they continue to run around in the web. It is clear that this discriminatory ability together with a tolerance toward the same species are necessary preconditions for any communal living (see chapter 9). The young spiderlings enjoy the advantage of being provided with much larger prey items than they could overpower themselves.

Among the cobweb spiders, *Theridion pictum* and *Achaearanea saxatile* also leave caught prey for their young (Hirschberg, 1969). When *Achaearanea* has caught prey, she performs stroking movements in the web to attract the young spiders. Conversely, plucking movements serve as warning signals that send the young ones scurrying to the retreat (Nørgaard, 1956). The cobweb serves here as a means of communication between mother and young spiders, just as does the sheet web in *Coelotes*.

The highest organization of brood care in spiders has evolved in a few species among the theridiids and the eresids. The mother feeds her young mouth-to-mouth with a special substance that she regurgitates. This kind of food transfer is well known from bees, ants, and termites but is extremely rare in spiders. *Theridion notatum*, *Theridion impressum*, and a few *Stegodyphus* species rank among these exceptions. The regurgitate consists of a mixture of predigested food and the mother's own intestinal cells, which are in a process of lysis (Nawabi, 1974). From time to time the mother exudes a drop of regurgitate at her mouth opening, and soon several spiderlings gather there to drink the liquid. Some spiderlings stroke their mother's legs, apparently to stimulate her to regurgitate. During the first days after hatching, the young are fed chiefly with regurgitate, and prey is offered only as an extra. In *Theridion impressum* the mother stops regurgitation feeding after the young have undergone their first molt. Prey animals, however, are still consumed communally (Kullmann, 1974). When the young spiders have grown further, they will be chased away by the mother (Hirschberg, 1969).

Regurgitation feeding has also been demonstrated by using radioactive markers such as labeled phosphorus (^{32}P). This method permits quantitative measurements of the regurgitated substances (Hirschberg, 1969; Kullmann and Kloft, 1969). In such studies the radioactive isotope (for example, $Na_2H^{32}PO_4$) is first fed to a fly. When the radioactively labeled fly is then given to the spider mother, she becomes radioactive herself (Fig. 177). After the young spiderlings have hatched from the cocoon, they are checked every day to see whether they have become radioactive or not. By measuring the amount of radioactivity in the spiderling, one can also determine how much food it has received from the mother. It was surprising to learn that regurgitation feeding already becomes

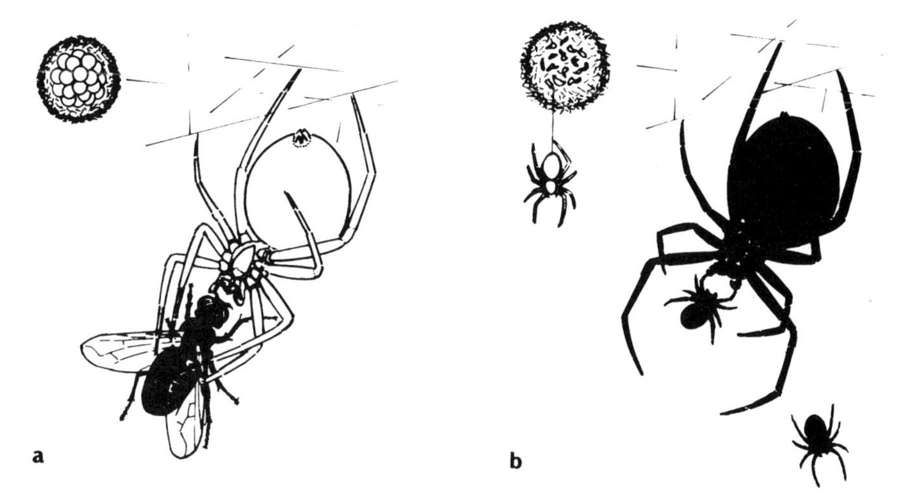

FIG. 177. Use of radioactive isotopes to demonstrate regurgitative feeding. (*a*) After having built the cocoon, the spider is fed with a radioactively labeled fly and becomes radioactive herself. (*b*) The newly emerged spiderlings are fed mouth-to-mouth and in turn become radioactive. Radioactivity is represented in black. (After Kullmann and Kloft, 1969.)

established on the first day after hatching. During the next days the young spiderlings feed additionally on prey animals. Thereby they quickly gain weight and are ready to molt after only one week.

The regurgitate is not absolutely necessary for the survival of the spiderlings. If they are kept in isolation and offered only the softened prey caught by their mother, they also grow up, although more slowly. Apparently the combination of regurgitate and the consumption of prey provides for the optimal development of the young spider (Hirschberg, 1969).

It is quite remarkable that regurgitation feeding is found in two completely unrelated families, namely, the ecribellate Theridiidae and the cribellate Eresidae. Probably, this highly developed brood care has evolved independently in both spider families (Kullmann et al., 1971; 1971/1972). It also seems plausible that regurgitation feeding represents a first step in the development of communal (or social) life. This mode of feeding requires not only close contact, but also tolerance and cooperation between conspecifics.

8

Development

The ontogeny of a spider can be divided into three periods: an embryonic period, a larval period, and a nympho-imaginal period (Vachon, 1957). The embryonic period encompasses development from the time an egg is fertilized until the typical shape of the spider's body is established. During the subsequent larval period the prelarva and larva still lack certain morphological characteristics, and hence are referred to as "incomplete stages" (Holm, 1940). The larval stages are unable to feed on their own and live solely on their yolk supply. Not until the nympho-imaginal period, when all the organ systems are finally present, do the young spiders become self-sufficient. The imago (adult) differs from the nymph (or juvenile) essentially by being sexually mature.

All stages from the prelarva to the adult are separated by a molting process. Growth occurs only during a molt, that is, between two consecutive stages, and thus takes place in graduated steps. Spider eggs are always enveloped by two layers, an inner *vitelline membrane* and an outer *chorion* (Fig. 151d). The chorion appears opaque and granular because the liquid that surrounded the egg during egg laying has dried onto it (Fig. 182a). Immersing an egg in paraffin oil renders the chorion transparent so that the internal egg structure and all developmental processes can be observed directly. Immediately beneath the vitelline membrane lies a thin plasmatic cortical layer. This peripheral zone gives off delicate extensions that radiate toward the center of the egg cell, where the nucleus is located. The main part of the egg, however, consists of yolk granules and "vitelline fluid." The yolk granules tend to be larger in the lower part of the egg; however, this does not reflect true polarity, because turning the egg upside down will reverse the yolk distribution (Holm, 1940).

Early Development

Fertilization takes plase in the uterus externus (see Fig. 156). One hour after oviposition the sperm nucleus is already moving toward the center of the egg. At that time the female pronucleus has just completed the second maturation division and is also on its way toward the center (Suzuki and Kondo, 1994a,b). The actual conjugation of the two nuclei occurs 1–2 hours later (Montgomery, 1908). The zygote begins a series of mitotic divisions called *cleavage*. A few hours after fertilization one can recognize two "cleavage nuclei" that shift their

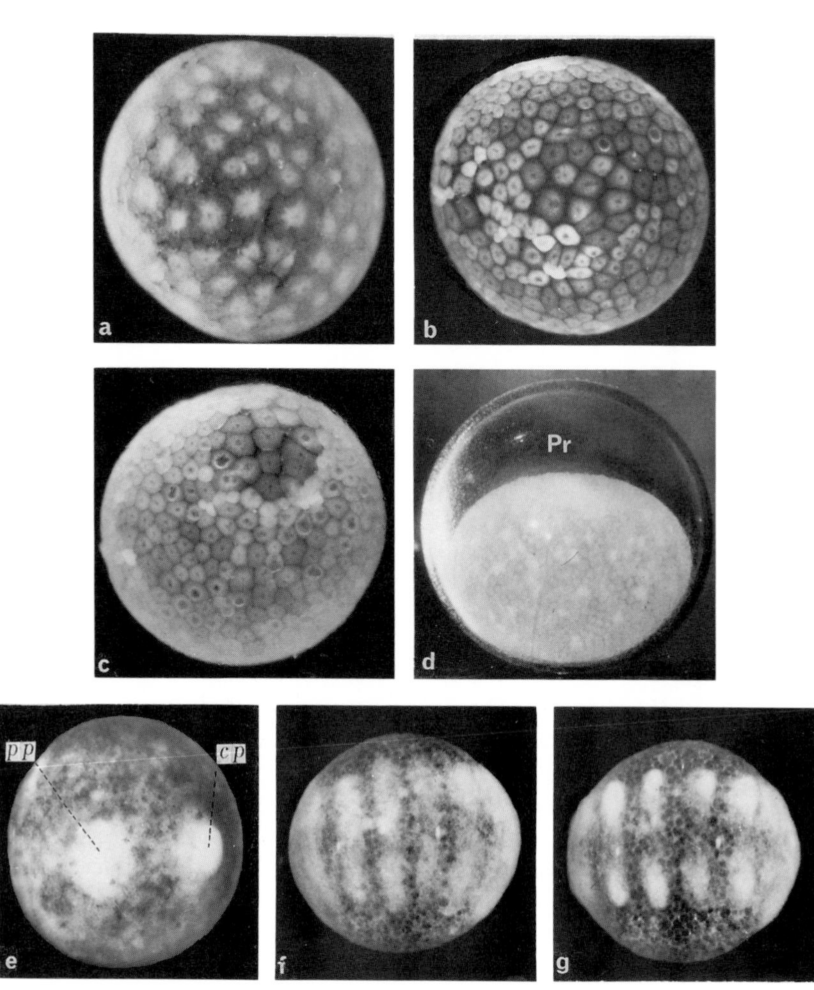

FIG. 178. Early embryonic stages of *Agelena labyrinthica*. (*a*) The cleavage cells spread within the cortical plasm (30 hours). (*b*) Blastula stage prior to contraction (48 hours). (*c*) Blastula stage after contraction. (*d*) Blastula of the tarantula *Ischnothele* in optical section. Pr = perivitelline space. (*e*) Formation of the primitive plate (pp) and of the cumulus (cp) (100 hours). (*f*) Beginning of segmentation of the prosoma (150 hours). (*g*) Primordia of the four pairs of legs (160 hours). (Photos: Holm.)

position from the center of the egg to the periphery; actual cell membranes have not yet formed. At the same time the cortical plasma divides into many polygonal patches (Fig. 178a–c). When the two cleavage nuclei have divided four more times (which for *Agelena* happens about 24 hours after the egg was laid) all 32 nuclei have moved to the periphery (Holm, 1952). Following three further nuclear divisions, each "cleavage cell" (nucleus) coalesces with from four to eight patches of the cortical cytoplasm, thereby establishing large, flattened cells. Thirty-five hours after egg laying, a thin *blastoderm* has formed by a process that can be called "superficial" cleavage, because the entire yolk mass, which does not divide, becomes concentrated in the center of the embryo. In this stage of development the embryo is called a *blastula*. The yolk granules now sink to the bottom part of the blastula, leaving behind a space, the blastocoel, that is filled with perivitelline fluid (Fig. 178d). The next stage is marked by the so-called contraction of the blastula (at 50–55 hours), which involves a transfer of the perivitelline fluid to the outside, so that the overlying blastoderm settles on top of the yolk mass.

The Metameric Phase

The polarity of the embryo first becomes apparent immediately after the contraction of the blastula, when one side of the flattened blastoderm is transformed into a columnar epithelium. Gradually a diffuse germinal disk, which consists of rather small cells, develops (at about 75 hours after egg laying). Ten hours later a slight indentation called the *primitive groove*, or blastopore, appears in the center of the germinal disk. It is at this spot that the migration of the mesendoderm cells into the interior takes place (*gastrulation*). The area, which is then underlaid by the mesendoderm cells, is referred to as the *primitive plate* (Fig. 178e).

The primitive plate is initially flat, but soon one portion of it bulges outward to form the primitive knot, or *cumulus* (at about 100 hours). The cumulus gradually shifts toward the margin of the germinal disk. By this time the primitive plate is several cell layers thick, because most of the mesendoderm has already migrated into the interior. Between the cumulus and the primitive plate lies an area of flat cells called the *dorsal field*. This sector of the gastrula enlarges laterally while the cumulus becomes reduced in size and finally disappears completely (130 hours after fertilization).

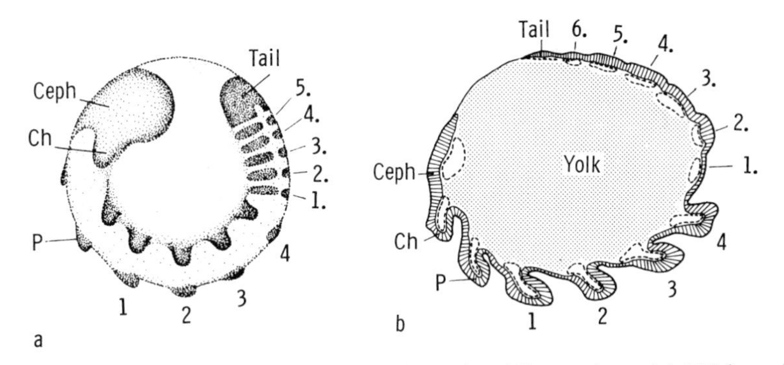

FIG. 179. (*a*) Germinal band shortly before inversion (*Cupiennius salei*, 130 hours). The abdomen is clearly segmented (1.–5.). On the prosoma the extremities differentiate. Ceph = frontal lobe, Ch = chelicerae, P = pedipalps, 1–4 = legs. (After Seitz, 1966.) (*b*) Longitudinal section of an embryo shortly before inversion (*Agelena labyrinthica*). Note the metameric coelomic cavities (dashed lines). Abbreviations as in (*a*). (After Wallstabe, 1908.)

When the sector of the dorsal field has spread half way around the developing embryo (180°), the first signs of segmentation appear in the form of two or three concentric furrows around the blastopore (Fig. 178f). At about 150 hours the germinal disk stretches into a broad *germinal band*, which consists of a cephalic lobe, a caudal lobe, and five intervening metameres (somites) that represent the respective segments of the pedipalps and the legs. The cephalic lobe soon gives off an additional segment, the cheliceral segment, while the caudal lobe splits to form the first abdominal segment. About the same time, the primordia of the extremities become visible as slight elevations on the prosomal segments; somewhat later (170–200 hours) similar budlike thickenings appear on the developing abdominal segments (Fig. 178g). Whereas the prosomal segments manifest themselves synchronously, the abdominal segments are added successively at the caudal end. In the interior of the body the coelomic cavities become established when the mesoderm separates (Fig. 179); the walls of these cavities later differentiate into muscle tissue.

Inversion of the Embryo

When the germinal band has reached its maximal length so that the cephalic and the caudal lobes almost touch each other (day 10), a longitudinal furrow appears in the midline of the germinal band (Fig. 180). This median furrow gradually widens, and the halves of the germinal band pull apart from each other laterally, again exposing the mass of yolk beneath the median furrow. The marginal areas of the embryo extend laterally until they meet in the dorsal midline; thus the originally convex germinal band straightens and finally curls in the opposite direction (*inversion* or *reversion*).

In some "primitive" spiders (the Mesothelae and certain Orthognatha) the embryo develops without such an inversion process (Crome, 1963; Yoshikura, 1975). During the early developmental phases a *postabdomen* composed of nine segments appears. It is sharply flexed toward the ventral side, and at this stage of development the embryo very much resembles those of some "primitive" arachnids (scorpions, for instance). Inversion is typical of most labidognath embryos and, from an evolutionary point of view, is considered a more advanced form of development.

The anlagen of the pedipalps and legs initially have only two segments, but these soon divide into four, and then into six or seven segments of equal length (Wallstabe, 1908). The anterior extremities of the abdomen gradually become smaller and move apart (Fig. 180d). Invaginations in the second and third abdominal segments represent the future book lungs and the tubular tracheae, respectively. The next abdominal extremities, on segments 4 and 5, shift toward the posterior end of the embryo and differentiate into spinnerets. A slight constriction of the first abdominal segment indicates the border between prosoma and opisthosoma. The initially distinct segmentation of the body becomes less apparent during later developmental stages. However, abdominal segmentation is often still noticeable in young spiders, either directly in the

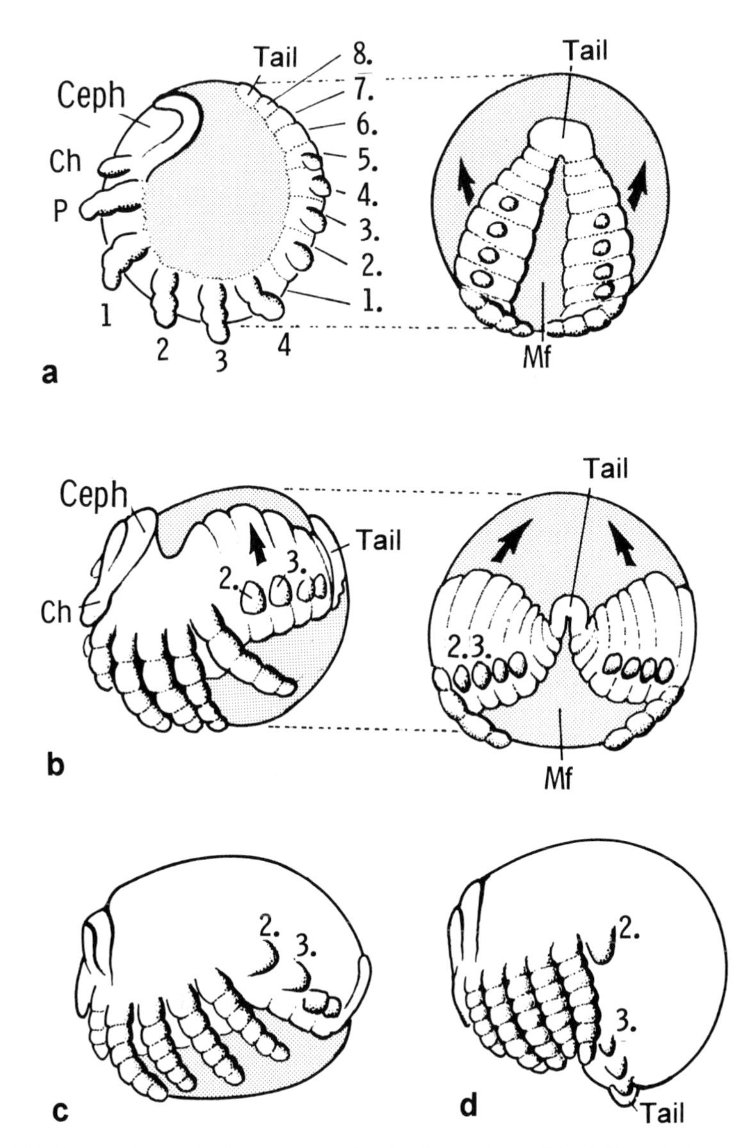

FIG. 180. Late embryonic development in *Agelena labyrinthica* (abbreviations as in Fig. 179). (*a*) Beginning of inversion, lateral view. Abdominal view of the same stage is seen on the right. The germinal band splits along the midline (Mf) and both parts expand laterally (arrows). The anlagen of the abdominal extremities appear on segments 2.–5. (*b*) In the middle of the inversion process the extremities consist of five segments and rest on top of the ventral yolk mass. (*c*) The lateral body walls have merged along the dorsal midline, and another segment has been added to the legs. (*d*) Completed inversion. The yolk mass is completely grown over by the embryo; prosoma and opisthosoma are flexed toward each other. The abdominal appendages move apart and differentiate into respiratory organs (segments 2. and 3.) and spinnerets. The extremities have attained their final number of segments (15 days after the egg was laid). (After Wallstabe, 1908.)

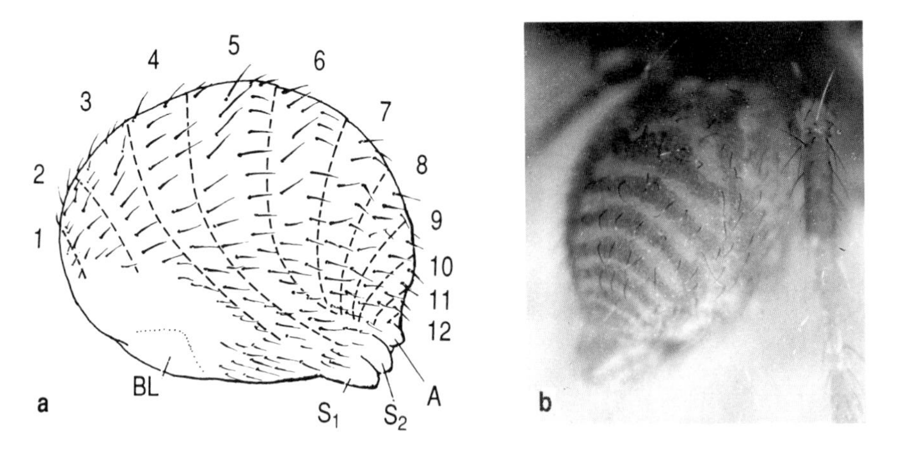

FIG. 181. (*a*) Lateral view of the opisthosoma of a young spider (*Nuctenea umbratica*). The embryonic segmentation (dashed lines) is reflected in the serial distribution of hair sensilla. A = anal tubercle; BL = book lung; S_1, S_2 = spinnerets; 1–12 = abdominal segments. (Tracing of a photograph by Holm, 1940.) (*b*) Abdomen of a young *Tegenaria*, dorsal view. The distribution of the dark pigment and of the hair sensilla corresponds to the original segmentation.

form of fine furrows or indirectly as a segmentally distributed pattern of abdominal hairs (Fig. 181). The typical shape of the spider's body emerges at the end of the inversion process. The embryo at this stage is then termed a *prelarva* by some authors (Vachon, 1957; Legendre, 1958).

The spider embryo has a marked capacity for regeneration, at least until segmentation begins. Even the loss of half of the embryo can be tolerated, and a normal embryo formed (Holm, 1952). The cumulus ostensibly acts as the organization center, for if it is extirpated, no dorsal field, and consequently no germinal band, develops. This observation has been challenged by experiments in which spider embryos whose cumuli had been destroyed by X-rays still developed normally (Seitz, 1970). However, it could be that some of the endodermal cumulus cells had survived the radiation and hence could still regulate the development of the embryo.

Organ Development

Most organ systems develop during the larval period from the cells of the three germinal layers. The nervous system, sensory organs, respiratory organs (book lungs and tracheae), poison glands, silk glands, foregut, and hindgut develop from the *ectoderm*. The *mesoderm* differentiates into the musculature, blood vessels, coxal glands, and reproductive organs. The *endoderm* provides the material for the midgut and the Malpighian tubules.

Development of the Nervous System

The primordia of the nervous system appear as medial, pillowlike thickenings of the ectoderm. Neuroblasts gather within each segment and form ganglia that sink into the interior of the embryo, where they aggregate to produce a coherent mass of nervous tissue. All abdominal ganglia migrate into the prosoma. There they combine with the ganglia of the extremities and form the large subesophageal ganglion. The cheliceral ganglion shifts anteriorly so that it comes to rest above the esophagus; topographically it is now a part (the tritocerebrum) of the supraesophageal ganglion. The proto- and deuterocerebrum of the supraesophageal ganglion develop as invaginations of the cephalic lobe and become connected with the subesophageal ganglion only secondarily (Dawydoff, 1949). The outgrowth of nerve fibers gives rise to a neuropil (Fig. 82) and the peripheral nervous system.

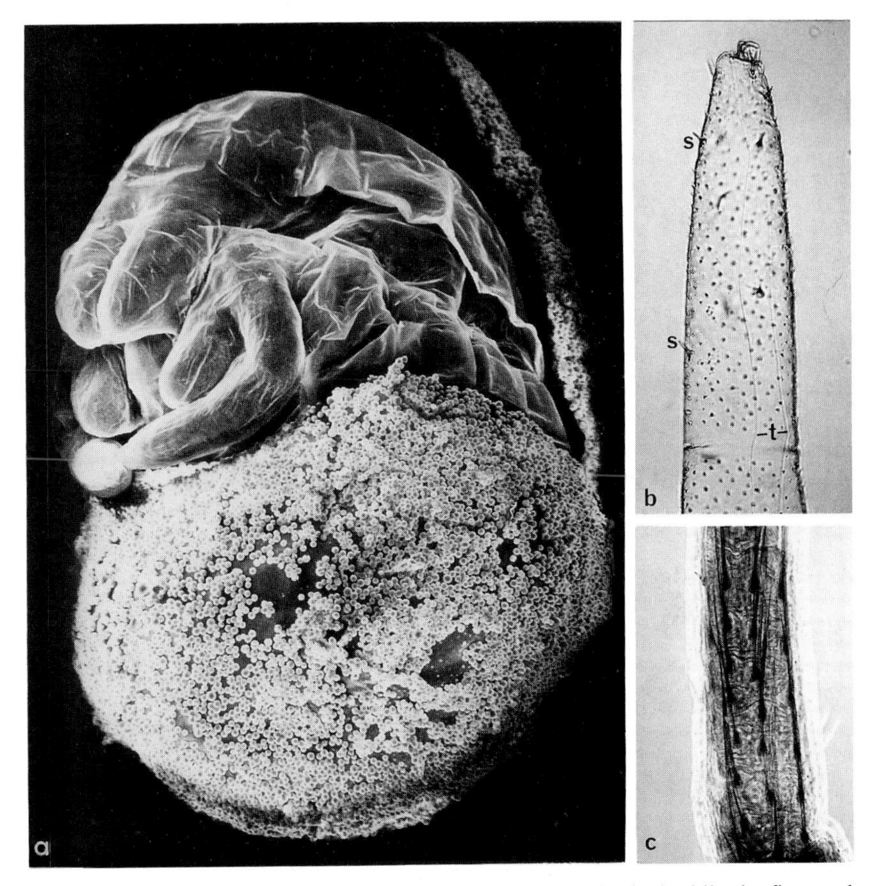

FIG. 182. (*a*) Hatching spider larva. The granular egg shell is shed while the first molt takes place. 90 ×. (*b*) Larval leg of a black widow (exuvium). Note the few short hair sensilla (S) and the two small claws that are moved by long tendons (t). 250 ×.
(*c*) First nymphal leg, still encased in the larval leg cast. Note the wrinkled surface of the new cuticle and the linear arrangement of hair sensilla. 280 ×.

Development of the Gut

The abdominal yolk mass becomes cleaved by intersegmental membranes which later divide further to form the prospective midgut diverticula. The median axis of the embryo remains unsegmented, however, and will turn into the glandular part of the digestive canal. The foregut and hindgut are ectodermal invaginations, and therefore possess a lining of cuticle. The union of the three independently formed parts of the intestinal tract does not take place until after hatching (Legendre, 1965).

In *Agelena* hatching occurs about 2 weeks after the egg was laid. Special cuticular denticles, or "egg teeth," which are situated at the base of the larval pedipalps, help to tear the chorion sheath of the egg. It is possible that transient hatching glands are also involved in perforating the egg membranes (Yoshikura, 1955). While these membranes are being shed, the spiderling molts for the first time (Fig. 182; Holm, 1940).

Classification of Developmental Stages

Normally it is the emergence from the egg that marks the transition from embryonic to postembryonic life. This can also be said of spiders (Holm, 1940); yet the stage at which they hatch may vary in different spiders. For instance, in most spiders it is the prelarva that sheds its egg membranes, but in some spiders this happens first at the larval stage. For this reason, Vachon (1957) has classified the different developmental stages without regard to when the spider hatches; instead, he defined the completed inversion process as the end of the embryonic period. The subsequent larval period begins with a spiderlike *prelarva*, which after molting develops either into another prelarva or into a more advanced *larva* (see Table 4). After one or two additional molts, the first complete stage, *nymph 1*, appears (Fig. 183). The next 5–10 nymphal stages (juveniles) differ mainly by their increasing size. Only with the last molt do the spiders reach sexual maturity and begin adult life. The main criteria for determining the different developmental stages are summarized in Table 4.

TABLE 4. Developmental Stages in Spiders. (After Vachon, 1957.)

	Larval: not self-sufficient (yolk)		*Nympho-Imaginal: self-sufficient (prey)*	
Characteristic	*Prelarva*	*Larva*	*Nymph*	*Adult*
Mobility	None	Very little	Complete	
Leg segmentation	Incomplete	Complete	Complete	
Hairs, spines	Neither	Both, undiff.	Large variety	
Claws	None	Simple	Differentiated	
Cheliceral claw	Undiff.	No poison canal	Differentiated	
Spinnerets	Barely diff.	Without spigots	Functional	
Sexual organs	Undeveloped	Undeveloped	Developed nonfunct.	Developed functional

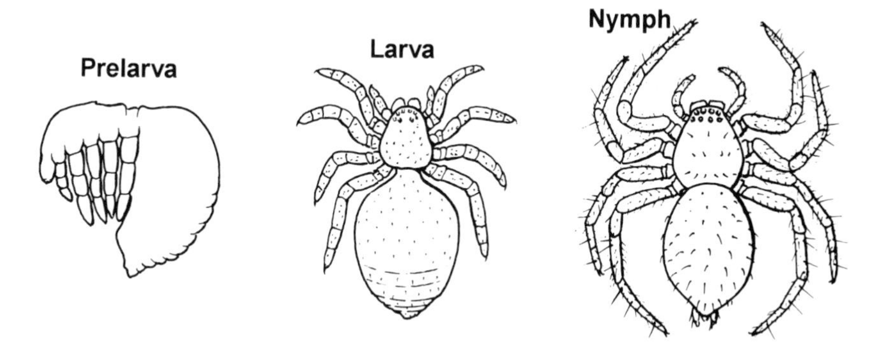

FIG. 183. Developmental stages in spiders. (After Vachon, 1957.)

The descriptions of embryonic development in spiders are actually rather confusing, because the various authors use different terms (e.g., prelarva, larva, pullus, postpullus, prenymph, nymph, etc.) for the embryonic stages. Although several authors have recently dealt with this problem (Canard, 1987; Downes, 1987; Canard and Stockmann, 1993), no general consensus has been reached so far. In general, there is now, again, a tendency to define *hatching* (i.e., the opening of the egg membranes) as the end of the embryonic period. The following immobile stages (1–3) are called *postembryonic* or *fetal*—until the first "true" molt occurs, in which a skin with extremities is shed. The following stages are then juvenile (nymphs) and they look like miniature spiders.

Growth and Molting

The rigid exoskeleton of an arthropod rather limits the growth of the body. In spiders only the soft abdomen can expand—the prosoma and the extremities, which are encased in hard exocuticle, cannot. Actual growth can thus occur only during a molt. The new cuticle lies wrinkled beneath the old body shell (Figs 182c, 188b) and can be stretched during and immediately after the molting process. It is this folding-extension mechanism that allows for a finite increase in size from one developmental stage to the next. In addition to an increase in size, some body proportions may also be changed, and certain sensory organs (such as hair sensilla) may increase in number or may appear for the first time.

Most obvious are the changes in sensory organs between the immobile larvae and the very agile nymphal stages (Wurdak and Ramousse, 1984). Larvae of *Araneus* have only a few short sensory hairs (on the distal leg segments), a single lyriform organ (on the metatarsus), and eyes without lenses. After molting into the first nymph, all types of sensory hairs are present (tactile and taste hairs, trichobothria, tarsal organs), each leg exhibits several single slit sensilla plus 14 lyriform organs, and the eyes are now equipped with bulging lenses. The legs have the three typical claws (instead of two simple hooks), the spinnerets bear functional spigots of various silk glands, and the mouth parts

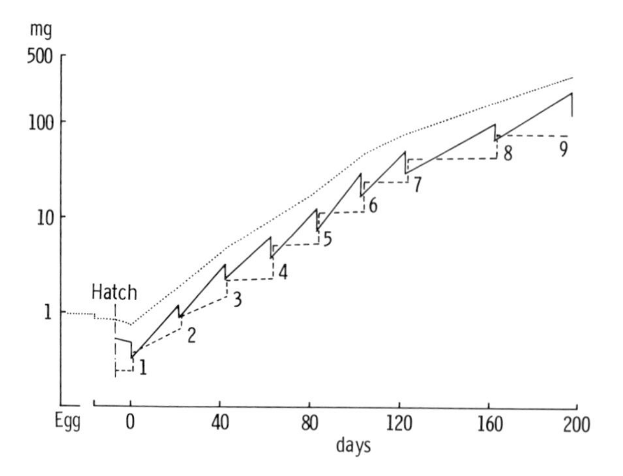

FIG. 184. Growth of a *Tegenaria agrestis* during nine molts (1–9). Dotted line = body growth, solid line = growth of opisthosoma, dashed line = growth of prosoma. Egg = time of egg laying (−145 days). The ordinate is represented logarithmically. (After Homann, 1949.)

are differentiated into cheliceral teeth, an open cheliceral fang, and a serrula as a cutting edge on the maxillae. Only quantitative changes of the sensory equipment will be oberved during all the following stages.

Early nymphal stages may molt every few days, but later instars need several weeks to prepare for the next molt (Eckert, 1967). The intermolt intervals are, of course, dependent on nutritional conditions. If provided with an ample supply of food, young *Tegenaria* molt about the time at which their weight has doubled (Fig. 184; Homann, 1949). The number of molts undergone depends on the ultimate body size. Small spiders need only a few molts (about 5), whereas large spiders pass through about 10 molts to reach the adult stage (Bonnet, 1930). The small males achieve maturity with 1–2 fewer molts than the (larger) females. In black widows, for instance, the male matures twice as fast as the female (Deevey, 1949; Forster and Kingsford, 1983). For most spiders the last molt marks the transition to sexual maturity; only in some exceptional cases do adult spiders still molt further (see below).

In general, young spiderlings molt independently of each other. However, in the colonial araneid *Eriophora bistriata*, where all the little orb webs are built closely together, the spiderlings molt at the same time (Fowler and Gobbi, 1988a). It is believed that this synchronization is due to a chemical communication (pheromones).

The Molting Process

An experienced arachnologist can usually predict when a spider is getting ready to molt. The animal withdraws into its retreat for several days and stops feeding. A closer examination of the spider's body reveals a darkening of the legs caused

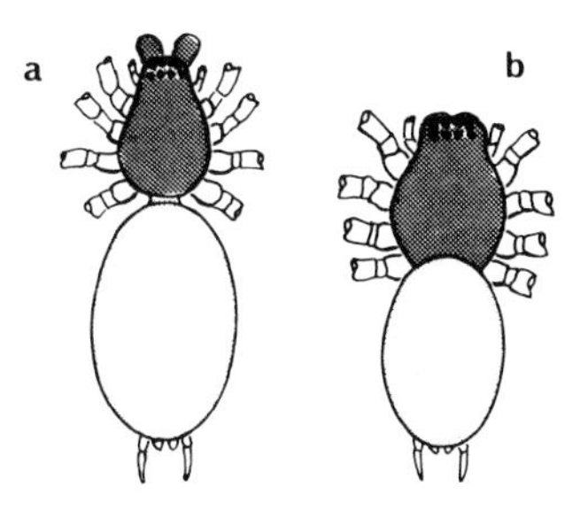

Fig. 185. Two young *Tegenaria agrestis* (*a*) immediately before and (*b*) after molting. Notice the change in the volumes of the prosoma and opisthosoma. (After Homann, 1949.)

by the new hair sensilla shining through the old cuticle. Another sign is the movement of the abdomen away from the prosoma so that the pedicel becomes visible from above (Fig. 185). The definite signal for the imminent molting process of most spiders is their suspension on a molting thread. "Tarantulas" do not use molting threads, but simply lie on their backs while they molt (Fig. 186; Melchers, 1964; Minch, 1977). In all spiders, however, three successive phases of the molting process can be distinguished: (1) lifting of the carapace; (2) liberation of the abdomen; and (3) extraction of the extremities.

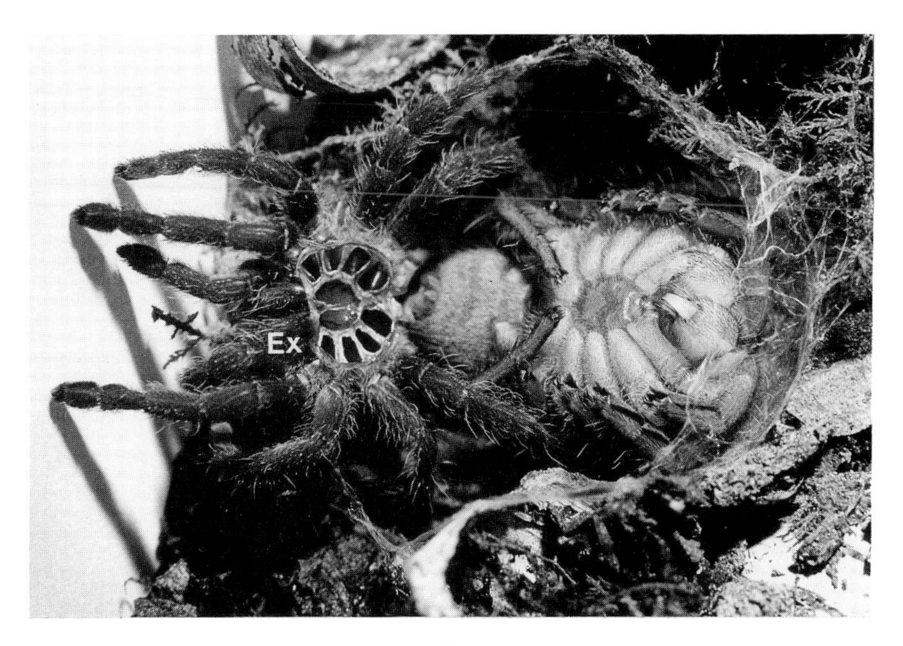

Fig. 186. "Tarantulas" lie on their back while molting. On the left the shed skin (Ex) can be seen, on the right the freshly molted animal. Note the lack of pigmentation and the legs pressed tightly to the body. A very delicate mesh of silk fibers encloses the spider.

Lifting of the Carapace

At the beginning of the molting process, the heart rate increases and more hemolymph is pumped into the prosoma. The weight of the cephalothorax increases by 80%, while the abdomen shrinks and loses about 30% of its weight. The normal hemolymph pressure of 150 mm Hg (about 17.5 psi) doubles because of the increased frequency of the heartbeat and the concomitantly higher stroke volume. Under normal conditions the cuticle can withstand pressures of 600–750 mm Hg before it will split. Shortly before molting, however, the cuticle loses about two-thirds of its rigidity, because at this time most of the endocuticle has been dissolved from the inside (see below). It has been shown experimentally that at this stage a pressure of 200–300 mm Hg is sufficient to rupture the cuticle of the cephalothorax (Homann, 1949).

The first tearing occurs along the frontal or lateral parts of the prosoma. The chelicerae, which normally point slightly forward, are moved back and forth, producing additional tension near the clypeus. The first rent is most likely to develop in this region and then extend laterally above the coxae until the entire carapace lifts off like the lid of a tin can (Fig. 187).

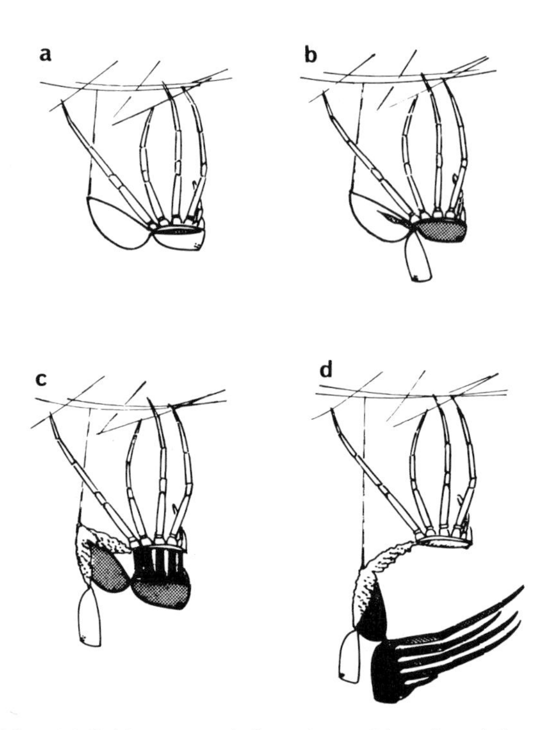

FIG. 187. Molting. (*a*) Spider suspended on the molting thread; lateral tearing of the cephalothorax. (*b*) Lifting off of the carapace and lateral fission of the abdominal cuticle. (*c*) Freeing of legs and abdomen. (*d*) Complete liberation from the exuvium. (After Bonnet, 1930.)

Liberation of the Abdomen

The two tears running along both sides of the prosoma soon spread to the opisthosoma. The old abdominal cuticle appears wrinkled because the volume of the abdomen has decreased, owing to the shift of the hemolymph anteriorly. Wavelike contractions of the abdominal musculature finally cause the abdomen to separate completely from the old skin (the exuvium). Just before the abdomen frees itself, the spinnerets attach a thread to the inside of the shed skin, and the spider descends a bit farther on this thread (Fischel, 1929).

Extraction of the Extremities

About the same time that the abdomen becomes free, the extremities are pulled out of their old cast. This is by far the most difficult part of the molting process, and consequently most complications arise during this phase. For instance, it may happen that a spider cannot extract all of its extremities from the old skin, a situation that is often fatal. Usually, however, the new cuticle is soft and sufficiently pliable that even rather bulky parts (like the bulb of the male palp) can be pulled through the narrow lumen of the old skin.

The increase in hemolymph pressure causes the folded leg cuticle to stretch somewhat and, by this action, the leg coxae are pushed out of the exuvium. The many hairs and spines arising from the new leg cuticle are actually advantageous, since they point distally and thus prevent the leg from slipping back into the old cast. However, the increase in hemolymph pressure is only partly responsible for the extraction of the legs. When the femora are halfway free, the cephalothoracic muscles become active. Several legs remain fixed in their positions and act as braces for those legs that are pulled out. This is usually done in a particular sequence; for example, legs 1 and 4 on one side stay in a fixed position while legs 2 and 3 are partially extracted on the other side. When the "new" legs are exposed down to the patellae, muscles of the legs themselves, the flexors, are activated. This leads to a shortening and, subsequently, to a jerky withdrawal of the legs from the old cuticle. Again, the hairs and spines of the legs help in this process, since they allow the legs to be moved only toward the body. As soon as the first legs are completely free, they aid in pulling out the remaining ones. Immediately after molting the legs usually hang down in a completely stretched fashion, but they are then bent abruptly. A period of "gymnastics" follows during which all extremities are flexed repeatedly. Presumably this helps to maintain elasticity of the joints while the cuticle begins to harden.

The largest force needed for extracting a leg (0.5 pond, or 5 mN) must be exerted at the beginning of the liberation process, because friction is greatest when the entire leg is still encased in the exuvium. Once the extraction is underway, the friction diminishes rapidly (Homann, 1949). The suspended position that most spiders assume during molting is not absolutely necessary but is physiologically very economical. Often the freshly molted legs would still be too limp to support the weight of the spider. Perhaps this also explains why some wandering spiders lie on their backs while they molt.

The entire molting procedure lasts only 10–15 minutes for small spiders but may take several hours for large "tarantulas." Some spiders start to groom themselves extensively after molting, pulling all their extremities through the chelicerae and mouth parts.

Physiology of Molting

In a strict sense molting comprises two different processes: (1) *apolysis*, the separation of the old cuticle from the hypodermal cells; and (2) *ecdysis*, the shedding of the entire old skin (*exuvium*), which corresponds to what most people think of as molting. Apolysis precedes ecdysis by about one week.

Cellular Events

The epidermal cells secrete certain enzymes (chitinases and proteases) into the gap that develops between the epithelium and the cuticle, the *exuvial space*. These enzymes gradually dissolve the endocuticle but attack neither the exocuticle nor the nerve fibers associated with the sensory hairs. The dissolved components of the endocuticle are largely resorbed, which means that very little material is actually lost in the molting process (Fig. 188). For a large "tarantula" weighing 80 g, the dry weight of the shed skin was found to be 2 g (Melchers, 1964).

About 4–5 days before the spider molts, the hypodermal cells secrete the first new layer of the cuticle (the *cuticulin*). The enzymes circulating in the exuvial fluid probably become activated only after the cuticulin layer has been deposited, that is, after the surface of the hypodermal cells is protected. Ecdysis begins when the endocuticle has largely been digested. That the prosoma tears first along the pleural areas (Fig. 187a) becomes quite understandable when one remembers that exocuticle is lacking in the joint regions and thus only the thin epicuticle remains there after apolysis.

The following sequence of events, which is very similar to that observed in insects, occurs when spiders molt:

1. Activation of the hypodermal cells (cell division and cell growth).
2. Secretion of the exuvial fluid and apolysis.
3. Secretion of the first new cuticle (cuticulin).
4. Activation of the enzymes in the exuvial fluid (dissolving of the endocuticle).
5. Absorption of the dissolved substances from the old cuticle.
6. Secretion of the new cuticle.
7. Ecdysis—the shedding of the old skin.
8. Stretching of the wrinkled new cuticle.
9. Sclerotization of the new exocuticle.

Subsequently, as further endocuticle is deposited during the weeks following the molt, the exoskeleton will increase considerably in thickness.

It seems particularly interesting that the sensory hairs remain functional

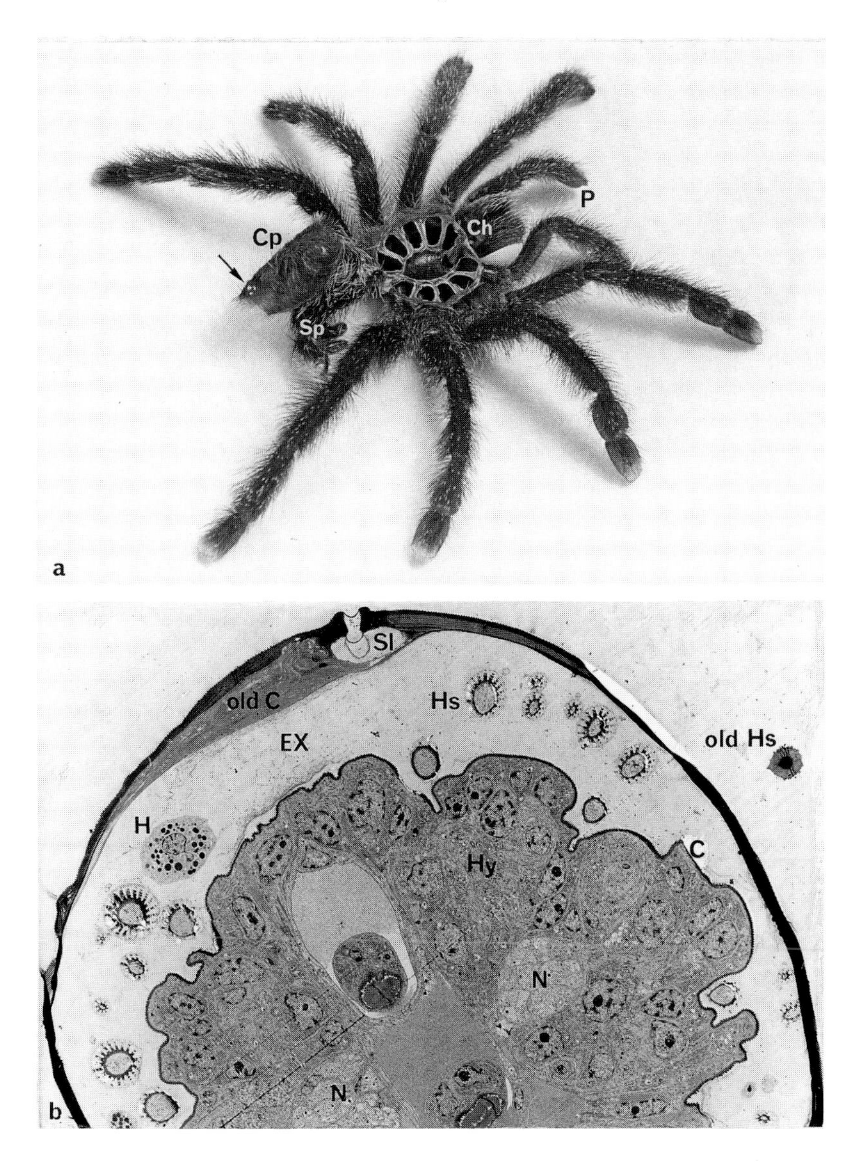

FIG. 188. (*a*) Exuvium of a young "tarantula". Ch = chelicerae, Cp = carapace with the cuticular lenses of the eyes (arrow), P = pedipalp, Sp = spinnerets. About natural size; weight of exuvium: 0.2 g. (*b*) Cross-section of a leg of *Hypochilus* shortly before molting. The old cuticle (old C) is separated from the hypodermis (Hy) by a large exuvial space (EX). The hypodermis cells have already produced a thin corrugated cuticle (C) as well as new hair sensilla (Hs). H = hemocyte, N = leg nerves, Sl = slit sensillum. 1,200 ×.

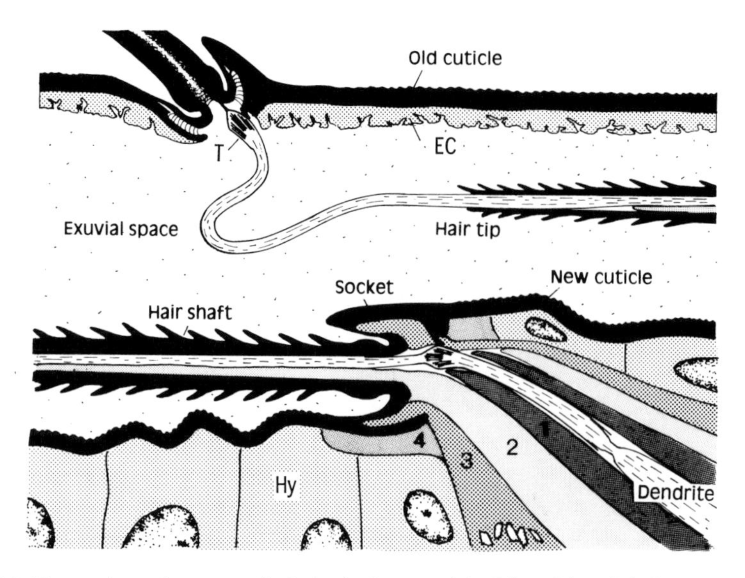

FIG. 189. Formation of new tactile hairs before ecydsis. The old cuticle is separated from the hypodermis by the exuvial space. The endocuticle (EC) has been largely dissolved. The dendritic connection between old and new hair sensillum is still continuous. T = dendritic terminals; 1–4 = sheath cells. (After Harris, 1977.)

during most of the molting process. The reason is that the innervation of the old sensilla is maintained while the new hairs develop on the surface of the hypodermis (Figs 188, 189). The long dendrites innervating each "old" hair sensillum lie freely in the exuvial space and become wrapped by a sheath cell (cell 2 in Fig. 189). This cell starts to secrete cuticle on its outside, thus producing the new hair shaft. Another sheath cell (cell 3 in Fig. 189) is responsible for the formation of the socket in which the hair shaft is articulated. About 24 hours before ecdysis all new hair sensilla are completely developed, and only then is the continuity between the old and the new sensory hair broken. When the exuvium is cast off, only the tips of the dendrites are lost, and the new hair sensilla are thus functional immediately after molting (Harris, 1977).

Hormonal Control

For spiders, molting is triggered and controlled by hormones, as it is for insects and crustaceans. An increase of the molting hormone ecdysone is found in the spider's hemolymph a few days before ecdysis (Eckert, 1967; Bonaric and de Reggi, 1977). By injecting β-ecdysone it is possible to shorten the intermolt periods, or even to induce ecdysis (Krishnakumaran and Schneiderman, 1968; Bonaric, 1976). Various experiments have demonstrated that ecdysone does not act in a species-specific way. For instance, when pieces of integument from a young *Tegenaria* are transplanted into a nymph of *Coelotes*, they molt synchronously with the host. Even when adult integument is used, ecdysis can be induced (Eckert, 1967). Very little is known about the production site of the molting hormone. A type of endocrine tissue that is diffusely distributed in the prosoma and that has been termed the "prosomal gland" (Legendre, 1958) might be responsible for producing ecdysone. This seems plausible, since the molting hormone in insects originates from an analogous gland, the prothoracic gland. Other endocrine tissues, such as the neurosecretory cells of the CNS or Schneider's organs, are also most likely to be involved in the control of molting.

It has been known for a long time that, besides the hypodermis, the hemolymph is also affected by the molting process (Browning, 1942). Shortly before molting, special blood cells, the leberidiocytes, which are characterized by a large fluid-filled vacuole, appear in the hemolymph. Presumably these cells participate in regulating the amount of water in the spider's body by absorbing water from the hemolymph (Seitz, 1976).

Autotomy and Regeneration

As a last resort in perilous situations, most spiders can "amputate" one of their own legs. This voluntary separation of an extremity following a dangerous stimulus is widespread among arthropods and is called *autotomy* (also termed *autotilly* or *autospasy* in a more narrow sense; Pieron, 1907; Roth, 1981; Roth and Roth, 1984). Luckily, though, lost legs can be replaced, at least as long as the animal still has to go through another molt. Regeneration of autotomized legs is thus restricted to juvenile spiders—except for those few cases where molting continues in the adult stage.

Autotomy

One can easily induce a spider to autotomize a leg by pulling or squeezing the femur of that leg. Tampering with the distal leg segments (the tarsus or metatarsus) does not usually lead to autotomy. It has been shown that anesthetized spiders cannot autotomize their limbs (Bonnet, 1930), so autotomy is indeed a voluntary act.

The site where a leg is autotomized lies usually between the coxa and the trochanter (Wood, 1926), rarely between the patella and the tibia (e.g., in linyphiids; Berland, 1932). An externally applied pulling force is not important for autotomy. More significant is the spider's own sudden jerking upward of the coxa while the femur maintains its position and acts as a brace. The trochanter tilts sharply, and the high tension that develops between the coxa and the trochanter causes the membrane of the joint to rupture dorsally (Fig. 190a). Further separation poses no problem, because the site of amputation is traversed by only one muscle (m. gracilis), which readily detaches itself from the trochanter and then withdraws into the coxal cavity. All other muscles insert on marginal thickenings (sclerites) of the joint membrane. These muscles aid in closing the open coxa by pulling the sclerites toward each other—a closing mechanism similar to that of certain garbage cans (Bauer, 1972). Additionally,

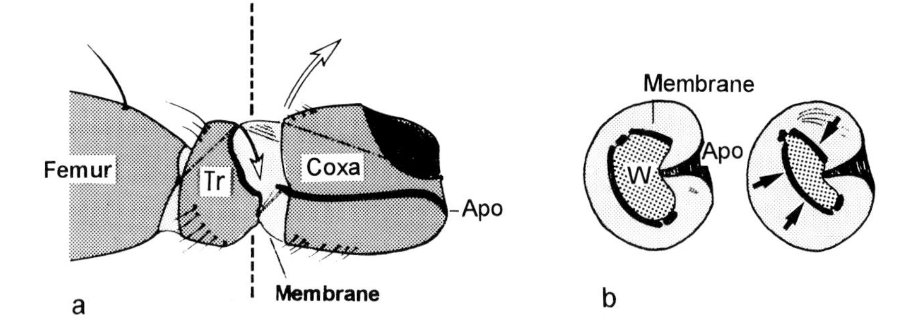

FIG. 190. (*a*) Site of autotomy in a spider leg (*Philodromus*), lateral view. A rapid jerk upwards (large arrow) causes a high tension in the joint membrane so that a tear (small arrow) develops dorsally near the edge of the trochanter (Tr). The articulation of the coxal apodeme (Apo) and the trochanter is shown already disrupted. (After Bauer, 1972.) (*b*) Closing of the wound after autotomy; frontal view of coxa. *Left*: The wound opening (W) is bordered by sclerites (black) of the joint membrane. *Right*: The large wound is sealed by a closing movement of the sclerites. (After Bonnet, 1930 and Bauer, 1972.)

the hemolymph pressure forces the joint membrane to bulge forward, which also helps to seal the wound.

Under natural conditions, autotomy is used mostly as a defense against an aggressor, often another spider, such as an attacking female. When bees or wasps get caught in a spider web, they sometimes manage to push their stinger into the soft joint membranes of the spider's legs. The afflicted leg is usually autotomized within seconds, before the deadly poison can spread into the body of the spider (Eisner and Camazine, 1983). Leg autotomy is also effectively used by *Filistata,* when grabbed by a scorpion (Formanowicz, 1990). Although the scorpion will eat the severed leg, the spider itself can escape. This is reminiscent of lizards, which readily autotomize their tail after being seized by a predator.

Another opportune use for autotomy may arise during molting, when an extremity cannot be freed from the old skin. Even if molting itself has been successful, injured or malformed legs can still be autotomized afterward. Such limbs are not discarded but are usually eaten by the spider (Bonnet, 1930).

Not all spiders show the same readiness to part with one of their limbs. Some spiders may autotomize at the slightest provocation (Philodromidae, Pholcidae, Selenopidae), others only as a last resort (Salticidae, Thomisidae, Uloboridae, Theraphosidae), and a few under no circumstances (*Tetragnatha, Metellina*; Vollrath, 1990). Under natural conditions most spiders lose one or two legs during their lifetime (Oppenheim, 1908). A careful inspection of field collections showed that between 5% and 20% of all spiders had at least one leg missing. Although this loss may affect the efficiency of prey capture, it is remarkable how well the spiders can compensate: orb weavers with only three or four legs left could still build fairly regular webs (Vollrath, 1987a, 1990).

Regeneration

When a juvenile spider loses a leg, it will often—but not always—be replaced at the next molt. Whether the leg is regenerated or not depends on the time of amputation: only if the leg has been severed within the first quarter of the intermolt phase will regeneration take place (Bonnet, 1930). If the leg is lost later, no regeneration occurs. Apparently, regeneration is an all-or-nothing process that is controlled hormonally.

Regeneration of an autotomized leg cannot be observed directly, because the leg grows inside the old coxa. The formation of the new leg is comparable to the differentiation seen during normal ontogeny. However, because of the limited space of the coxal cavity, the new leg will be highly folded (Fig. 191). After ecdysis the regenerated leg appears a bit shorter and thinner than the original leg, but all leg segments are present in the right proportions. Even claws, spines, and sensory organs are regenerated, although the sensory organs are not always complete. The tarsal organ, for instance, is lacking in the first regenerate (Blumenthal, 1935), and the number of sensory hairs is usually less than normal. However, these are rather quickly supplemented in the course of further molts.

Whether a regenerated leg really replaces the original completely is a question that cannot be answered fully. Aside from the sensory organs, all the muscles and their innervation must be formed entirely anew. An analysis of regenerating "tarantula" legs has shown that the new legs do indeed have all 30 leg muscles—and all of them are inserted at the proper sites (Ruhland, 1976). However, the number of muscle fibers is less than normal. The pattern of innervation in the new leg muscles is also the same as it was in the original. Yet it is rather odd that newly regenerated legs are often stretched out and held away from the body, and are not used for walking at all. If they are occasionally used for walking, their movements are not well coordinated with the other legs. This impaired function is presumably caused by undifferentiated neuromuscular contacts in the regenerated leg.

A spider's ability to regenerate a missing part is not restricted to legs; pedipalps, chelicerae, endites, the labium, and even the spinnerets can be regrown as well (Bonnet, 1930). If only the distal segments of these extremities are severed—and, of course, if the leg is not autotomized—the missing segments are regenerated directly at the site of injury. This is for instance the case in black widows, yet interestingly, they cannot regenerate a *completely* autotomized leg (Randall, 1981). This suppression of regeneration (if autotomized at the coxa–trochanter joint) has also been noticed in some orb web spiders (*Nephila, Tetragnatha, Uloborus, Zygiella*; Vollrath, 1990).

The potential for regeneration in spiders is enormous. If several legs are amputated early in an intermolt phase, they will all regenerate by the time of the next molt. Indeed, in one drastic experiment, all eight legs were induced to autotomize, and all were nevertheless regenerated simultaneously (Bonnet, 1930). This was, however, possible only because the torso of the spider was fed artificially. If a regenerated leg is amputated again, it can also be replaced at

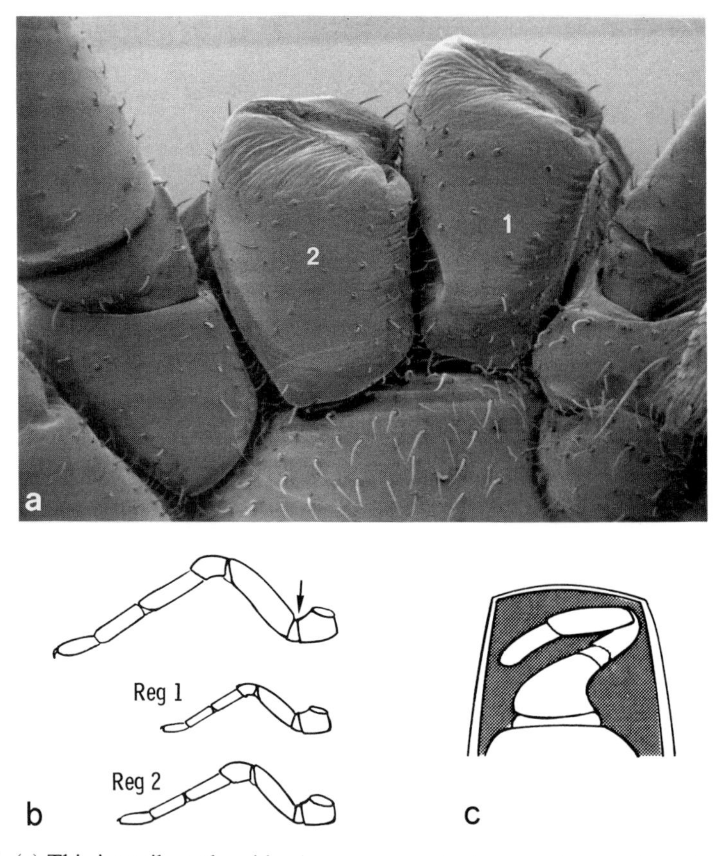

Fig. 191. (*a*) This juvenile crab spider (*Misumenops*) had lost both right front legs and only the coxal stumps (1,2) remained. After the next molt complete legs of 8 mm length had regenerated within the short coxae (1 mm!). (*b*) Comparison of the legs of a "tarantula" before autotomy (above) and after regeneration (Reg 1, 2) in the subsequent two molts. The arrow indicates the site of autotomy. (Tracing of a photograph by Ruhland, 1976.) (*c*) The strongly folded regenerate developing inside the old coxa. (After Bonnet, 1930.)

the next molt (Bonnet, 1930; Ruhland, 1976). After the final molt a spider can still autotomize its legs, but no signs of regeneration will be detectable in the remaining coxa.

Life Cycle and Longevity

The exact course of the life of a spider has been investigated in only a few spiders. In the temperate zones of central Europe, the main reproductive period is in May (Tretzel, 1954), and the spiderlings hatch during the summer. Some species may reach the adult stage in autumn, but most overwinter as nymphs.

If the egg sac is constructed during late fall, as it is for most species of *Araneus*, the hatched spiderlings will remain inside the cocoon until spring.

Only females of the European spider *Tegenaria atrica* can be found in spring. They lay their eggs in April and the spiderlings hatch 21 days later. By the end of August most of the spiderlings have molted nine times and are then adults. In late summer there are thus two generations of females (a young and an old one), but only one generation of males (a young one). The reproductive period lasts from August to October, and shortly thereafter the males die. The young females overwinter; their ovaries start to grow as early as November (Collatz and Mommsen, 1974).

Most spiders of temperate regions do live only one year, but some may live for two. The "primitive" spiders are renowned for their longevity. The orthognath purse-web spider *Atypus* (Fig. 107) can survive 7 years and large "tarantulas" over 20 (Berland, 1932; Canard, 1986). Another spider noted for its long life is the cribellate *Filistata*, which can live to be 10 years old. As a rule, only female spiders have a high life expectancy; most males die shortly after mating.

9

Ecology

The interactions between spiders and their environment have been investigated systematically only within the past few decades. During this time, however, information about the ecology of spiders has burgeoned to such an extent that it is impossible to include all aspects in the scope of this book. For more detailed and up-to-date discussions, especially of field experiments, the reader is referred to *Spiders in Ecological Webs* by David Wise (1993), and to *Ecophysiology of Spiders*, edited by W. Nentwig (1987).

Occurrence and Distribution of Spiders

Spiders are found almost everywhere on earth, from Arctic islands to dry desert regions. They are particularly abundant in areas of rich vegetation. (Shaking bushes over an opened umbrella or sweeping through low vegetation with a butterfly net are thus very efficient methods of collecting large quantities of spiders.) However, spiders are also found in rather barren environments such as sand dunes, tidal zones (Lamoral, 1968), or mountain tops. It is no exaggeration to say that spiders have conquered all possible ecological niches on land (Turnbull, 1973).

Such widespread distribution leads to the question of how spiders manage to populate very remote areas, such as islands, for instance. One explanation is based on the ability of many young spiders to float through the air on their own threads. This *ballooning*, as it is called, is actively undertaken by the spider. The spiderling stands on "tiptoe," facing the wind, and inclines its abdomen upward while exuding a silk thread. This strand is caught by the slightest current, which can easily lift the spider. Usually the voyage ends after a rather short distance. Under favorable wind conditions, however, these aeronauts may attain quite remarkable heights and distances. Spiders, as a kind of "aerial plankton," have been caught from airplanes flying at altitudes of several thousand meters (Glick, 1939; Gertsch, 1949). In 1832 Darwin noted in his diary that innumerable small spiders had been blown into the rigging of *The Beagle* 100 km off the coast of South America. Yet, as Berland (1932) has rightly pointed out, spiders cannot really bridge large distances by aeronautical means. East Africa and Madagascar, for instance, are only 400 km apart, yet each has a quite different spider fauna. The problem of bridging large distances is even greater when one considers the Pacific islands, which are often thousands of kilometers from the mainland. It is more likely that driftwood provided a route for settlement of spiders there.

Formerly it was believed that only *young* spiders could fly on their own threads, but later many small males (Richter, 1971; Blanke, 1973a) and even some medium-sized females (1 cm body length; Wickler and Seibt, 1986) were found to do so as well. Young spiderlings are real "featherweights" (0.2 mg; Coyle et al., 1985; Greenstone et al., 1987), which can take off at the slightest breeze. Adult females of 100 mg can still be lifted off, if they release several loops of thread into the air. In the linyphiids, adult males and females can take to the air—often thousands of them at a time (Bristowe, 1939; Van Wingerden and Vugts, 1974; Vugts and Van Wingerden, 1976). Whereas the gossamer threads of young spiderlings are seen in Indian summer, the mass flight of linyphiids occurs during fall and winter. A prerequisite for such aeronautical behavior is a sudden rise in temperature (Duffey, 1956). If the weather shifts to warm, sunny days after a long period of cold, the temperature on the ground will rise rapidly. This causes an updraft of air, and the spiders are easily lifted away, especially small ones. Of course, how far they will travel depends mainly on the wind—under favorable conditions up to several hundred kilometers (Okuma and Kisimoto, 1981). The fact that the islands of the Great Barrier Reef in Australia have been colonized by seven different spider families, but not by any of the (large) mygalomorphs, can only be explained by an aerial dispersal (Main, 1982).

Only a few of the aeronauts will survive; many fall victim to birds (such as swallows), or they land on open water or in some other wholly unsuitable environment. However, since the number of spiders transported is so great, there are always a few that will survive to invade new habitats. Less than a year after the catastrophic explosion of the volcano Krakatoa in 1883, the first settler on the barren island was found to be a web spider (linyphiid), and 50 years later more than 90 species of spiders were recorded there (Bristowe, 1931, 1934; Thornton and New, 1988). Most likely they had drifted in by winds from the neighboring islands Java and Sumatra. Similarly, only 2 years after the eruption of the volcano Mount St Helen in 1980 over 40 spider species were found, which had colonized the previously dead land by air. Their takeoff points must have been up to 30 km away (Edwards, 1986).

The "primitive" spiders (Mesothelae, Orthognatha) disperse only exeptionally by air (Coyle, 1983; Coyle et al., 1985). Their mode of life is strictly terrestrial and bound to their silk-lined burrows. When the young ones leave their mother's home, they will establish small new tubes of their own very close by (Decae, 1987). A first step toward ballooning is seen in some wandering spiders. They let themselves drop from one leaf down to the next, until they have found a suitable habitat (Barth et al., 1991). This "drop-and-swing-dispersal" can lead to true aerial flight, when strong side winds break the safety thread of the spiderlings, carrying them aloft.

The small spiderlings which usually live together during their first period of life gain two advantages by their aerial dispersal: (1) they become rapidly distributed over the available habitats so that overcrowding is minimized; and (2) they escape the gradually emerging cannibalistic behavior of their siblings (Tolbert, 1977).

Habitat

Most spiders live in strictly defined environments. The limitations are set by physical conditions, such as temperature, humidity, wind, and light intensity, and also by biological factors, such as the type of vegetation, the food supply, competitors, and enemies. Ecologically, vegetation can be classified into four vertical layers (Duffey, 1966): (1) a soil zone, consisting of leaf litter, stones, and low plants up to 15 cm in height; (2) a field zone, consisting of vegetation from 15 to 180 cm; (3) a bush zone of shrubs and trees of 180–450 cm in height; and (4) a wood zone of trees and treetops over 450 cm in height. Each zone has its characteristic microclimate, various niches for retreats, and a different spectrum of prey animals. Accordingly, we often find a corresponding "stratification" of different spider species (Toft, 1976, 1978).

Among wolf spiders, *Lycosa pullata* (now *Pardosa pullata*) lives in the lowest zone (0–5 cm), whereas *Lycosa nigriceps* is dominant in the lower field zone (20–30 cm; Duffey, 1966). Even within the same zone the microclimatic conditions can vary considerably, and these may cause an ecological separation of different species. For instance, *Pardosa pullata* and *Pirata piraticus* live together in bogs where sphagnum moss is found. However, *Pardosa* keeps almost exclusively to the surface of the moss, whereas *Pirata* prefers the more secluded stem region of the moss (Nørgaard, 1951). Continuous monitoring shows that the temperature fluctuates a great deal on the surface of the moss layer, but varies little inside (Fig. 192).

However, it cannot be stated as a general rule that only a single species will live in a given microhabitat. The American wolf spiders *Lycosa carolinensis* and *Lycosa timuqua* live in the same habitat, hunt the same type of prey, and are both active at night (Kuenzler, 1958). The same is true for two linyphiids (*L. triangularis* and *L. tenuipalpis*), which share exactly the same habitat on *Calluna* heaths; they can even be manipulated to accept each others webs (Toft, 1987, 1990). Other species, particularly among the jumping spiders, migrate back and forth between different vegetative zones (Luczak, 1966). Generally, small spiders usually stay close to the ground (Tretzel, 1955).

A vertical distribution of various spider species can be studied nicely on tree trunks. It is interesting that the composition of the spider fauna changes not only with respect to the height of a tree, but also depends on the structure of the bark, the microclimate, the prey availability and even on the age of a particular tree (Wunderlich, 1982; Simon, 1991). Many web-building spiders are vertically distributed within a vegetation zone. The two orb weavers *Argiope aurantia* and *Argiope trifasciata* build their webs at different heights during the period from May to September (Fig. 193). Only during the adult stage do their habitats overlap (Enders, 1974, 1977; Brown, 1981). An explanation for this

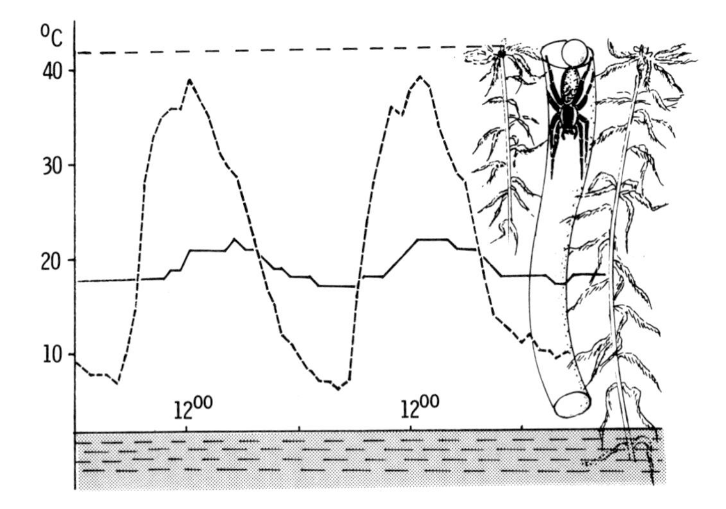

FIG. 192. Temperature fluctuations in the habitat of the wolf spider *Pirata piraticus* during two summer days. On the surface of the sphagnum moss the temperature changes between 8° and 39°C (dashed line). In contrast, the temperature stays near 20°C inside the moss layer (solid line). During sunny periods the female exposes her egg sac to the sun at the entrance of her silken tube. (After Nørgaard, 1951.)

vertical stratification might be found in the distribution of prey animals. Spiders that build their webs higher up can also include larger flying insects among their prey. This becomes evident when comparing the prey of the cross spider *Araneus marmoreus* with that of *Argiope bruennichi*. The webs of *Araneus*, built 50–70 cm above the ground, trap mostly small dipterans, whereas the lower webs (0–50 cm) of *Argiope* capture rather large grasshoppers (Pasquet, 1984).

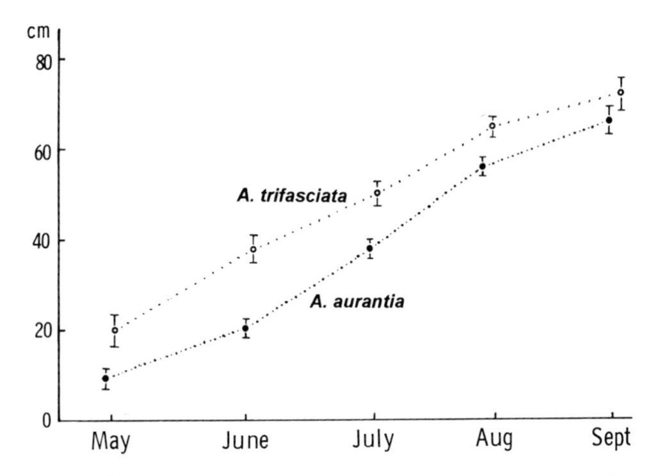

FIG. 193. Vertical stratification of the orb webs of *Argiope aurantia* and *A. trifasciata*. Throughout the summer the height of the webs above the ground is different for both species. Only in September, when both species are adult, do the heights of their webs overlap. (After Enders, 1974.)

In general, a specific spatial distribution of spider species would seem to be an adaptation to interspecies competition—that is, a strategy that aims to avoid such competition (Tretzel, 1955). Later investigations of different species of wolf spiders showed, however, that the distribution pattern was determined more by environmental factors than by interspecific competition (Schaefer, 1972, 1974). Ecological separation may also be achieved by different reproductive periods, as is shown for certain species of *Lycosa* (now *Pardosa*) in Fig. 194 (Tretzel, 1955; Kronestedt, 1990). The same habitat may be used at different times of day by different species. Such differences in activity are known for many species of wandering spiders, but also for some web spiders, which exhibit different daily periods of web building. It should be noted that not only different species, but even different genera and families may compete for the same habitat (Wise and Barata, 1983). For instance, *Hypochilus, Achaearanea* and *Coelotes* occupy the same rocky outcrops in the Appalachian mountains and they have been called "ecological equivalents" (Riechert and Cady, 1983). Analyzing a desert spider community of 90 species, however, led to the conclusion that most related species are separated by spatial and temporal differences (Gertsch and Riechert, 1976).

The habitats of web spiders not only require specific microclimates but also must meet certain spatial demands. The environment must provide plenty of attachment sites for the scaffolding of the web (Uetz et al., 1978). In addition, sufficient open space must be available, either vertically (for orb webs) or horizontally (for sheet webs). Web spiders are thus not distributed at random in their environment, and their population density is necessarily limited. It is likely that the space that a web spider occupies is about as large as is necessary to provide for the animal's minimal energy requirements (Riechert, 1974).

Web spiders are sit-and-wait predators, and they probably expend less energy for prey capture than do wandering spiders. In addition, web spiders often catch several prey items in their webs at once and thus can maintain a cache of food. In general, web spiders also kill larger prey than errant spiders

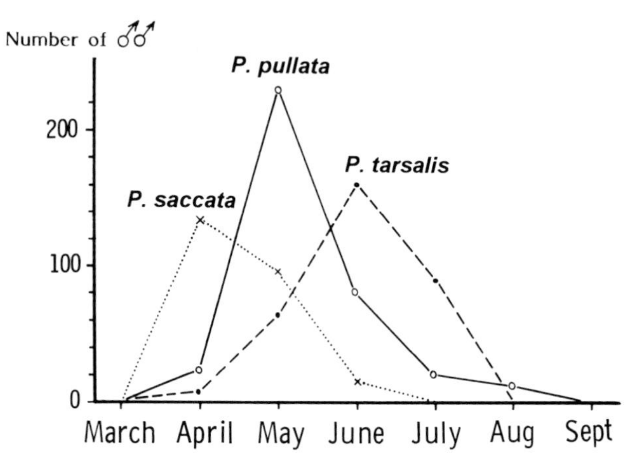

Fig. 194. Seasonal separation of the reproductive period in different species of wolf spiders (Pardosa). (After Tretzel, 1955.)

do. Apparently, catching prey with a web is more efficient than direct attack (Enders, 1975). This view, however, has been challenged following the demonstration that several species of wandering spiders can also overpower rather large prey (Fig. 195; Rovner, 1980c). For instance, some salticids subdue dragonflies that are two or three times their own size (Robinson and Valerio, 1977) and crab spiders are well known to tackle even bumblebees successfully.

The web itself can be considered the spider's territory, since it is usually defended against invaders. Sometimes the web is conquered by a conspecific, or even an unrelated spider (Hoffmaster, 1986). Such web robbery may be committed by conspecific males, who normally do not build webs of their own after their final molt (Eberhard et al., 1978). In all other cases web robbery can be interpreted as a competition for the same habitat, because such incidents happen more often during periods of high population density. Wandering spiders rarely show any territorial behavior. Only for the wolf spider *Pardosa* has a "mobile territory" been described: an area of 10 cm in diameter in the immediate vicinity of the spider is defended against invaders (Vogel, 1971, 1972).

A special type of web invasion is found in the pirate spider *Ero* and the small theridiid *Argyrodes* (formerly *Conopistha*). *Ero* sneaks into the web of another spider and lures the owner by carefully plucking the threads (*aggressive mimicry*). When the web spider moves toward the source of vibration, it is suddenly attacked and succumbs to the poisonous bite. *Argyrodes* parasitizes

FIG. 195. (*a*) A nocturnal huntsman spider (Sparassidae) has captured a large hawkmoth (*Coelonia brevis*; 5 cm body length) while it was hovering at blossoms. (Photo: Wasserthal.) (*b*) A small crab spider (*Misumena vatia*), well-camouflaged between the blossoms of an orchid (*Platanthera bifolia*), has overpowered a honey bee; note the typical "neck grip" of the victim. (Photo: Gasser.)

in the webs of various orb weavers (e.g., *Argiope, Cyrtophora, Nephila*), where she stealthily removes previously caught prey (Wiehle, 1928; Kullmann, 1959). She builds her irregular web close to the orb web and attaches "signal lines" at the hub. All her movements are very slow and deliberate so that she will not be detected. The signal for stealing prey is always triggered by the orb weaver, or more specifically, by her wrapping the prey (Vollrath, 1979). Only those movements stimulate *Argyrodes* to climb slowly toward the hub, where the prey package is normally stored. It is then localized with slow circular motions of the front legs and carefully cut out. Usually, *Argyrodes* is just stealing a prey, but it has also been observed that she may eat the host spider, or her cocoon, or the spiderlings (Exline and Levi, 1962; Wise, 1982; Whitehouse, 1986).

Often several *Argyrodes* parasitize in one orb web; the maximal number of parasites per web was 45! (Vollrath, 1987b). Such a high pressure may cause the host spider to search for another web site (Larcher and Wise, 1985; Rypstra, 1981). Occasionally several species of *Argyrodes* (e.g., *A. elevatus* and *A. caudatus*) coexist in the same web. Interspecific competition is apparently reduced by occupying different regions in the web and by being active at different times (Vollrath, 1976).

Several interesting ecological investigations have focused on the question of whether spiders occupy a certain habitat more or less by chance, or whether they select it specifically. The funnel-web spider *Agelenopsis* and the orb weaver *Nephila* have been shown to build their webs not only at sites where the microclimate is favorable, but also where prey is rather abundant (Riechert, 1974, 1976; Rypstra, 1985). For instance, if the spider constructs its web in bushes that are in bloom, it will have a better chance of catching insects there. Some spiders have been observed to change the sites of their webs if they do not catch enough prey (Turnbull, 1964; Olive, 1982). On the other hand, the abundance of prey seems to have no effect on some orb-web spiders in their selections of web sites (Enders, 1976). However, orb weavers build larger webs when hungry, and thus probably enhance their foraging success (Higgins and Buskirk, 1992; Sherman, 1994).

Prey Animals

Insects are by far the largest part of the diet of a spider. Other arthropods, such as sowbugs or millipedes, are also on the list—as are spiders themselves! It is rather unusual to find vertebrates among a spider's prey (McCormick and Polis, 1982), although tadpoles or small fish may fall victim to certain lycosids or pisaurids. The large desert spider *Leucorchestris* (3 cm body length, 5 g weight) can capture small geckos of about her own size (Henschel, 1990b), and the raft spider *Dolomedes* overpowers fish that outweigh her 4–5 times (Fig. 196). However, reports that claim that large "tarantulas" have been feeding on birds, lizards, or snakes are rarely scientifically verifiable.

Among the insects, flies and the wingless Collembola contribute to the bulk of the spider's diet. Because Collembola occur in huge numbers, they are very

FIG. 196. The raft spider *Dolomedes* has caught a small fish under water. She is now pulling her prey ashore to feed on it. (Photo: Bleckmann.)

important to many small spiders (Bristowe, 1941). Beetles, grasshoppers, and butterflies are another abundant source of prey. However, certain insects, such as stinkbugs, ants, and wasps, as well as certain beetles, moths, and caterpillars are generally avoided by most spiders. Either these insects actively use chemicals to defend themselves (as do ladybugs) or they may have an unpleasant taste.

The webs themselves can be regarded as selective filters for the potential prey. Insects with well-developed flying abilities and acute eyesight (like the hoverflies, the Syrphidae) can apparently avoid spider webs. Pollinating insects (bees, wasps, certain flies, and beetles) are rarely found trapped in the space webs of theridiids and linyphiids, but plant-sucking insects (Aphidae, Homoptera, and Thysanoptera) are caught extremely often, that is to say, selectively (Nentwig, 1980). On the other hand, all kinds of "aerial plankton" become stuck in spider webs: dust particles, salt crystals (Fig. 197), algae, fungal spores, and pollen grains (Linskens et al., 1993). Pollen adhering to the sticky threads of orb webs is most likely ingested when the spider eats ("recycles") her old web. It seems that young spiderlings may even derive some energy from this unusual food source (Smith and Mommsen, 1984). Even more surprising is the observation that some male crab spiders regularly drink nectar from blossoms and apparently exploit its sugar content (Pollard, 1993). This is not to say that these spiders are vegetarians!

Most spiders are not particular about the type of prey they feed on. Such spiders are called *polyphagous*, that is, they are generalists with respect to their prey. A good example is *Linyphia triangularis*, which, in one experiment, accepted 150 of 153 offered species of prey (Turnbull, 1960). Quite in contrast are *stenophagous* spiders, which are specialized for feeding on only a very few species of prey, such as ants (*Zodarium, Callilepis*) or other spiders (*Ero*).

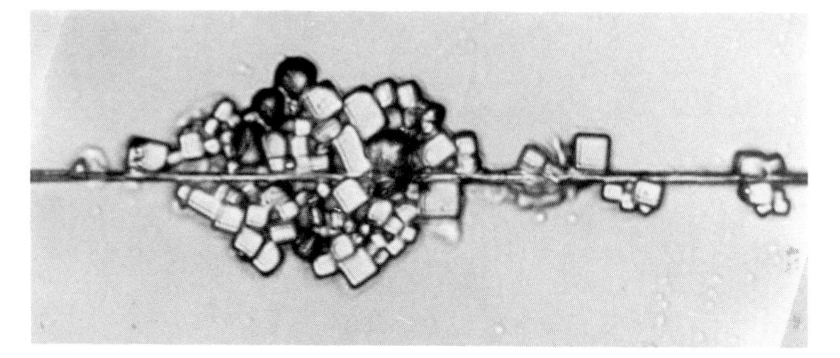

Fig. 197. Spider webs as aerial filters. Salt spray from the nearby ocean forming tiny crystals of sodium chloride on the silk threads of an orb web (Zygiella). One wonders whether such salty webs can possibly be eaten ("recycled") by the spider before building a new one. 210 ×.

One would expect that predation by spiders would severely limit insect populations. Indeed, we have impressive hypothetical calculations (Turnbull, 1973) based on an estimate of the population density of spiders ($130/m^2$) and on an assumption that each spider consumes 0.1 g of prey daily. The conclusion is that all the spiders living on 1 ha of land would devour $1.3 \times 10^6 \times 0.1 \times 365 = 4.75 \times 10^7$ g, or 47,500 kg of prey per year. Although 0.1 g/spider/day is certainly an inflated estimate, even half that amount would be astounding. Bristowe (1958) gave a less quantitative yet less abstract estimate when he stated that "the weight of insects consumed annually in Britain exceeds that of the human inhabitants."

More recent estimates from wood areas and agricultural land in central Europe are much more modest, assuming only 100 kg (or perhaps only 1 kg!) of insect prey/hectare/year (Kirchner, 1964; Nyffeler and Benz, 1979, 1987). The results of such studies are certainly strongly dependent on the habitat of the spiders. In natural grassland, but also in orchards (Mansour et al., 1985; Mansour and Whitcomb, 1986), spiders are quite numerous and will have a measurable effect on the insect population—we only need to consider the high density of various wolf spiders in the litter zone, or of agelenids, linyphiids, and araneids in undisturbed grass and wood areas. In the modern agricultural landscape, in contrast, the absolute numbers and species diversity of spiders are rather low—and so is the effect of spiders as a means of pest control. One way to improve this situation would be to provide a less monotonous environment, for instance by creating more extensive border regions between fields and woods, or by leaving pockets of uncultivated land as "ecological cells." Such an approach would quickly lead to a marked increase in the number of spider species, as has been shown in recent experiments (Nentwig, 1986). Equally important, however, is a reduction in the use of herbicides and pesticides (Riechert and Lockley, 1984).

Overall, spiders hardly play a major role in the control of insect pests. As we have seen, most spiders are generalists with respect to their diet but, for efficient pest control, spiders would need to specialize in eating such pests. Furthermore, spiders generally do not form social colonies (as ants or bees do), so their populations cannot become very dense. Despite those restrictions, spiders may have an important "buffer effect," for instance, during the early development of an insect population, when growth is exponential (Clarke and Grant, 1968). A good example is the spiders living in the litter zone of forests, where densities of 50–200 individuals/m^2 have been recorded (Nyffeler, 1982).

Many spiders can adjust to the available food supply by eating more prey when it is abundant. This maximal energy uptake allows the spider not only to grow but also to mature more quickly (Miyashita, 1968; Ward and Lubin, 1993). If a long period of food deprivation then follows, such well-fed spiders have better chances of surviving it. (A similar strategy of an "optimal food uptake" is also found in a completely different group of animals, the snakes.) Of course, small spiders take in much less "biomass" than large spiders do. The small wolf spider *Pardosa* eats about 3.5 mg of insects daily, equivalent to 12% of its body weight (Edgar, 1970). Comparable values were found for the larger wolf spider *Trochosa* (3–12%; Breymeyer, 1967; Breymeyer and Jozwik, 1975) and the sheet-web weaver *Linyphia* (10–25%; Turnbull, 1962).

Some spiders produce relatively more eggs when the food supply is abundant (Blanke, 1974b; Kessler, 1973a; Wise, 1975). In general, however, the number of eggs is mainly a function of the spider's size: the larger the spider, the more eggs she will lay (Petersen, 1950; Kessler, 1971a). On the other hand, the number of eggs decreases with an increased level of brood care.

The astonishing ability of spiders to survive several months without food is primarily the result of their low rate of metabolism (Fig. 198; Anderson, 1970). The amount of oxygen a "tarantula" consumes at rest is only one-fifth of the oxygen a cold-blooded vertebrate needs, and only one-hundredth of that being used in a warm-blooded animal (Paul, 1990).

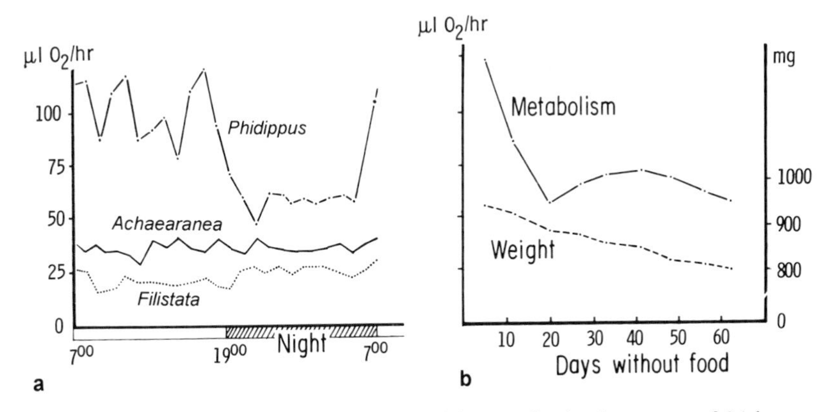

FIG. 198. (*a*) Oxygen consumption in three spider species in the course of 24 hours. Note the much higher metabolism of the day-active jumping spider (*Phidippus*) as compared to the sedentary theridiid *Achaearanea* and to the tube-dwelling *Filistata*. (After Anderson, 1970.) (*b*) Decrease in body weight and metabolism in *Lycosa lenta* during starvation. (After Anderson, 1974.)

Under conditions of starvation, metabolism is reduced drastically—in extreme situations by more than 80% (Collatz and Mommsen, 1975). Wolf spiders, which have a life expectancy of only 300 days, can survive up to 200 days without food (Anderson, 1974). Although such experiments were conducted in the laboratory, long periods of starvation no doubt occur naturally, for example, while the spider overwinters. In autumn the house spider *Tegenaria* deposits mainly fat, but also some carbohydrates and proteins in its body; this "fuel" is then metabolized during the winter, when no sources of food are available. After 50 days of starvation, 60% of the stored carbohydrates, 47% of the fatty acids, and 9% of the original proteins have already been metabolized (Collatz and Mommsen, 1974).

Enemies

Even the most skilled predators can fall victim to other hunters—and this applies likewise to spiders and their many foes. The main enemies of spiders are spiders themselves (Jackson, 1992). Many wandering spiders will attack each other, and web spiders, while defending their territories, often duel successfully with invaders. The pirate spiders (Mimetidae), which feed exclusively on other spiders, have already been discussed (see chapter 6). Another unusual example of a spider that prefers to attack other spiders is the cryptic jumping spider *Portia* (Jackson and Blest, 1982; Forster and Murphy, 1986). *Portia* has a large visual field and excellent eyesight (Fig. 199). She approaches prey—mostly other

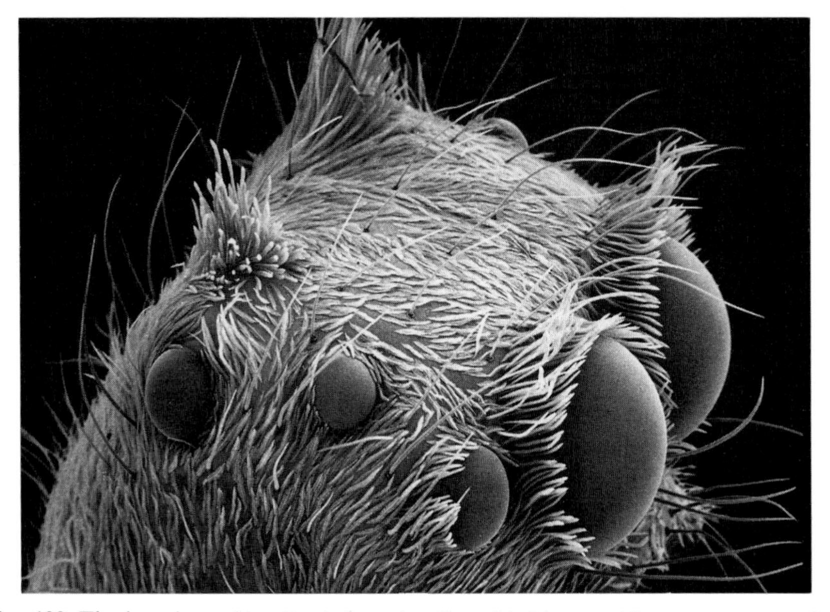

FIG. 199. The jumping spider *Portia* is a visually guided hunter. The arrangement of the smaller secondary eyes on the prosoma provides a huge visual field, while the large main eyes have the highest acuity of any spider eyes. 20×.

spiders—very slowly and in a robot-like fashion. Even other jumping spiders with similarly acute vision do not notice this stealthy approach. When quite close (1–2 cm), *Portia* suddenly lunges and bites (Fig. 200b). The poison seems very potent against spiders: it paralyzes them within 10–30 seconds, whereas it takes about 4 minutes to be effective against insect prey (Jackson and Hallas, 1986). When stalking web spiders, *Portia* uses an "aggressive mimicry" tactic, i.e., she lightly plucks the threads to lure the victim. The approaching web spider has hardly any chance to launch an attack and, even if she manages to get a hold of a leg, *Portia* will autotomize immediately. *Portia's* success as a spider hunter thus lies in a clever combination of camouflage, slow motion, excellent vision, a rapid lunge, a highly effective poison, and various prey-specific capture strategies.

Other, less sophisticated spider hunters may also be quite successful. The fragile-looking pholcids, for instance, are capable of overpowering rather large spiders such as *Tegenaria* or araneids (Fig. 200a). They can also use "aggressive mimicry" by imitating prey vibrations. When the web owner rushes out, it is quickly wrapped in silk threads using the long hindlegs. This is a rather safe method because *Pholcus* always keeps a good distance to its victim (Jackson and Brassington, 1987).

Among the nematodes (roundworms) several species are known as parasites of spiders (Müller, 1983; Poinar, 1987). Early juvenile stages of the nematode enter the spider's opisthosoma and feed on its tissues (Fig. 201). Effects of the parasite are a reduction of the muscles, the midgut, and the reproductive system,

Fig. 200. Spiders as spider enemies. (*a*) The pholcid *Holocnemus pluchii* attacks the araneid *Cyrtophora citricola*, which has just caught prey. (From Blanke, 1972.) (*b*) The salticid *Portia fimbriata*, having delivered a fatal bite to a large theridiid.

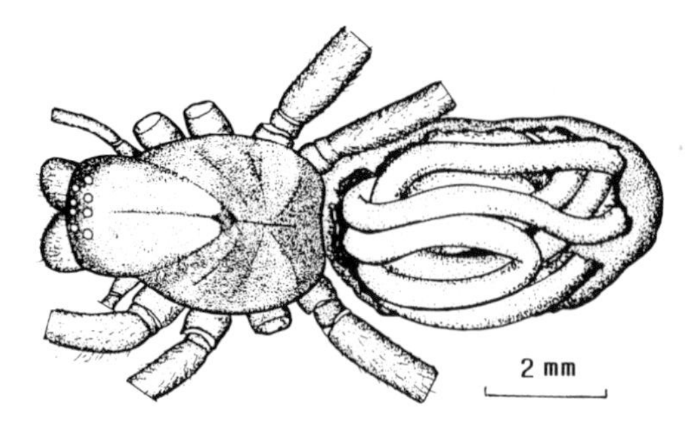

Fig. 201. A funnel-web spider (*Coelotes*), parasitized by a nematode. The roundworm occupies almost the entire abdomen of the host. (After Müller, 1983.)

leading to parasitic castration. The host's behavior may also change, for instance a migration toward a source of water is commonly observed, which is apparently important for the emerging nematode. Most of the infected spiders die just before or after the emergence of the parasite.

Among the insects, certain types of wasps are the arch enemies of spiders (Fig. 202). The spider wasps (Pompilidae) hunt spiders exclusively, whereas the mud daubers (Sphecidae) pursue insects as well as spiders. The fate of the spider is always the same: it is paralyzed by the wasp's sting and an egg is laid on its abdomen. When the wasp larva hatches, it feeds on the living tissue of the paralyzed spider until the spider succumbs.

Different species of pompilid wasps use different strategies (Rathmayer, 1966). Some species paralyze the spider only temporarily, deposit an egg on it, and fly away. Other species first dig a hole into which they drag the paralyzed spider. When the spider recovers after a few hours, it finds itself locked inside a narrow tube where it cannot move. Most pompilids, however, paralyze the spider permanently, or at least for several weeks, and enclose their victim in an underground chamber. The wasp's poison acts very rapidly (in less than 1 second), via the hemolymph, on the motor system of the spider. Apparently neuromuscular transmission is blocked rather than the muscles themselves (Rathmayer, 1978).

In general, pompilids do not hunt one particular species of spider, although they may have a preference for certain species. The wasp *Episyron rufipes*, for instance, goes mostly after araneids, whereas *Pompilus cinctellus* searches for salticids. Some spider wasps (for example, *Pseudagenia*) do not dig holes in the ground but build tiny cells of clay into which they carry the paralyzed spiders. Some other wasp species (such as those of the genus *Anoplius*) do not even bother to hunt their own spiders; instead they attack other pompilids and take paralyzed spiders away from them.

It is very curious that spiders scarcely defend themselves when they are attacked by pompilid wasps. As soon as the wasp appears, the spider appears to be terrified and tries to escape (Bristowe, 1941). The spider's panic flight is probably caused by a scent that the excited wasp exudes (Petrunkevitch, 1926).

FIG. 202. Wasps as spider enemies. (*a*) The pompilid wasp *Anoplius fuscus* has paralyzed a wolf spider and is now carrying it away. (Photo: Krebs.) (*b*) The pompilid *Episyron funerarius* stings a *Cyrtophora citricola* in the abdomen. (From Blanke, 1972.) (*c*) The araneid *Cyclosa conica* with a larva (L) of an ichneumonid wasp attached to its carapace. (Photo: Blanke.)

Some wolf spiders (such as *Geolycosa*), however, do make some attempts to defend themselves, either by pushing the wasp away, or by maintaining a plug of sand between itself and the wasp (Gwynne, 1979; McQueen, 1979). One spider, the cribellate *Amaurobius*, pays no heed to pompilid wasps. If a pompilid gets ensnared in the hackle band of the catching threads, *Amaurobius* simply overpowers it and eats it (Bristowe, 1941).

Even large "tarantulas" frequently fall victim to pompilids, and only rarely do they counterattack. A famous example is the huge wasp *Pepsis*, with a body length of 8 cm and a wingspan of 10 cm. It persecutes tropical "tarantulas" and does not even hesitate to follow them into their burrows. In its choice of prey, *Pepsis*, incidentally, is strictly species specific. If it is put in a cage with the "wrong" species of "tarantula," it does not attack and, in fact, may be killed by the spider. In order to determine the "right" species, the wasp crawls over the entire spider, with its antennae vibrating. Only after a thorough olfactory inspection and positive identification of the spider does *Pepsis* set its stinger into the soft pleural areas near the spider's coxae (Petrunkevitch, 1952).

Mud daubers (Sphecidae) sting mainly web spiders (Rehnberg, 1987), which they drag along the ground or, if these are light enough, carry off in flight. The immobilized spiders are either enclosed in natural cavities (in soil or in wood) or they are sealed off in vertical mud tubes that look like little pipe organs (Coville, 1987). Even the very poisonous black widow spider is not safe from wasp attacks: *Chalybion cyaneum* regularly preys on black widows (Irving and Hinman, 1935). A single sphecid wasp may have a considerable effect on a spider population, as one female catches 100–300 spiders in the course of one summer (Bristowe, 1941). The tropical *Trypoxylon* species may take an even higher toll, since they provision each clay cell with an average of 20 spiders, taken in less than a day (Coville and Coville, 1980). It should be stressed that the spiders serve as a food source only for the developing larvae and not for the adult wasps, which feed mostly on plant juices (nectar and honey dew).

Members of the wasp family Ichneumonidae also plague spiders. They attach their eggs to the spider's body or deposit them in the spider's egg sacs. *Polysphincta* raids orb weavers and fastens its eggs to their abdomens (Fig. 202c). The larva lives as an ectoparasite on the spider's body, ingesting its body fluid (Nielsen, 1923, 1932). The hatched larvae of other ichneumonids, *Gelis*, for example, feed on spider eggs inside the cocoon (see Fig. 171d; Horstmann, 1970; Kessler, 1971b; Kessler and Fokkinga, 1973). Certain dipteran flies are also known to act as cocoon, or egg, parasites. Well-known examples are members of the families Chloropidae, Sarcophagidae, and Asilidae. It is quite astonishing that these flies manage to deposit their eggs in the spider cocoons, since the cocoon wall is usually tough and is often well camouflaged.

Certain "spider flies" (Acroceridae) attack the spider directly and are true endoparasites (Schlinger, 1987). The young fly larva attaches to a spider's leg, crawls slowly onto the abdomen and penetrates between the lamellae of a book lung. There it may spend several months or, as in the case of large "tarantulas," several years. The fourth instar, or mature larva, is the active feeding and destructive stage. Although it will last only 1–2 days, this period is deadly for the spider, because all its body tissues will be devoured by the parasite. The enormous larva will then break through the dorsal abdomen, attach itself to the spider's web and pupate within a few days. In general, a host spider is parasitized only by a single acrocerid larva, but in large "tarantulas" up to 14 larvae have been found in one abdomen. It is quite interesting that the more primitive "spider flies" (Subfamily Panopinae) are always associated with the likewise primitive "tarantulas," whereas the more advanced Acrocerinae parasitize the higher evolved Araneomorphae—which makes a good case for coevolution.

Finally, vertebrates are also enemies of spiders. Many fish, particularly trout, eat spiders that happen to fall onto the surface of the water. Among amphibians, toads are considered to feed most often on spiders; spiders constitute about 5% of a toad's diet (Bristowe, 1941). Lizards and geckos occasionally eat spiders, but in general reptiles were thought to have little effect on the spider population. However, observations in the Bahamas and Caribbean showed that islands with lizards (*Anolis*) had 10–30 times less spiders than those without lizards (Schoener and Toft, 1983; Pacala and Roughgarden, 1984; Schoener and Spiller, 1987). The influence of birds as a factor controlling spider populations is generally overestimated. It is true, however, that spiders are a major prey for many birds during winter (Askenmo et al., 1977; Hogstad, 1984) and in the spring spiders are also fed to the nestlings. Yet, spiders are relatively poorly perceived, and hence little harassed, by birds for several reasons: (1) many spiders hide in retreats; (2) most web spiders sit motionless in their webs; (3) many species are camouflaged; and (4) most spiders are active at night. Birds, in contrast, are visually oriented and usually active during the day. Thus, it is no surprise that the spiders found in the stomachs of birds were always rather conspicuous species, whereas cryptic forms, such as *Cyclosa*, were absent. Only a few spider enemies are found among the mammals. In shrews spiders make up 1–2% of their diet (Bristowe, 1941). Bats also prey on spiders (Kolb, 1958; Shiel et al., 1991) and, as recent analyses of bat feces have shown, they can pick up ground spiders as well as orb weavers (Wolz, 1992). Monkeys feed on all kinds of arthropods and apparently do not spare spiders. The South American woolly monkeys *Lagothrix* have been observed to eat the poisonous brown recluse spider, *Loxosceles* (Fig. 39c), without any adverse effects (Macedo, pers. comm., 1980).

Not all enemies of spiders threaten the life of the spider directly. Some steal only their prey, as do the kleptoparasitic spiders of the genus *Argyrodes* (see p. 239) and *Mysmenopsis* (Coyle et al., 1991). The scorpionfly *Panorpa* also invades spider webs in order to steal prey (Thornhill, 1975; Nyffeler and Benz, 1980) as do certain wasps and dragonflies (Fincke, 1984). Many small birds, especially hummingbirds, were watched snatching silk material and wrapped prey (Young, 1971). Cribellate silk is used as a kind of Velcro to link their nesting material to twigs and leaves, or to attach lichen to the outside of their nests. Regular silk threads, mostly from cocoons, are utilized by several birds (e.g., New World flycatchers) to suspend their nests like a hammock between forked branches (Hansell, 1993).

Some tiny flies and gnats may partake in a spider's meal but owing to their small size have hardly any effect on food intake (Robinson and Robinson, 1977). A bit more of a parasite is the symphytognathid spider *Curimagua bayano*, which shares in the prey of the rightful web owner (Vollrath, 1978). This tiny spider (1–2 mm body length) rides on its host, a large (40 mm) *Diplura*, in a way reminiscent of young wolf spiders sitting on top of their mother. When *Diplura* is feeding on its prey, *Curimagua* climbs down the large chelicerae of its host and imbibes the liquefied prey tissue. In fact, *Curimagua* seems unable to capture any sort of food on its own.

Adaptations

Some spiders can tolerate a certain fluctuation in environmental factors and are referred to as *euryecious*. In contrast, *stenecious* species are less flexible to changing conditions. The distribution and population density of spiders in their habitat are therefore not accidental but are functions of a whole range of graded factors within a given biotope. These factors are not merely of a physical nature (like temperature, humidity, or wind conditions), but are more related to the type of vegetation present in the spider's habitat. Quite often, certain spiders associate with certain plant species. For instance, in tussocks of broom sedge (*Andropogon virginicus*) live five typical spider genera: *Phidippus, Oxyopes, Eustala, Clubiona* and *Mangora*; the population density of these spiders is about the same in all tussocks, independent of the biotope in which this grass may grow (Barnes and Barnes, 1955). Of the many specific adaptations of spiders to their environment, I shall select only those that show how well spiders can cope with rather extreme environmental conditions.

Thermoregulation

The ambient temperature is certainly important for determining the activity of a spider. Many web spiders, for instance, cease building their webs when the temperature drops below a critical point. Although spiders belong to the poikilothermic animals, it would be incorrect to believe that their body temperature simply follows the ambient air temperature. Spiders can adjust their behavior to keep their body temperature at a higher or lower level than that of the environment. Such thermoregulation is well known in lizards, yet a comparable ability in spiders has only recently been observed (Lubin and Henschel, 1990).

For example, the Australian wolf spider *Geolycosa godeffroyi* lives in a burrow in the soil but often comes up to the surface to warm its body in the sun (Humphreys, 1974, 1975, 1987). Its body temperature can then climb quickly to 40°C, although the ambient air temperature may be only between 15 and 25°C (Fig. 203). The spider avoids overheating by returning to its cool burrow; this situation is reversed at night, when the burrow provides insulation from the cold outside. In winter the spider has been found to maintain its body temperature at +1.8°C inside the burrow while the outside temperature was −2.5°C. Throughout the year, the spider maintains a body temperature that is, on average, 4–5°C above the ambient air temperature. An even larger temperature difference has been noted inside the egg sacs of *Geolycosa*, which the spider periodically exposes to the sun at the entrance of its burrow (see Fig. 192).

Behavioral thermoregulation has also been observed in various web spiders. Some araneids (*Nephila, Cyrtophora*) of tropical and subtropical regions are often exposed to extreme solar radiation when they sit in the center of their webs. The spider reacts by adjusting the axis of its body to make it parallel to the incident sun rays, thus keeping the exposed body area to a minimum

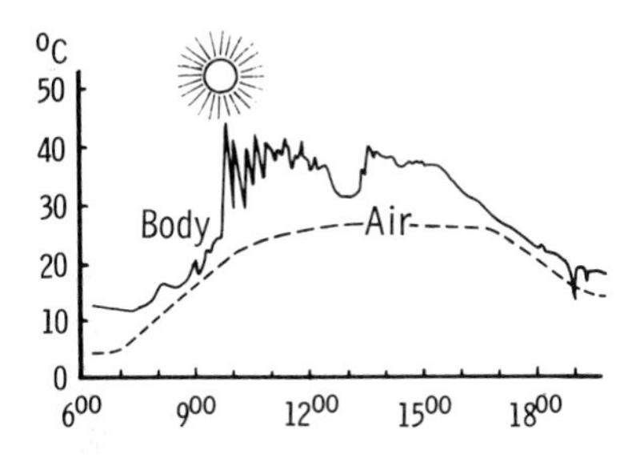

Fig. 203. Changes in the air temperature and body temperature of the wolf spider *Geolycosa* in a 24-hour period. Early in the morning (6:00), when the spider is still inside the burrow, its body temperature is 8°C higher than the ambient air temperature. Exposure to the sun causes a rapid increase in body temperature (first peak). The following fluctuations in body temperature result from the spider's shuttling between the warm outside and the cool burrow. (After Humphreys, 1974.)

(Krakauer, 1972; Robinson and Robinson, 1974). The light-reflecting guanine crystals of the abdomen may also play a role in preventing overheating. If the body temperature increases to above 40°C, the spider withdraws into the shade (Blanke, 1972); such high temperatures cause an anesthesia-like state in the spider, and such stupors can easily be induced experimentally.

Some web spiders can also adjust their body temperatures to cold conditions. In Florida, for instance, *Nephila* may experience rather low temperatures during winter. In response, these spiders orient their webs in an east–west direction; the spider's body thus becomes fully exposed to the sun's rays and can reach 7°C above the ambient air temperature (Carrel, 1978). In addition to temperature there are also other factors (such as wind and light) that play a role in determining the orientation of a web. In response to wind, *Araneus gemmoides* builds her web parallel to the prevailing wind direction, whereas *Araneus diadematus* simply reduces the area of the exposed web. Both strategies tend to prevent mechanical load and possible damage of the web (Hieber, 1984).

Behaviorally mediated thermoregulation entails one disadvantage for spiders: a higher body temperature also means a higher transpiration rate and a correspondingly greater loss of water. If a spider loses more than 20% of its body weight because of transpiration, it will die (Cloudsley-Thompson, 1957). It seems, however, that most spiders can take up water from moist soil, provided that the humidity of the substrate is more than 12%. Normally, early morning dew is quite sufficient for that purpose.

A rather curious example of a "passive thermoregulation" is provided by the parasitic pompilid wasp that attacks the eresid spider *Stegodyphus lineatus* in the hot Negev desert (Ward and Henschel, 1992). The spider is paralyzed inside its aerial silk tube and a single egg is laid on its abdomen. If it were left inside the tube, the immobile spider would heat up to 40°C and die within a few days. Instead the wasp moves the spider to the cooler nest entrance—incidentally the same position the spider normally adopts in response to the heat of the day. Wind convection at the nest entrance causes a considerable drop in temperature and the host is thus kept cool long enough for the wasp larva to develop into a pupa.

Overwintering

In the temperate zones, poikilothermic animals have to adjust to the harsh cycles of the seasons. Spiders have developed several adaptations to survive such adverse conditions as cold, dampness, flooding—and naturally, lack of food. Spiders meet these challenges by colonizing the appropriate microhabitats, by increasing their resistance to cold, and by reducing their metabolic rate. They are thus well prepared for overwintering and, as a consequence, mortality during the cold winter months is surprisingly low.

About 85% of the spider fauna overwinter in the soil, mainly in leaf litter, which is a good insulator against the cold. During this time most spiders assume a rigid posture, with the legs drawn close to the body so that the exposed body surface is kept to a minimum. The microhabitat of the leaf-litter zone protects not only from extreme temperature fluctuations but also from desiccation (Edgar and Loenen, 1974). Even heavy snow cover is by no means lethal for spiders. On the contrary, the insulating properties of a layer of snow ensure a rather steady temperature of about 0°C (Buche, 1966). Thus, ambient air temperatures of −40°C, which have been recorded, for example, in Canada, have little effect on the spiders beneath the snow (Aitchison, 1978, 1987).

Spiders of the temperate zones can be put into one of five groups, based on their annual cycles or periods of maturation (Schaefer, 1976a,b, 1977): (1) *eurychronous* species which take a long time to mature and overwinter in various developmental stages (these spiders constitute 23% of the species investigated); (2) *stenochronous* species whose reproductive period lasts from spring until summer and which overwinter as nymphs (45%); (3) stenochronous species whose period of reproduction is during the fall and whose young spiderlings remain inside the egg case during winter (7%); (4) *diplochronous* species, having two separate reproductive periods (spring and fall) and overwintering in the adult stage (4–13%); and (5) stenochronous species with their reproduction period in winter, the winter-active spiders (9%). Some members of the linyphiids (erigonids), lycosids, clubionids, thomisids, and tetragnathids belong to this group. They can build webs even below 0°C (Fig. 204) and also catch prey (e.g., collembolans) (Hågvar, 1973; Aitchison, 1984).

The winter-active spiders of the temperate zones are mostly found among the family Linyphiidae (Fig. 204). They are not particularly resistant to cold but simply most active at rather low temperatures. Below −4°C they become as rigid as other spiders do, and below −7°C they die. Those spiders that overwinter passively are much more cold resistant. Most garden spiders (*Araneus* species) can withstand temperatures of −20°C, even in unprotected locations (Kirchner, 1973). It is not quite clear how these spiders achieve their remarkable resistance to cold. It is known that the spider's hemolymph contains glycerol, which acts as an antifreezing agent, and that the glycerol content is

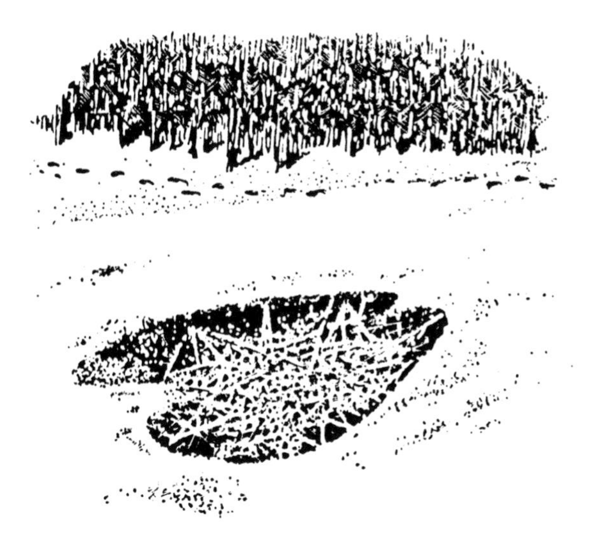

FIG. 204. Web of the small linyphiid *Lepthyphantes* in a deer footprint in snow. (After Buchar, 1968.)

markedly higher in winter than in summer. Nevertheless, it seems unlikely that the glycerol alone can account for such resistance to cold, since it has been found that this chemical lowers the freezing point of the spider's hemolymph by only 1°C (Kirchner and Kestler, 1969). The freezing point (or melting point, to be more precise), however, is not equivalent to the much lower undercooling point (i.e., just before ice crystals are formed) that is measured in live spiders (Kirchner, 1987). This is apparently due to certain proteins of the hemolymph, which can lower the freezing point by 20°C (Duman, 1979; Husby and Zachariassen, 1980; Zachariassen, 1985). It remains still a bit mysterious, however, how cold-resistant spiders manage to survive long and cold winters (Kirchner and Kullmann, 1975).

Many spider eggs can also withstand temperatures as low as −24°C (Schaefer, 1976a,b, 1977). One would expect that the egg stage would thus be an ideal form to survive the winter, yet only 7% of our spiders actually use this strategy. Perhaps it is more favorable for a young spider to be self-sufficient by early spring.

Special Adaptations: Camouflage and Mimicry

More impressive than the physiological adaptations of spiders toward the changing physical conditions of their environments are adaptations that protect spiders from their enemies. These protective measures range from simple camouflaging colors to complex behavior, including mimicry. Most spiders are rather drab and are not very noticeable within their environment. Even such conspicuously green species as *Micrommata virescens* or *Araniella cucurbitina* are difficult to detect in nature, because they live mostly on leaves. More interesting, however, are those spiders that can actively change their coloration to match the substrate they live on: the crab spiders *Misumena vatia* and *Thomisius onustus*, for example, change the color of their white or yellow abdomens to correspond to the white or yellow blossom on which they are sitting. This process usually takes a few days (Weigel, 1941). The advantage of this color adaptation is twofold: first, insects visiting the blossoms do not perceive the camouflaged spider and can be taken by surprise, and second, the spider is not recognized by its enemies (Bristowe, 1958). We should also bear in mind that many insects can see UV light and therefore their visual impressions may be quite different from ours. White and yellow crab spiders, for instance, were found to absorb UV light and are probably inconspicuous to insect eyes, especially when sitting in the middle of blossoms (Sato, 1987).

Aside from this slow color change, a rapid color change, as a kind of startle response, can also occur in some spiders. The tropical araneid *Cyrtophora cicatrosa* drops immediately out of its web when disturbed; at the same time, the color pattern of its abdomen changes rapidly so that the animal blends with its background (Blanke, 1975a). This rapid color change probably results from a quick reduction of the light areas (that is, the guanine deposits) in the abdomen. The return to normal coloration takes only a few minutes.

Startle reactions among spiders such as dropping suddenly on a thread, are quite common. When the spider falls to the ground, it often pulls its legs close to its body and "plays dead," a behavior that is termed *catalepsy*. Some web spiders, *Argiope*, for instance, react by violently shaking their webs if they are disturbed by sudden vibrations or by shadows passing overhead. The most striking response occurs in daddy-longlegs spiders (Pholcidae), which can tremble so rapidly in their webs that they become too blurred to be seen (Jackson, 1990; Jackson et al., 1990). Many wandering spiders quickly assume defensive postures when threatened. The front legs are jerked upward, the chelicerae are widely spread, and the whole prosoma is lifted up (Tretzel, 1959). "Tarantulas" may remain in such threatening postures for several minutes.

Perhaps the most impressive adaptations are the various kinds of mimicry. Spiders are renowned for their ant mimicry: several hundred species copy the shapes and body colors of ants, and also move in a way that is very similar to that of ants. Striking examples are found in certain salticids and clubionids, whose short and sturdy forelegs look like the antennae of ants. This optical illusion is even more convincing when one watches a spider in motion: the forelegs are rarely used for walking but instead are raised above the body, quivering in an antlike fashion. The other legs are thin, and the prosoma (or opisthosoma) may be constricted so that the typical shape of an ant's body is achieved (Fig. 205). Of course, morphological and behavioral adaptations vary considerably among different spider species (Jackson, 1986a; Elgar, 1993). Some species are only superficially antlike and still walk with all eight legs, whereas others are detailed copies of a specific ant species (Reiskind, 1977). The Central American *Castianeira rica* (Clubionidae) has one of the most interesting types of mimicry because it exhibits multiple mimicry (Reiskind, 1970). Females of this species look like a rather generalized version of the ant subfamily Ponerinae, but the bright orange males resemble ants of the genera *Atta* and

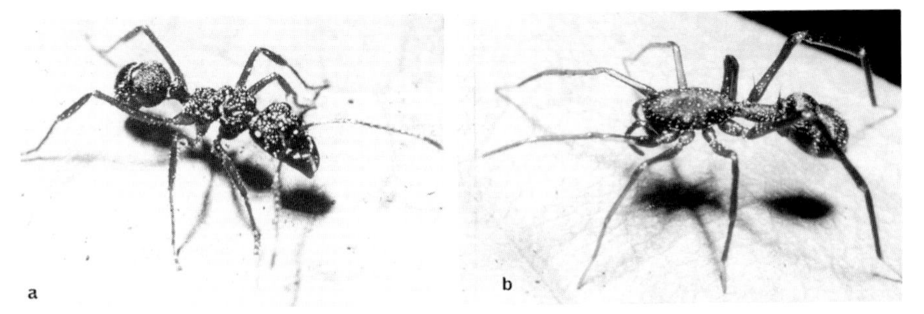

FIG. 205. Ant mimicry. (*a*) The "model," the tropical ant *Ectatomma ruidum*. (*b*) The "copy," the clubionid spider *Mazax pax*, which lives in the same habitat. Note the close resemblance in body size, shape, and coloration. (Photos: Reiskind.)

Odontomachus. In contrast, the second and third nymphal stages of *C. rica* are black and shiny like ants of the subfamily Myrmicinae, while the fourth and fifth nymphs are yellow–orange, mimicking leaf-cutter ants (*Atta*).

The crucial question of ant mimicry, namely, whether the spider derives any advantage from its resemblance to ants, has been discussed quite often (Berland, 1932); certainly, there is little experimental evidence that would demonstrate a biological meaning for this mimicry. It has been proved, however, that birds cannot distinguish at least one ant mimic (*Synageles*, Salticidae) from the actual ant (Engelhardt, 1971). Furthermore, it is known that birds usually avoid eating ants, presumably because they are distasteful. In one experiment, when a *Synageles* was fed to a bird by hand, it was readily accepted; however, when *Synageles* was offered in a dish together with ants (*Lasius niger*), the bird ate neither. This speaks clearly in favor of protective mimicry (Batesian mimicry), for which a prerequisite is that the model and the mimic occur in the same habitat. This is indeed true for most cases of ant mimicry in spiders. Furthermore, those spiders that live together with ants almost never feed on ants (Jackson and Willey, 1994). The only example of an "aggressive mimicry" is the spider *Aphantochilus* (Oliveira and Sazima, 1984) and perhaps the antlike crab spider *Amyciaea*, which lures ants by her movements before seizing them (Hingston, 1927; Mathew, 1954).

Overall, we must conclude that ant mimicry in spiders has mainly a protective function. This is also supported by the observation that the antlike *Myrmarachne* species rarely become victims of sphecid wasps, in contrast to the "normal"-looking jumping spiders (Edmunds, 1978, 1993). Other, more indirect evidence that ant mimicry is self-protective for spiders is provided by the ant-mimicking salticids *Synageles venator* and *Peckhamia picata*, which lay only 3–4 eggs: this low number of offspring must have a relative good chance for survival. One might object that the low number of eggs is simply a result of their small and narrow abdomen. Among the different *Myrmarachne* species, for instance, only the small ones lay few eggs (2–6), whereas the larger ones have about 25—which is close to the usual 25–35 eggs found in the cocoons of "regular," middle-sized jumping spiders (Bristowe, 1939).

It is interesting that not only do spiders mimic insects, but insects mimic spiders. Some fruit flies (*Rhagoletis*, *Zonosemata*) have a conspicuous pattern of dark stripes on their wings that resembles the legs of certain jumping spiders (*Salticus, Phidippus*). When they lift and lower their wings it gives the impression of a moving spider (Greene et al., 1987; Mather and Roitberg, 1987). A *Rhagoletis* fly gives such a display when encountering a *Salticus* spider. The jumping spider apparently mistakes the fly for a threatening conspecific and carefully withdraws. However, in experiments in which the stripes on the wings had been removed, the spider recognized the fly and attacked immediately. This protective mimicry is rather specific and works only with jumping spiders. Other visual hunters such as lynx spiders or preying mantis are not fooled by this mimicry.

Communication

As mentioned previously in chapters 4 and 7, spiders have developed various means to communicate with each other. Especially during courtship, mechanical, chemical, or visual signals may play important roles in communication. The exchange of chemical signals was probably the first type of communication in spiders to evolve (Weygoldt, 1977), because contact chemoreceptors permit mutual identification of the sexual partners. Also important are receptors that perceive various mechanical signals, especially the vibrations transmitted by a web (Barth, 1982). It is quite likely that the vibrational signals emitted by web spiders during courtship are species specific and thus also serve to identify potential mates (see Fig. 160). Wandering spiders also generate vibratory signals during courtship; the vibrations may be transmitted by a solid substrate (such as the ground or leaves) or through the air as sound. The following discussion of communication in spiders is restricted to the exchange of acoustical signals, because the other modes of communication have been described in chapter 7.

Like certain insects, several species of spiders produce sounds. They can use three methods: (1) drumming with parts of the body (palps, legs, or abdomen) against the substrate; (2) *stridulation*, that is, generating a sound on the spider's body by moving a "scraper" over a "file"; and (3) *vibration* of legs or abdomen (Rovner, 1980b; Rovner and Barth, 1981). Drumming is seen in many male wolf spiders during courtship (see chapter 7). Since females react to it, the communicative function of drumming is clear (Bristowe and Locket, 1926; Harrison, 1969). The situation is somewhat more complex for stridulation. Many types of stridulatory organs have been well described structurally, yet any communicative function ascribed to them has been met with skepticism (Berland, 1932; Legendre, 1963). However, investigations of the cobweb spider *Steatoda bipunctata* have shown that stridulation is indeed involved both in courtship and in agonistic encounters (Gwinner-Hanke, 1970).

In *Steatoda bipunctata* only the adult male possesses stridulatory organs. These consist of ridged areas (1 mm^2) on the prosoma that are apposed to strong cuticular spurs on the opisthosoma (see Fig. 206a). When the spider vibrates

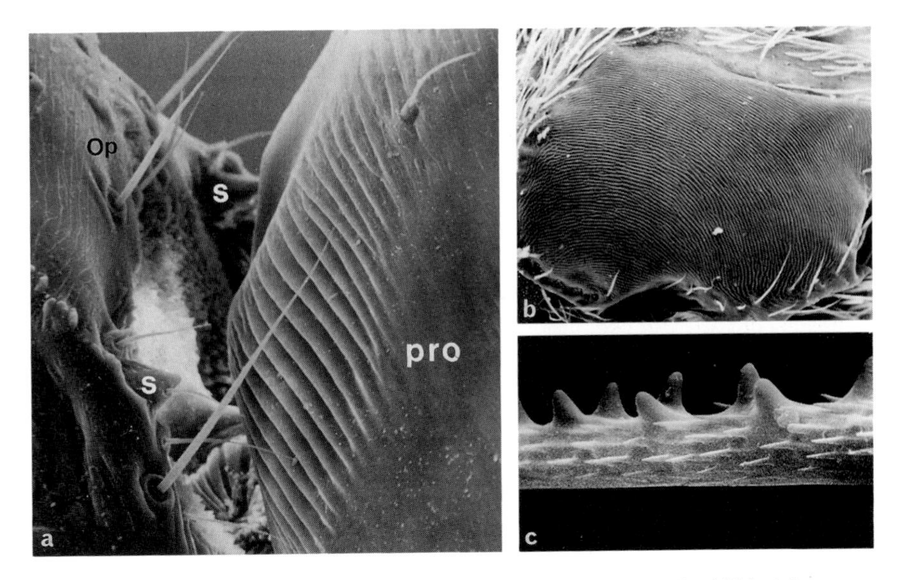

Fig. 206. Stridulatory organs. (*a*) The kleptoparasite *Argyrodes* (Theridiidae) has strong cuticular spines (S) on its opisthosoma (Op). The spines are scraped across the rippled area on the prosoma (pro). The hair sensilla projecting from the opisthosoma may function as proprioreceptors. 420 ×. (Photo: Vollrath and Foelix.) (*b*, *c*) The male wolf spider *Pardosa fulvipes* has spiny bristles on the coxae of legs 4. These are rubbed over the rills on the cover of the book lungs. (*b*) 70 ×; (*c*) 1,300 ×. (Photos: Kronestedt.)

its abdomen vertically, the organs produce a faintly audible sounds as does a fingernail run along the teeth of a comb. The main frequency of the sound is around 1,000 Hz, or, as a music connoisseur would say, a C‴. The male stridulates only in the presence of a female or if a competing male is nearby. In such a case, only the dominant male will stridulate (Lee et al., 1986). Proof that the stridulatory organs have a communicative function is seen from the female's reactions: she waves her forelegs, plucks the threads of her web, vibrates her entire body, and makes searching movements. The male's direct contact with the female's web is not necessary, and this is evidence that the stridulatory signals are transmitted through the air as sound. When the stridulatory organs are put out of action by covering them with glue, the males do not start courtship at all. Courtship is probably elicited by chemical substances that adhere to the female's threads. These threads alone suffice to trigger the male's courtship; they fail to do so after the substances have been washed off with ether. In the related theridiid *Steatoda grossa*, however, it is not yet clear whether a communicative function can be assigned to the stridulatory organs, since the females do not show any definite reactions toward stridulating males (Gwinner-Hanke, 1970).

How can the female spiders differentiate between the stridulations of males belonging to different species? How can the female wolf spider *Schizocosa ocreata* recognize when a "wrong" male (e.g., *Schizocosa rovneri*) is stridulating nearby? The answer lies most likely in the different way the sound is produced: *Schizocosa ocreata* stridulates for several seconds at 800 Hz, whereas *Schizocosa rovneri* produces short klicks (0.2 s) at 500 Hz, which are separated by long pauses (Uetz and Stratton, 1982). Although both species use the same "instruments" (stridulatory organs on the palps), they apparently play different "melodies."

Morphologically, the stridulatory organs are always paired "files" that lie close to cuticular teeth on an adjacent body part. Eight types of stridulatory organs, occurring in over 100 species of 23 families, have been described (Legendre, 1963, 1970). In addition to the type described for *Steatoda* (opisthosoma–prosoma), stridulatory organs may lie between the chelicerae and the pedipalps (as they do in *Scytodes*) or between the opisthosoma and the fourth legs (Fig. 206b,c; Hinton and Wilson, 1970; Kronestedt, 1973). Many of the large "tarantulas" have stridulatory organs in both sexes, not just in males. When one of these spiders is threatened, it emits a hissing noise that is quite like the defensive hissing of a snake. In general, then, stridulatory organs may serve a defensive purpose when present in both sexes and are used for identification when restricted to the male of the species (Meyer, 1928).

Stridulation is quite common among crustaceans (in fiddler crabs, for example) and in insects. Crickets, grasshoppers, and cicadas use stridulation mainly to identify their own species. Many beetles, roaches, bugs, and butterflies also possess stridulatory organs, which in many cases serve to startle predators. In leafcutter ants, stridulation functions as an alarm system: when an ant is in trouble, buried under sand or caught in a spider's web, it will stridulate vigorously; this attracts other ants in the vicinity that quickly come to the aid of the victim (Markl, 1967).

Social Spiders

Anyone who has collected live spiders knows that they have to be kept apart, or they are likely to kill each other. The reputation spiders have for being solitary and intolerant is certainly well deserved. Yet among the 34,000 different species of spiders, about 20 species are known to live together in peaceful colonies; these are the social spiders.

What exactly is meant by "social" spiders? There are many different interpretations in the literature (Kullmann, 1972b; Krafft, 1985; D'Andrea, 1987) and they are all derived from the concept of sociality in insects (see E. O. Wilson, 1971). The most recent discussion of social spiders calls simply for a "cooperation among mutually tolerant individuals" (Downes, 1995). This definition abandons an earlier postulated *interattraction* ("a tendency to aggregate..."; Kullmann, 1972b) as a necessary precondition for sociality. Although there are a number of spiders that live together communally, only few (about 20) fulfill the main criterion of sociality, i.e., *cooperation* in prey capture and in brood care. These spiders belong mainly to web spiders, namely the agelenids (*Agelena consociata,*

A. republicana), the eresids (five species of *Stegodyphus*), the theridiids (four species of *Anelosimus*, two species of *Achaearanea*) and two dictynids (*Mallos gregalis, Aebutina binotata*; Avilés, 1993). It was long held that only web spiders could be social, because the threads of their webs transmit vibrations and thus serve as the necessary means of communication. However, in recent years a social crab spider (*Diaea socialis*; Main, 1988), a social huntsman spider (*Delena cancerides*; Rowell, 1985), and a social lynx spider (*Tapinillus* sp.; Avilés, 1994) have been reported. It should be pointed out that all these spiders have not quite reached the *eusocial* level of termites, ants, bees, or wasps, because they have never developed a caste system; yet it still seems permissible to call them social, or at least quasi-social (E. O. Wilson, 1971).

One of the first social spiders studied was the African funnel-web spider *Agelena consociata* (Fig. 207a). Several hundred, or even a thousand, individuals may live together in one web without showing any animosity toward each other. This tolerance must no doubt be based on mutual recognition, because social spiders are just as aggressive toward their prey as are their solitary relatives. Most likely, a spider is recognized as a member of the same species by the typical vibrations it generates in its web. When a spider is made to vibrate artificially, its conspecifics attack it immediately. However, they quickly release it after they briefly touch it with their first tarsi. Sometimes they also use their pedipalps for a closer inspection, but a spider of their own species will never be bitten. Probably, certain chemical substances (pheromones) help the spiders recognize a conspecific, since spiders washed with a mixture of ether and alcohol are bitten (Krafft, 1971).

The chemical signals are apparently not very specific, because it is possible to exchange spiders from different web communities and even of different species without eliciting any aggressive reactions. For example, the solitary *Agelena labyrinthica* can be peacefully introduced into the web of the social *Agelena consociata* (Krafft, 1975), and similarly, populations of social *Stegodyphus* species (*S. mimosarum, S. dumicola*) can be mixed together (Seibt and Wickler, 1988a). The social spiders thus differ remarkably from the "closed" societies of certain ants, where members of the same species from different nests often fight each other; the open societies of social spiders are more comparable to the anonymous schools of fish or flocks of birds. However, some kind of hierarchical system seems evident in *Agelena consociata*—at least, for example, when one of the spiders is to carry prey from the web to the retreat.

Even more impressive than the mutual tolerance of social spiders is their cooperation. Web building, prey catching, feeding, and brood care are often performed communally. Of course, cooperative web building is not particularly coordinated and is dictated mainly by necessity. Several animals of different age classes spin largely independently of each other when repairing or extending the existent web structure.

Prey catching in *Agelena consociata* is also communal (Darchen, 1965; Krafft, 1971). When prey falls on the sheet web, its movements attract the attention of all spiders in the vicinity. If the prey is small (less than 1 cm) only a single adult spider rushes in, grasps the victim, bites it, and carries it off to

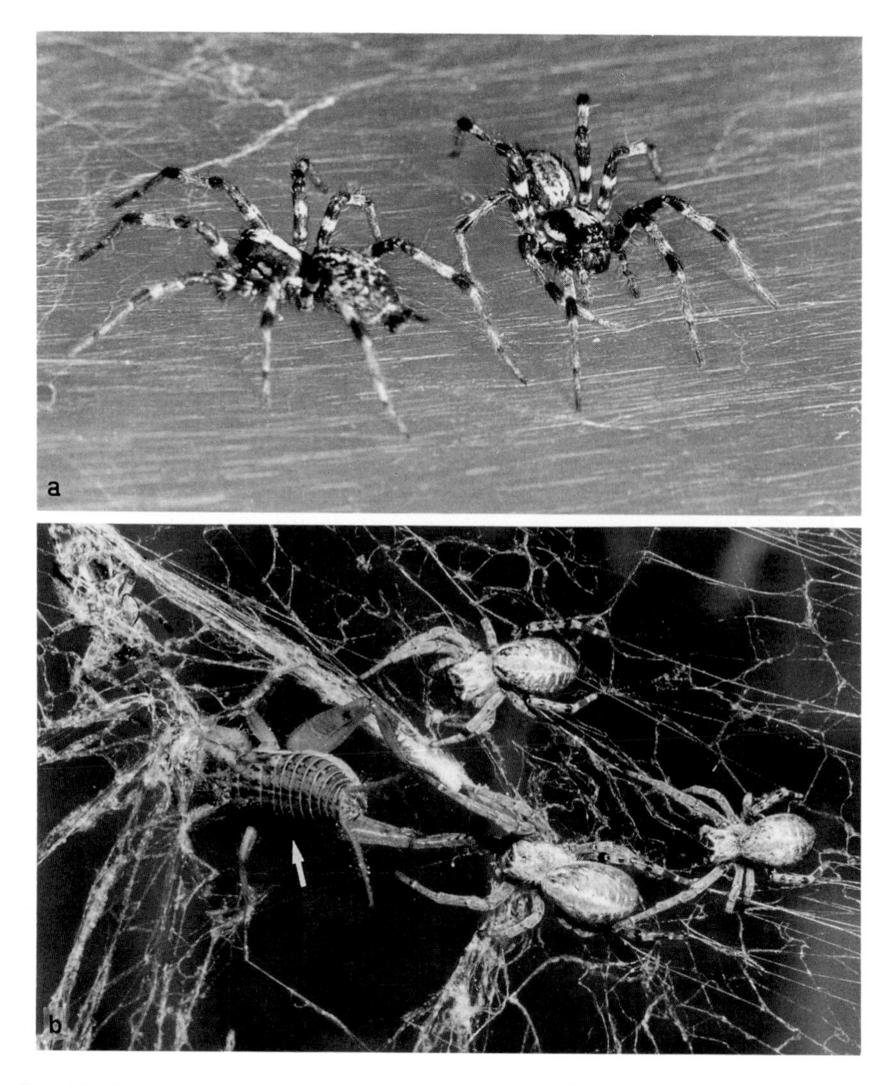

FIG. 207. Social spiders. (*a*) Two females of the African funnel-web spider *Agelena consociata* in close contact on the communal sheet web. (Photo: Krafft.) (*b*) Several *Stegodyphus dumicola* females attacking a cricket (arrow) together in their cribellate web. (Photo: Bar Shahal.)

the common retreat. There the actual feeding begins and as many other spiders participate as can find a place on the prey. If a larger prey animal has blundered into the web, several, mostly adult spiders attack it. Some hold fast to the legs or wings while others climb on the victim's back to inflict the poisonous bite. Nevertheless, only one spider carries off the prey to the retreat. The spiderlings do not participate in hunting but wait until food is brought to them by an adult spider. During the entire course of hunting and feeding, aggression is never displayed between conspecifics.

This lack of aggression is apparently not typical for all social spiders. The social *Stegodyphus* species (Fig. 207b), for instance, also hunt and feed together, but not as peacefully as it may appear at a first glance (Seibt and Wickler, 1988b,c). Transportation of large prey is not really well coordinated but is rather a "tug-of-war" of several individuals that eventually ends in the retreat. Similarly, communal feeding is not free of conflicts, but includes mutual leg kicking and dominance of the larger spiders (Ward and Enders, 1985). Apparently there is a competition for food, especially in larger colonies. It must also be pointed out that no active recruitment of new spiders occurs during prey capture—as is typically seen in social insects. Thus we do not find a true division of labor in these social spiders.

It is quite curious that sociality is not obligatory in certain spider species. For instance, in *Stegodyphus dumicola* many individuals live singly in small nests outside the colonies and, in fact, reproduce at a higher rate than their social siblings (Henschel, 1992). Nevertheless, a social mode of life gives a species an important advantage: A group of cooperative spiders can overpower prey larger than a single spider could (Christenson, 1984; Nentwig, 1985; Rypstra, 1990). In addition, sociality has other advantageous effects (group effects) that cannot be explained solely on the grounds of better nutrition: (1) a higher rate of metabolism and more rapid growth; (2) a higher life expectancy; (3) a lower rate of mortality; and (4) a smaller number of eggs.

The fact that social spiders produce fewer eggs than their solitary relatives (*Agelena consociata*, 12; *Agelena labyrinthica*, 50–100) seems paradoxical at first. However, it is a general trend that a higher level of sociality correlates with a lower number of eggs laid. As in other animals, a relatively small number of offspring will suffice to continue the species if brood care is highly developed. The longer life of a spider within a group has been demonstrated in experiments where social spiders were kept singly and their life spans were compared to those of their grouped siblings (Darchen, 1965). Similar observations of social insects are well known: bees kept in isolation survive only for a few days, although inside the hive they live longer than a month.

The Origin of Sociality

Social spiders have been found in quite different families, such as Dictynidae, Eresidae, and Desidae (Downes, 1994) among the cribellate spiders, and Agelenidae, Theridiidae, and Thomisidae among the ecribellates. Since these families are hardly related, sociality in spiders must have evolved independently several times. How did this sociality originate?

Only a few of the approximately 20 species of social spiders have members that live together on a permanent basis, while all the other species are only "periodically social;" that is, their individuals congregate only temporarily (Kullmann, 1972b). Furthermore, although some species (of Araneidae and Uloboridae, for example) form web colonies, each member retains its individual snare (Buskirk, 1975b; Lubin, 1980b; Burgess and Uetz, 1982; Uetz, 1986). One might refer to such loosely organized web colonies, which lack any cooperation, as *parasocial.* Those species in which the temporary social bond is closely linked with brood care may be termed *subsocial.* It is now generally assumed that evolution from a solitary to a permanently social mode of life has proceeded through various subsocial stages (Fig. 208). Quite likely, any higher form of sociality originated from a mother–offspring relationship (Markl, 1971). This relationship seems to be the basis of social systems of insects and spiders (Buskirk, 1981) as well as of the highly organized social systems of certain mammals. As an alternative explanation *neoteny* has been proposed, i.e., the persistence of juvenile behavior patterns (e.g., mutual tolerance) into adulthood (Burgess, 1978; Kraus and Kraus, 1988).

The decisive point on the route to sociality in spiders was probably that the young ones did not disperse but stayed in their mother's web, even after reaching maturity. As a consequence inbreeding and group selection became effective, as is seen in the skewed sex ratio of social spiders: each colony has 5–10 times as many females as males. This is of advantage for the size of a colony, because resources are invested into reproductive females rather than in "unnecessary" males (Avilés, 1986, 1995; Avilés and Maddison, 1991; Lubin, 1991; Vollrath, 1986b).

The evolutionary origins of sociality in spiders must probably be sought among the solitary *web* spiders, since their threads certainly play a central role in communication between individual spiders. It even seems that their webs

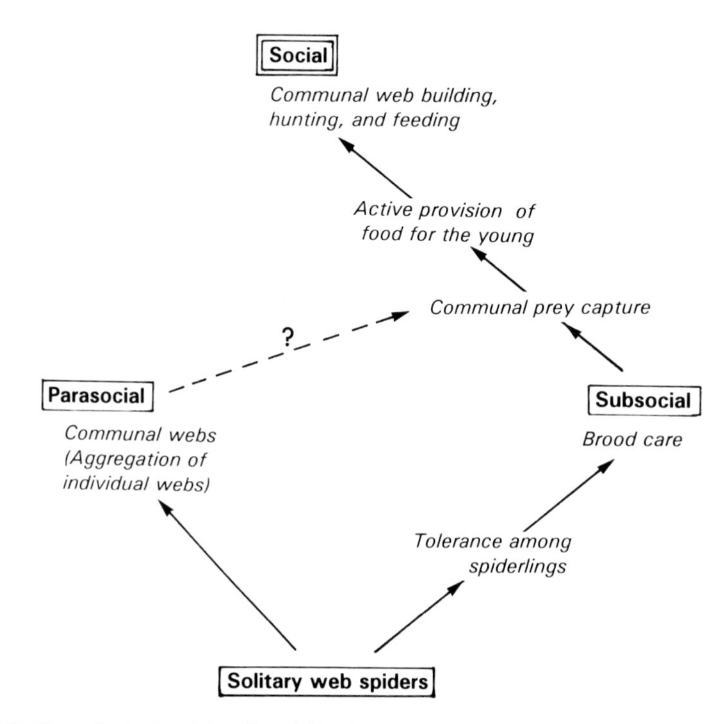

FIG. 208. Hypothetical origin of sociality in spiders. (After Shear, 1970.)

needed to be *irregular*, because only those can be used by several individuals simultaneously. Among the freely roaming wandering spiders we find very few social species, probably because the "connecting wires" are lacking. However, even in the case of social crab spiders there are many threads holding their communal leaf nest together (Main, 1986; 1988) and the social lynx spiders actually build a communal web (Avilés, 1994).

Within some spider families (such as Eresidae and Theridiidae), we can see several different "periodically social" stages that ultimately may have led to

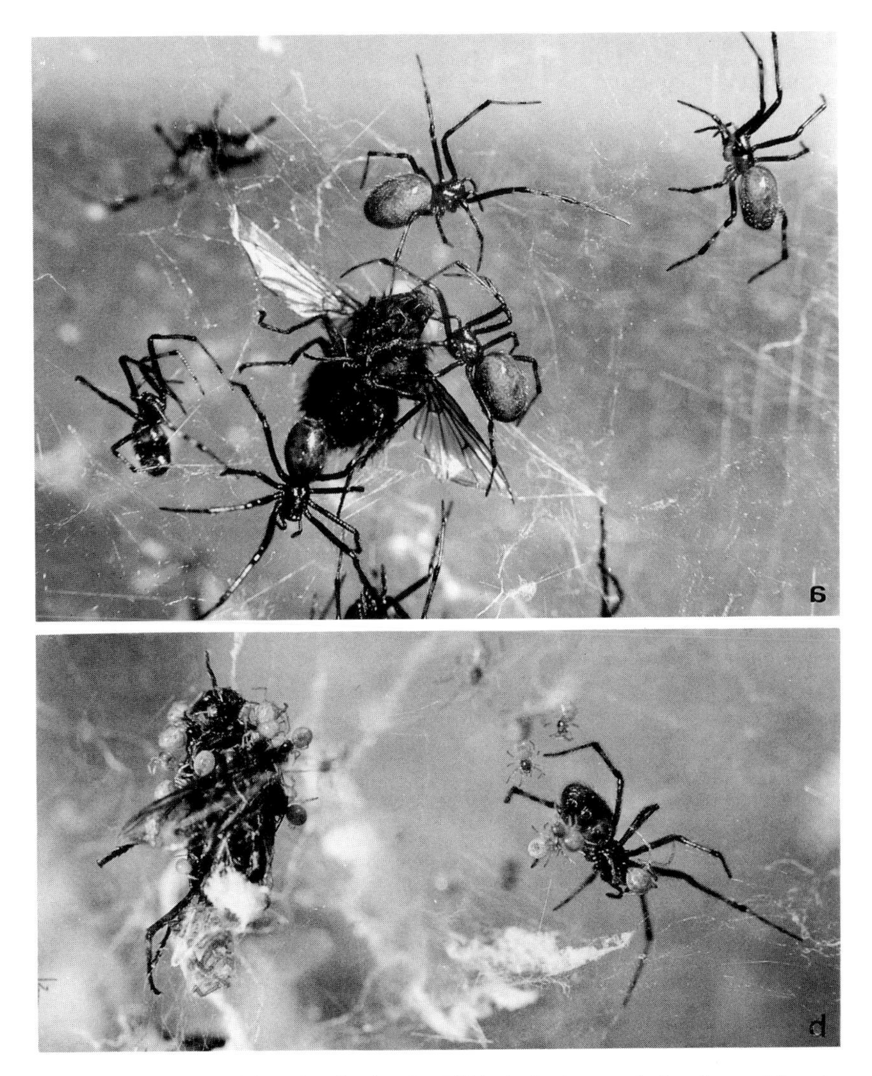

FIG. 209. (*a*) Communal hunting in the theridiid *Anelosimus eximius*. Several females fling sticky threads over a fly that became entangled in the mesh of the cobweb. (*b*) Young spiderlings gather around a female to be fed mouth-to-mouth (left) or they may feed communally on the prey killed by the adults (right).

permanent sociality (Kullmann, 1972b). Among theridiid species gradually improved methods of brood care can be observed, starting with the mere guarding of the egg case, followed by passive provision of the brood with food (e.g., the dead mother may serve as their nourishment, as in *Coelotes*), and finally, actively supplying food for the spiderlings. Examples for the latter are regurgitative feeding, carrying prey to the young, and communal prey hunting and feeding. Each increase in the complexity of brood care denotes a higher level of sociality.

A comparison between the two social theridiids *Anelosimus eximius* and *Anelosimus studiosus* is enlightening in this regard (Brach, 1975, 1977). *A. eximius* can be regarded as a typical social spider, since these spiders demonstrate tolerance, interattraction, and cooperation (Fig. 209) almost as well as *Agelena consociata*. In *A. studiosus*, however, tolerance is not permanent among the members of a colony. After the death of the colony founder, the mother, the colony remains together peacefully only until one sibling female has reached maturity. This female then tolerates only juvenile spiders and adult males, but chases any other adult females away. This aggression toward conspecific females precludes the formation of large colonies in this species. In contrast, up to 1,000 individuals may live together peacefully in one web of the consistently placid *A. eximius*.

A gradual development of sociality may also be noted among the Dictynidae, for instance from the communal yet territorial *Mallos trivittatus* to the communal, nonterritorial *Mallos gregalis* (Jackson, 1978). The colonial orb-weaver *Eriophora bistriata* could be placed between parasocial and quasi-social, because—despite building individual webs—a communal capture and feeding behavior can be observed when handling larger prey (Fowler and Diehl, 1978; Fowler and Gobbi, 1988b).

10

Phylogeny and Systematics

Now a book on biology is hardly the place to insert a chapter on classification.

W. S. Bristowe, 1938

Despite Bristowe's warning it seems necessary to cover the descent and classification of spiders briefly. I must admit, however, that our real knowledge of the phylogeny of spiders is very scanty, and hence to present any reliable pedigree is quite impossible.

Fossils

Fossils of spiders are rare and have been found only from widely separate geological eras. Although several fossils of spiders (*Arthrolycosa, Protolycosa*) have been preserved from the Carboniferous period of the Paleozoic era 300 million years ago, only very few remained from the entire Mesozoic era (Eskov, 1984, 1987; Selden, 1989; Eskov and Zonshtein, 1990). Only from the Tertiary period do we find fossils of spiders again, mostly in the form of well-preserved amber spiders (Petrunkevitch, 1942). For instance, more than 300 species have been described from the Baltic amber (about 40 million years old), and more than 100 species from the Dominican amber (20 million years; Wunderlich, 1986, 1988). These spiders from the Tertiary period, however, resemble recent spider families so closely that they provide hardly any clues as to their phylogeny (Fig. 210).

The first spiders probably appeared during the Devonian period, almost 400 million years ago (Selden et al., 1991). Definite spinnerets, which presumably belonged to a mesothele or mygalomorph spider of the mid-Devonian period, were recently discovered in North America (Fig. 211; Shear et al. 1989). The arachnids were quite likely at their peak during the Paleozoic era, since 4 of the 16 known arachnid orders were already extinct by the Carboniferous period (Petrunkevitch, 1955). During that time the evolution of insects was still in its infancy and flying was just about to be "invented" (Bristowe, 1958). The ancient spiders (Araneae) of the Carboniferous period were all segmented, quite similar to the Mesothelae (Liphistiidae) that still live today (Fig. 212). It is for this reason that the liphistiids are considered the most primitive of all living spiders

FIG. 210. (*a*) Fossilized amber spider (Hersiliidae) from the Dominican Republic, about 20 million years old. The typical long spinnerets (arrows) can easily be recognized. (*b*) Even silk threads from spider webs have been preserved in amber, as this capture thread from an orb weaver demonstrates; arrows indicate small glue droplets. (From Wunderlich, J.: Spinnenfauna gestern und heute. Bauer Verlag, Wiesbaden, 1986.)

and are often referred to as "living fossils." The Paleozoic spiders *Arthrolycosa* and *Protolycosa* were certainly closely related to the liphistiids of today (Petrunkevitch, 1933, 1955). Yet probably even during the Carboniferous period there were other spiders with segmented abdomens not closely related to the liphistiids (such as the Archaeometidae). Remnants of abdominal segmentation are still clearly present in some extant spider families, for instance in the cribellate Hypochilidae as well as in several orthognath spiders of the Antrodiaetidae family.

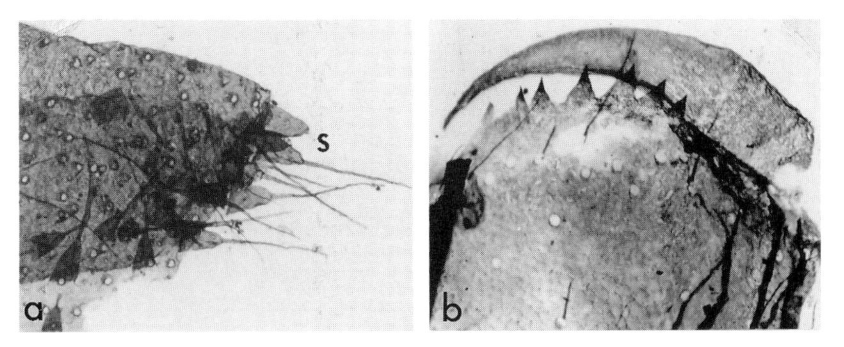

FIG. 211. The oldest fossil spider is *Attercopus fimbriungis* from the Devonian period of North America (380 million years old). (*a*) The spinnerets bear about 20 spigots. 180 ×. (*b*) Cheliceral fang opposing several cheliceral teeth. 55 ×. (Photos: Selden.)

There are hardly any fossil spiders known from the Permian period (260 million years) and even from the Mesozoic era (250–70 million years) only few findings have been reported. It is assumed that the first three-dimensional webs for capturing flying insects evolved during the Permian period, although definite proof is missing (Shear and Kukalova-Peck, 1990). Such webs existed most likely in the Jurassic period (about 160 million years), as we can conclude from findings of *Juraraneus* (Araneoidea) in Siberia (Eskov, 1984). From the

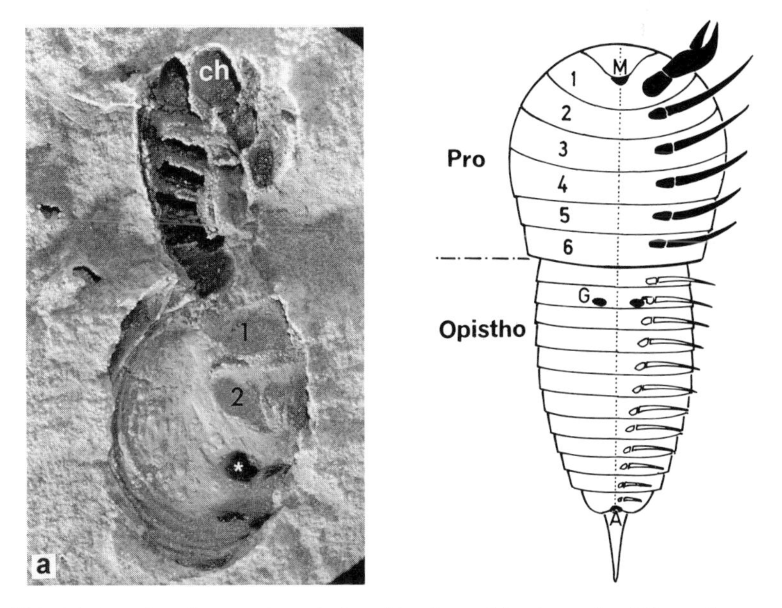

FIG. 212. (*a*) A carboniferous mesothele spider from Montceau-les-Mines, France (about 290 million years old). Note the imprints of the chelicerae (ch), two booklungs (1,2) and the spinnerets (*). (Photo: Almond). (*b*) Hypothetical archetype of a chelicerate from which the various arachnid orders may be derived. M = mouth opening, G = genital opening, A = anus, 1–6 = prosomal segments. (After Størmer, 1955.)

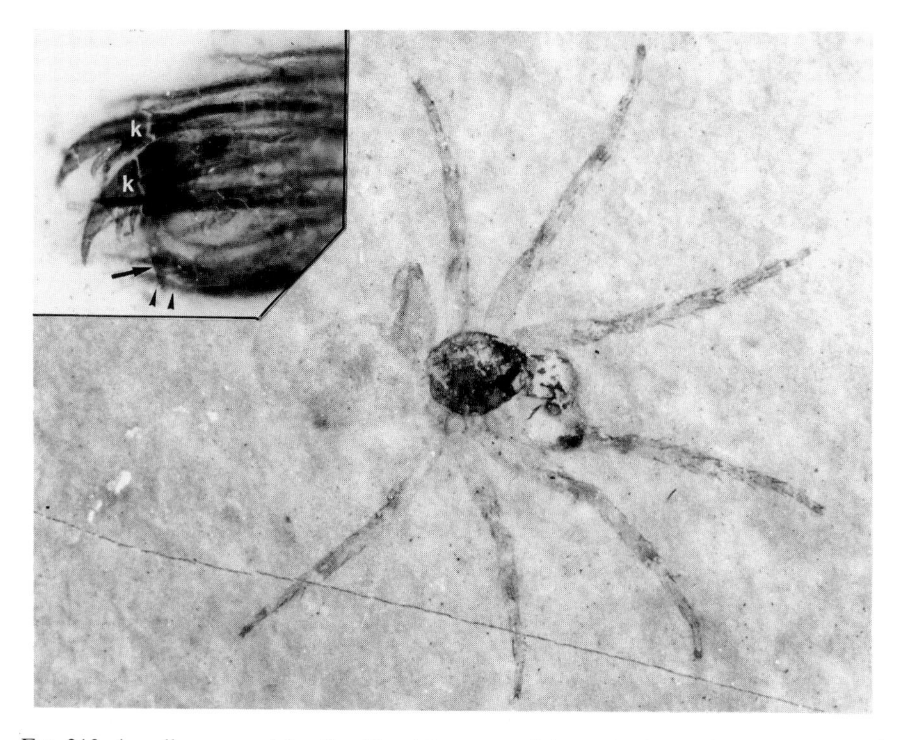

Fig. 213. A well-preserved fossil spider (*Cretaraneus*) from the lower Cretaceous period in Spain (130 million years). 15 ×. Inset: The tarsus has two combed main claws (k), a smooth middle hook (arrow), and serrated bristles (arrow heads), all of which are typical for orb weavers (cf. Fig. 17). 480 ×. (From Selden, P. A.: Orb-weaving spiders in the early cretaceous. Nature 340, 1989, p. 711.)

Cretaceaous period (about 100 million years) we know already orb weavers (Araneidae) with such typical features as three claws and serrated bristles on their tarsi (Fig. 213; Selden, 1989, 1990).

Evolutionary Trends

Because the paleontological record of fossilized spiders is scant, we must apply indirect methods to elucidate—at least partly—their phylogeny. Comparative studies of the spider's closest living relatives, the whip spiders and whip scorpions (Weygoldt and Paulus, 1979; Shultz, 1990), and of spider embryology are particularly suited for this purpose. A reconstruction of the Paleozoic archetype of a spider would exhibit the following characteristics (Fig. 212b; Bristowe, 1958; Savory, 1971):

- 6 prosomal and 12 abdominal segments consisting of tergites and sternites.
- Body segment 8 bearing the genital opening.
- Body segments 8 and 9 with respiratory organs.

- Body segments 10 and 11 carrying spinnerets.
- Prosoma and opisthosoma initially broadly joined together, later constricted at the pedicel.
- Prosoma consisting initially of 6 segments, later of a single plate (carapace).
- Chelicerae of 2 or 3 segments, formerly chelate (pincers).
- Palps resembling legs.

During the course of evolution this archetype acquired several new characteristics that can be explained primarily as adaptations to a terrestrial mode of life (Grasshoff, 1978). For instance, preoral digestion, typical of arachnids, is possible only on land. Concomitantly, more efficient mouthparts were needed and apparently the palpal coxae were best suited, and thus were preadapted, for that purpose. Also, the chelicerae may have proved to be more efficient tools than pincers, once their terminal segment was modified into a movable claw and the basal segment became mainly a site of muscle attachment. Further improvements were certainly the development of cheliceral poison glands and a second, postcerebral pump (the sucking stomach), within the intestinal tract. Other changes involved: (1) a constriction of the seventh body segment, which led to the formation of the typical pedicel; (2) the hydraulic extension of certain leg joints (see p. 22); (3) a direct transfer of sperm with modified palps instead of deposited spermatophores; (4) a concentration of all ganglia into a solid CNS within the prosoma; and finally, (5) the development and refinement of the spinning apparatus for use in prey catching. In general, spiders probably became smaller and lived shorter lives, while their entire behavioral repertoire became more complex.

Classification

A "natural system" of animal classification should mirror the relationships among different species. Since direct kinships of spiders can not be proved, because of the lack of any continuous fossil records, spider systematics is based almost entirely on comparative morphology. The procedure commonly used is to regard certain characters as primitive (*plesiomorphic*) and others as derived (*apomorphic*). For instance, the orthognath position of the chelicerae is considered plesiomorphic, whereas the labidognath position represents an apomorphic state (although recently both orthognath and labidognath have been considered as derived from an older *plagiognath* state; Kraus and Kraus, 1993). If within two different groups the same (homologous) character has been changed in the same way, it is referred to as synapomorphic, and a direct interrelationship between the groups ("sister groups") is deduced. This mode of classification implies that the direction, or polarity, of the evolutionary pathway is known.

In practice, however, it is not always simple to decide whether a certain character state is primitive or derived. One possible method of deciding is to assume that a character is apomorphic if it represents an adaptive improvement of a basic construction. Since evolution is probably directed toward an improvement of efficiency, any changes of a phylogenetic character may be thought of as a part of a continuous process of adaptation. Thus the polarity of a character transformation can be determined, and—at least hypothetically—the evolutionary pathway can be reconstructed (Grasshoff, 1975). To date, all spider systems are based on similarities, and depending on which and how many characters are being considered—and how they are appraised—taxonomists arrive at widely different classifications.

From the viewpoint of a cladistic analysis the most important synapomorphies of the Araneae (which would indicate their monophyly) are the following (Coddington and Levi, 1991): (1) abdominal extremities transformed into spinnerets; (2) abdominally located spinning glands; (3) chelicerae with poison glands; (4) male palp modified for sperm transfer; and (5) loss of abdominal segmentation, except in Mesothelae and some Mygalomorphae (yet even there the segmentation is only external).

Suborders

Most of the present classifications of spiders are derived from the system of the French arachnologist Eugène Simon, and a modified form of this system has been presented in chapter 1. Systematicists already disagree about the various suborders, especially about the position of the Mesothelae (Liphistiidae). Although some authors (Millot, 1949; Glatz, 1973) group them with the "tarantulas" into one suborder Orthognatha, others assign the Mesothelae a much more isolated position (Platnick and Gertsch, 1976): they are thought to be a sister group of all other recent spiders, which are then classified as the suborder Opisthothelae. Similarities that no doubt exist between Liphistiidae and some orthognath families (such as an abdominal segmentation in the Antrodiaetidae) are not considered synapomorphic but symplesiomorphic, and thus do not indicate direct relationships. In most current systems of classification, the order Araneae (Araneida) is divided into three suborders of equal rank, namely the Mesothelae, the Mygalomorphae (Orthognatha) and the Araneomorphae (Labidognatha).

The Mesothelae (40 species in two genera, or one family, resp.) have several ancient ("primitive") characters such as a segmented abdomen and four pairs of ventrally located spinnerets. However, the male palps are rather complex (Kraus, 1978; Haupt 1979) and therefore the question arises whether this is a plesiomorphic character for all spiders or whether this was a single modification that happened only in the Mesothelae (*autapomorphy*).

The Mygalomorphae comprise over 2,200 species in 260 genera, or 15 families, respectively. The anterior median spinnerets—which are still present in the Mesothelae—are completely lacking, and the anterior lateral spinnerets are strongly reduced. Raven (1985) offered a cladogram for all mygalomorphs based on 39 characters (Fig. 214). He distinguishes two groups, namely the Tuberculotae (with an oblique thoracic region and eyes raised on tubercles) and the Fornicephalae (with an arched head region). Among the latter group, the Atypidae and Antrodiaetidae are marked by the common trait "very few tarsal trichobothria," which seems to be an apomorphy. All the other families of the Fornicephalae share the following common characters: a rastellum (a reinforced part of the chelicera adapted for digging), and tubular burrows with trapdoors (Fig. 141).

The Araneomorphae refer to all "true spiders", i.e., 32,000 species in 2,700 genera or 90 families, respectively. They contain both cribellate and ecribellate spiders (see details below on the former classification in Cribellatae and Ecribellatae). The cribellate families always have rather fewer species in comparison to their ecribellate sister groups. Among the Haplogynae (see below) there are only about 90 cribellate species (Filistatidae) yet more than 2,400 ecribellate species; likewise, among the orb weavers, we find about 300 cribellate species (Uloboridae, Deinopidae) but more than 10,000 ecribellate Araneoidea.

The largest systematic unit within the Araneomorphae, the Araneoclada, shares several synapomorphies, the most important being the transition of the posterior book lungs into tracheae. The Araneoclada can be divided into a

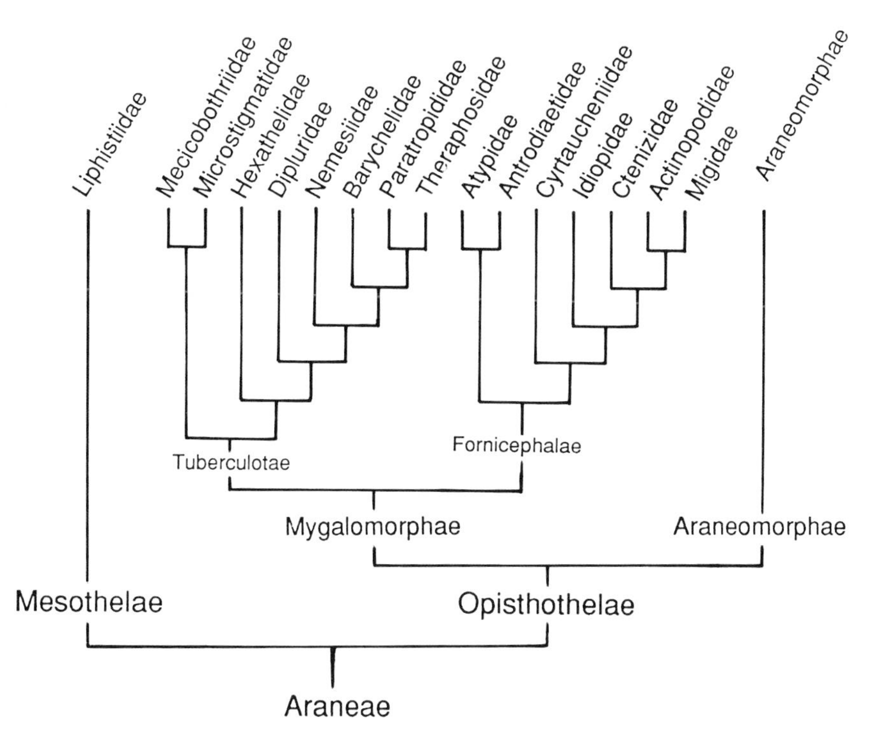

FIG. 214. Cladogram of the monophyletic main groups of spiders: Mesothelae, Mygalomorphae, Araneomorphae. Only the families of the Mygalomorphae are listed here; the much larger infraorder Araneomorphae is shown in Fig. 215. (After Raven, 1985; Coddington and Levi, 1991.)

small group, the Haplogynae, and a much larger group, the Entelegynae. Haplogynae are defined by their simple female genital organs, mainly the lack of fertilization ducts. For quite some time they were regarded as an artificial group (Cooke, 1970; Brignoli, 1975, 1978; Lehtinen, 1975), but recent cladistic studies tend to re-establish the "classic" Haplogynae as a monophyletic group (Platnick et al., 1991). Entelegynae (about 70 families) have complex female genital organs with paired fertilization ducts that end in a sclerotized platelet, the epigyne (Figs 156, 216). The former subdivision of the Entelegynae in Dionycha (two claws) and Trionycha (three claws), however, can no longer be maintained, because even within the same family some genera may have two claws, and others three. The most likely explanation is that the Dionycha evolved from Trionycha by reducing the third (middle) claw. It is noteworthy that in some spiders (*Ctenus, Phoneutria*) the nymphs exhibit three claws whereas the adults have only two (Homann, 1971). The third claw was probably convergently reduced in several spider families during the transition from life in a web to free hunting. The third claw thus is not a useful taxonomic character, although it can be quite convenient for identification. Whether the Dionycha are monophyletic and how the sister groups should be arranged (Fig. 215) still seems debatable.

At the base of the Entelegynae one can separate some primitive, cribellate families (e.g., Oecobiidae, Eresidae, Hersiliidae) from the higher evolved families; the former have eyes with a simple reflecting tapetum (Homann, 1971), the latter have derived and complex tapeta as well as special spinning glands for building cocoons (Coddington, 1989). Only a few examples of the higher Entelegynae are cited here: the Lycosoidea comprise 10 families with 235 genera and 3,700 species (Griswold, 1993). Their classic synapomorphy is a grate-shaped tapetum in the secondary eyes (Homann, 1971). The large group of orb weavers (Orbiculariae: Araneoidea and Deinopoidea) have been mentioned previously (cf. p. 145). New interpretations, based on cladistic analysis, consider the orb web as primitive and cribellate (Coddington, 1986, 1989; Coddington and Levi, 1991). Only in a few cribellate representatives (mainly in Uloboridae) did the plesiomorph characters persist (cribellum/calamistrum, tracheae, distribution of trichobothria), while the cribellum disappeared in the sister group of the "higher" Araneoidea. In some families (Linyphiidae, Theridiidae) it must be assumed that even the orb web disappeared. In other words, their sheet and frame webs would not be precursors of an orb web, as was postulated by Kullmann et al. (1971/72; Fig. 123), but would be derived from orb webs (see below).

The Araneoidea make up about one-third of all spiders (10,000 species in 740 genera or 11 families, respectively). Their most important synapomorphy is the elastic sticky thread—although, in some families a secondary loss of this thread must be postulated. The position of the Linyphiidae and of the Theridiidae among the Araneoidea is still somewhat uncertain; both are considered to be highly derived groups. Compared to former spider classifications several significant changes can be noted in Fig. 215, for instance the separation of the Tetragnathidae and Theridiosomatidae from the Araneidae.

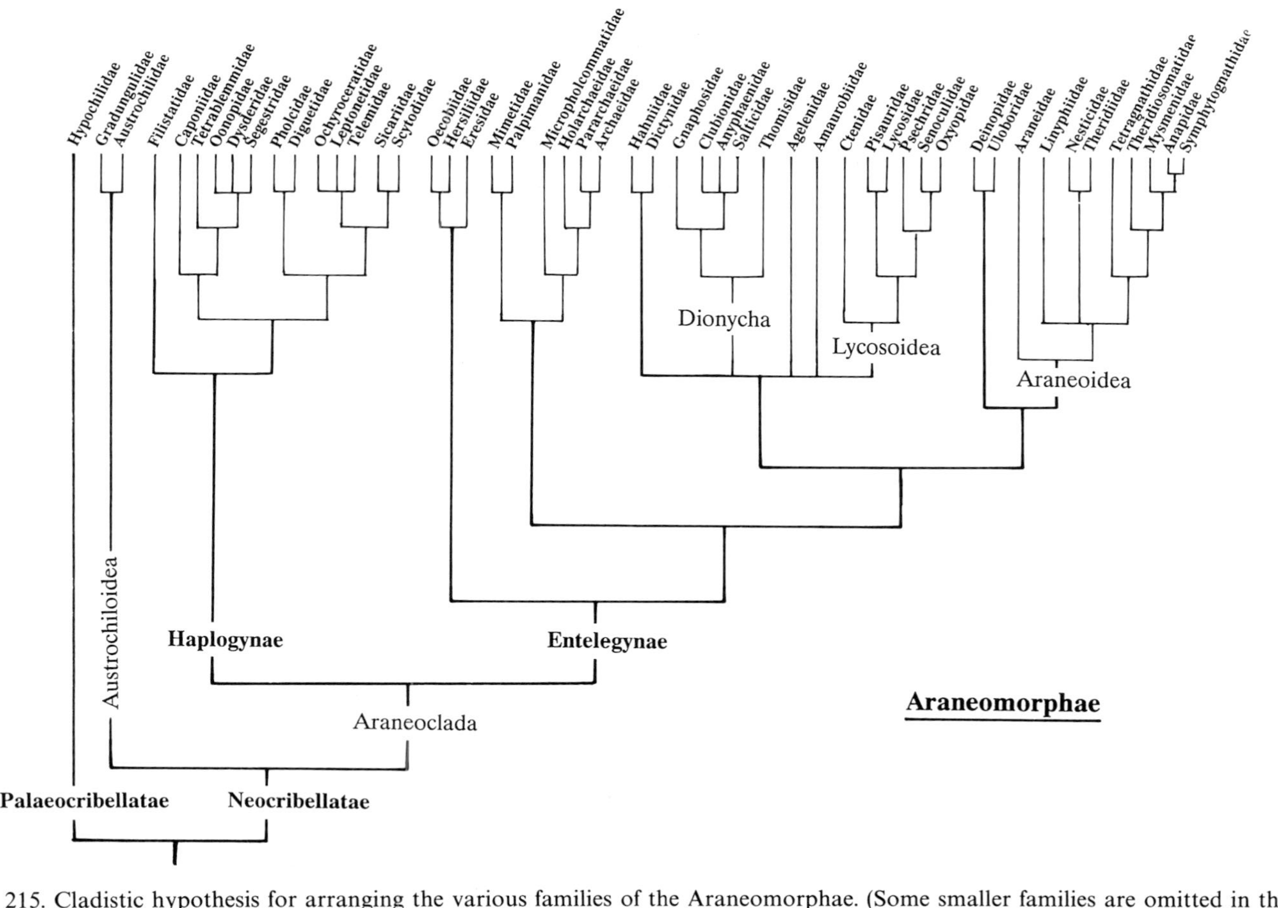

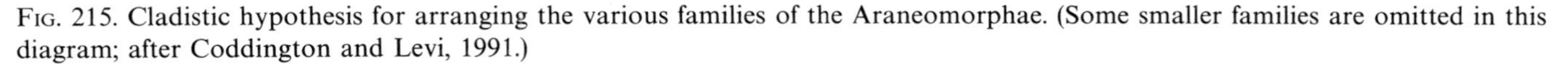
Fig. 215. Cladistic hypothesis for arranging the various families of the Araneomorphae. (Some smaller families are omitted in this diagram; after Coddington and Levi, 1991.)

The Cribellate Problem

The idea that all orb weavers (Orbiculariae) are monophyletic is rather new and is not accepted by all arachnologists. Since it is a basic issue in arachnology, it seems appropriate to discuss this "cribellate problem" briefly here.

Formerly the labidognath spiders (Araneomorphae) were divided into two suborders, the Cribellatae and the Ecribellatae, yet this classification has been questioned repeatedly (Lehtinen, 1967; Platnick, 1977). In general, cribellate forms are considered primitive (plesiomorphic), because the cribellum corresponds to the anterior median spinnerets of the Mesothelae. The ecribellate forms lack both a cribellum and anterior median spinnerets; instead they may possess a vestigial structure with no apparent function, the colulus. Based on embryological studies, however, anterior median spinnerets, cribellum, and colulus are believed to be homologous structures.

Now, what relationships exist between the Cribellatae and the Ecribellatae? Did they evolve independently from a common ancestral group? Or is one derived from the other? Is there any fundamental difference at all between cribellate and ecribellate spiders?

Most arachnologists agree that cribellate spiders are monophyletic, that is, that they are all descendants of a common ancestor. It is argued to be highly improbable that a complex character combination like the cribellum–calamistrum could have evolved independently several times. However, many ecribellate spiders hardly differ from some cribellate forms except that they lack such a cribellum–calamistrum complex. One could deduce then that ecribellates may have evolved from cribellates by having lost their cribellum and calamistrum. To put it pointedly, ecribellates are in fact cribellate spiders, which have simply lost their cribellum during their evolution. But have all ecribellate spiders passed through a cribellate stage? If so, such a transition would have taken place more than 30 times, independent of each other, during the spider's phylogeny. An alternative explanation would assume that, through convergent evolution, similar spider families developed among the cribellates and the ecribellates—comparable to the parallel evolution of marsupials and the other mammals.

Both interpretations may be valid. Transformations from a cribellate to an ecribellate form probably occurred between the Oecobiidae and the Urocteidae (Kullmann and Zimmermann, 1976)—or from the Psechridae to the Lycosidae (Homann, 1971). In contrast, the similarities between the ecribellate orb weavers Araneidae and cribellate Uloboridae could be the result of convergence (Kullmann et al., 1971/72). Superficially the orb webs of both are strikingly similar, but a closer investigation reveals distinctly different catching threads (araneids use glue, whereas uloborids use cribellate wool) that originate from different, nonhomologous silk glands (Kullmann, 1972a; H. M. Peters, 1984, 1987). It is thus difficult to imagine that the araneid orb web evolved from a uloborid orb; more likely, both webs could be merely analogous snares (Kullmann et al., 1971/72).

Family Relationships

The number, position, relative size, and internal structure of the eyes are very important characters for classifying spider families. In ancient spiders, all eight eyes were probably the same size but then developed differently during their phylogeny. The lateral eyes, which can be separated into three groups according to the type of tapetum present, are systematically quite useful, since their synapomorphies can be established rather well (Homann, 1971).

Probably all complex lateral eyes with a grated tapetum are of monophyletic

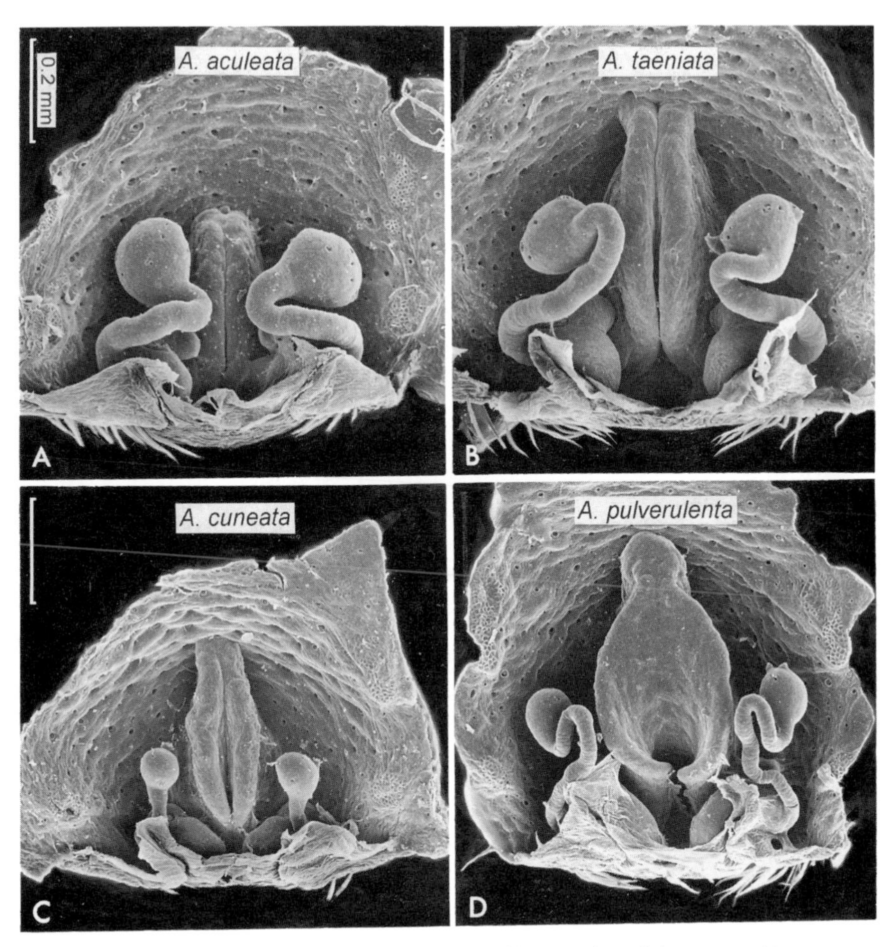

FIG. 216. Epigynes of four closely related wolf spider species of the genus *Alopecosa* (dorsal view, cf. Figs 156, 157). *A. aculeata* (A) and *A. taeniata* (B) show only minor differences in the structure of their epigynes, yet differ distinctly in their courtship behavior. (From Kronestedt, 1990; Copyright © 1990 by The Norwegian Academy of Science and Letters.)

origin; at the same time they represent an apomorphic character. This means that all spiders with a grated tapetum are related to one another, for example, the Psechridae, Zoropsidae, Lycosidae, Oxyopidae, and Senoculidae. It seems noteworthy that these families contain both cribellate and ecribellate forms. As mentioned previously, the Lycosidae are presumably derived from the Psechridae by loss of the cribellum. Like the reduction of the third claw, the loss of the cribellum may be correlated with the transition from life within a web (Psechridae) to free hunting (Lycosidae).

Whereas genital structures are used mainly for the separation of species (Fig. 216), the classification of families relies more on the structure of the spinnerets, chelicerae, tarsal claws, and the labium. The divergence in spider classification is caused less by the different opinions of systematicists about the observable "facts" than by their interpretation of them. At present many "classical" families are radically rearranged and may even group cribellate and ecribellate forms within the same genus (Lehtinen, 1967; Forster and Wilton, 1973).

Instead of the classical pedigree of spider families (Bristowe, 1938), a modern spider phylogeny is presented here based on cladograms (Coddington and Levi, 1991; Fig. 215). Although many data and standardized methods have been used to compile it, we can not expect this to be the final word on spider phylogeny. Nonetheless, we should be glad that Coddington and Levi had the courage to update the systematic relationship of more than 100 spider families. Critical counter proposals can be anticipated in the near future.

Bibliography

Adams, M. E., E. E. Herold, V. J. Venema: Two classes of channel-specific toxins from funnel web spider venom. J. comp. Physiol. A 164 (1989) 333

Aitchison, C. W.: Spiders active under snow in southern Canada. Symp. Zool. Soc. London 42 (1978) 139

Aitchison, C. W.: Low temperature feeding by winter-active spiders. J. Arachnol. 12 (1984) 297

Aitchison, C. W.: Feeding ecology of winter-active spiders. In Nentwig, W., ed.: Ecophysiology of Spiders. Springer Verlag, Berlin 1987 (p. 264)

Alberti, G.: Comparative spermatology of Araneae. Acta Zool. Fennica 190 (1990) 17

Alberti, G., C. Weinmann: Fine structure of spermatozoa of some labidognath spiders (Filistatidae, Segestriidae, Dysderidae, Oonopidae, Scytodidae, Pholcidae; Araneae; Arachnida) with remarks on spermiogenesis. J. Morphol. 185 (1985) 1

Alberti, G., B. A. Afzelius, S. U. Lucas: Ultrastructure of spermatozoa and spermatogenesis in bird spiders (Theraphosidae, Mygalomorphae, Araneae). J. submicr. Cytol. 18 (1986) 739

Anava, A., Y. Lubin: Presence of gender cues in the web of a widow spider, *Latrodectus revivensis*, and a description of courtship behaviour. Bull Brit. arachnol. Soc. 9 (1993) 119

Anderson, J. F.: Metabolic rates of spiders. Comp. Biochem. Physiol. 33 (1970) 51

Anderson, J. F.: Responses to starvation in the spiders *Lycosa lenta* Hentz and *Filistata hibernalis* (Hentz). Ecology 55 (1974) 576

Anderson, J. F.: Energy content of spider eggs. Oecologia 37 (1978) 41

Anderson, J. F., K. N. Prestwich: The fluid pressure pumps of spiders (Chelicerata Araneae). Z. Morph. Tiere 81 (1975) 257

Anderson, J. F., K. N. Prestwich: The physiology of exercise at and above maximal aerobic capacity in a theraphosid (tarantula) spider, *Brachypelma smithii* (F. O. Pickard-Cambridge). J. comp. Physiol. B 155 (1985) 529

André, J., C. Rouiller: The ultrastructure of the vitelline body in the oocyte of the spider *Tegenaria parietina*. J. biophys. biochem. Cytol. 3 (1957) 977

Angersbach, D.: Oxygen pressures in haemolymph and various tissues of the tarantula *Eurypelma helluo*. J. comp. Physiol. 98 (1975) 133

Angersbach, D.: Oxygen transport in the blood of the tarantula *Eurypelma californicum:* PO2 and pH during rest, activity and recovery. J. comp. Physiol. 123 (1978) 113

Anton, S., F.G. Barth: Central nervous projection pattern of trichobothria and other cuticular sensilla in the wandering spider *Cupiennius salei* (Arachnida, Araneae). Zoomorph. 113 (1993) 21

Askenmo, C., A. von Broemssen, J. Eckman, C. Jansson: Impact of some wintering birds on spider abundance in spruce. Oikos 28 (1977) 90

Aspey, W. P.: Agonistic behaviour and dominance–subordinance relationships in the wolf spider *Schizocosa crassipes*. Proc. 6th Int. Arachnol. Congr. 1974. Amsterdam 1975 (p. 102)

Austadt, S. N.: Evolution of sperm priority patterns in spiders. In: Smith, R. L., ed.: Sperm Competition in the Evolution of Animal Mating Systems. Academic Press, New York 1984 (p. 223)

Avilés, L.: Sex ratio bias and possible group selection in the social spider *Anelosimus eximius*. Amer. Nat. 128 (1986) 1

Avilés, L.: Newly-discovered sociality in the neotropical spider *Aebutina binotata* Simon (Dictynidae?). J. Arachnol. 21 (1993) 184

Avilés, L: Social behaviour in a web-building lynx spider, *Tapinillus* sp. (Araneae: Oxyopidae). Biol. J. Linn. Soc. 51 (1994) 163

Avilés, L: Causes and consequences of cooperation and permanent sociality in spiders. In: Choe, J., B. Crespi, eds: Social Competition and Cooperation in Insects and Arachnids. II. Evolution of sociality. Princeton University Press, New Jersey 1995 (in press)

Avilés, L., W. Maddison: When is the sex ratio biased in social spiders? Chromosome studies of embryos and male meiosis in *Anelosimus* species (Araneae: Theridiidae). J. Arachnol. 19 (1991) 126

Babu, K. S.: Anatomy of the central nervous system of arachnids. Zool. Jb. Anat. 82 (1965) 1

Babu, K. S.: Certain histological and anatomical features of the central nervous system of a large Indian spider *Poecilotheria*. Amer. Zool. 9 (1969) 113

Babu, K. S.: Histology of the neurosecretory system and neurohaemal organs of the spider *Argiope aurantia* (Lucas). J. Morphol. 141 (1973) 77

Babu, K. S.: Postembryonic development of the central nervous system of the spider *Argiope aurantia* (Lucas). J. Morphol. 146 (1975) 325

Babu, K. S., F. G. Barth: Neuroanatomy of the central nervous system of the wandering spider *Cupiennius salei* (Arachnida, Araneida). Zoomorph. 104 (1984) 344

Babu, K. S., F. G. Barth, N. J. Strausfeld: Intersegmental sensory tracts and contralateral motor neurons in the leg ganglia of the spider *Cupiennius salei* Keys. Cell Tissue Res. 241 (1985) 53

Baccetti, B., R. Dallai, F. Rosati: The spermatozoon of arthropoda. VIII. The 9+3 flagellum of spider sperm cells. J. Cell Biol. 44 (1970) 681

Bachmann, M.: Das Gift der orthognathen Spinne *Pterinochilus* spec. Isolierung und teilweise biochemische und biologische Charakterisierung eines Neurotoxins und einer Hyaluronidase. Doctoral thesis, University of Basel, 1976

Baerg, W. J.: The life cycle and mating habits of the male tarantula. Quart. Rev. Biol. 3 (1928) 109

Balogh, J. I.: Vorläufige Mitteilungen über radnetzbauende Pachygnathen. Fol. Zool. Hydrobiol. 6 (1934) 94

Baltzer, F.: Ueber die Orientierung der Trichterspinne *Agelena labyrinthica* (Cl.) nach der Spannung des Netzes. Rev. Suiss. Zool. 37 (1930) 363

Barnes, R. D., B. M. Barnes: The spider population of the abstract bromsedge community of the southeastern piedmont. Ecology 36 (1955) 658

Barnes, W.J.P., F.G. Barth: Sensory control of locomotor mode in semi-aquatic spiders. In: Armstrong, D.M., B.M.H. Bush, eds: Locomotor Neural Mechanisms in Arthropods and Vertebrates. Manchester University Press, Manchester, New York 1991 (p. 105)

Barrows, W. M.: The reactions of an orbweaving spider, *Epeira sclopetaria* Clerck, to rhythmic vibrations of its web. Biol. Bull. 29 (1915) 316

Bartels, M.: Sinnesphysiologische und psychologische Untersuchungen an der Trichterspinne *Agelena labyrinthica* (Cl.). Z. vergl. Physiol. 10 (1929) 527

Bartels, M.: Ueber den Fressmechanismus und chemischen Sinn einiger Netzspinnen. Rev. Suisse Zool. 37 (1930) 1

Bartels, M., F. Baltzer: Ueber Orientierung und Gedächtnis der Netzspinne *Agelena labyrinthica.* Rev. Suiss. Zool. 35 (1928) 247

Barth, F. G.: Ein einzelnes Spaltsinnesorgan auf dem Spinnentarsus: Seine Erregung in Abhängigkeit von den Parametern des Luftschallreizes. Z. vergl. Physiol. 55 (1967) 407

Barth, F. G.: Die Feinstruktur des Spinnenintegumentes. 1. Die Cuticula des Laufbeins adulter häutungsferner Tiere (*Cupiennius salei* Keys.). Z. Zellforsch. 97 (1969) 137

Barth, F. G.: Die Feinstruktur des Spinnenintegumentes. II. Die räumliche Anordnung der Mikrofasern in der lamellierten Cuticula und ihre Beziehung zur Gestalt der Porenkanäle (*Cupiennius salei* Keys., adult, häutungsfern, Tarsus). Z. Zellforsch. 104 (1970) 87

Barth, F. G.: Der sensorische Apparat der Spaltsinnesorgane (*Cupiennius salei* Keys., Araneae). Z. Zellforsch. 112 (1971) 212

Barth, F. G.: Die Physiologie der Spaltsinnesorgane. 1. Modellversuche zur Rolle des cuticulären Spaltes beim Reiztransport. J. comp. Physiol. 78 (1972a) 315

Barth. F. G.: Die Physiologie der Spaltsinnesorgane. II. Funktionelle Morphologie eines Mechanoreceptors. J. comp. Physiol. 81 (1972b) 159

Barth, F. G.: Microfiber reinforcement of an arthropod cuticule. Laminated composite material in biology. Z. Zellforsch. 144 (1973a) 409

Barth, F. G.: Bauprinzipien und adäquater Reiz bei einem Mechanoreceptor. Verh. dtsch. zool. Ges. 66 (1973b) 25

Barth, F. G.: Sensory information from strains in the exoskeleton. In: Hepburn, H. R., ed.: The Insect Integument, Elsevier, Amsterdam 1976 (p. 445)

Barth, F. G.: Spiders and vibratory signals: sensory reception and behavioral significance. In: Witt, P. N., J. S. Rovner, eds: Spider Communication, Mechanisms and Ecological Significance. Princeton University Press, Princeton, New Jersey 1982 (p. 67)

Barth, F. G.: Slit sensilla and the measurement of cuticular strains. In: Barth, F. G., ed.: Neurobiology of Arachnids. Springer Verlag, Berlin 1985a (p. 162)

Barth, F. G.: Neuroethology of the spider vibration sense. In: Barth, F. G., ed.: Neurobiology of Arachnids. Springer Verlag, Berlin 1985b (p. 203)

Barth, F. G.: Vibrationssinn und vibratorische Umwelt von Spinnen. Naturwiss. 73 (1986) 519

Barth, F.G.: Sensory guidance in spider pre-copulatory behaviour. Comp. Biochem. Physiol. 104A (1993) 717

Barth, F. G., W. Libera: Ein Atlas der Spaltsinnesorgane von *Cupiennius salei* Keys. Chelicerata (Araneae). Z. Morph. Tiere 68 (1970) 343

Barth, F. G., P. Pickelmann: Lyriform slit sense organs. Modelling an arthropod mechanoreceptor. J. comp. Physiol. 103 (1975) 39

Barth, F. G., J. Stagl: The slit sense organs of arachnids. A comparative study of their topography on the walking legs. Zoomorphology 86 (1976) 1

Barth, F. G., H. Bleckmann, J. Bohnenberger, E.-A. Seyfarth: Spiders of the genus *Cupiennius* Simon 1891 (Araneae, Ctenidae). II. On the vibratory environment of a wandering spider. Oecologia 77 (1988) 194

Barth, F. G., S. Komarek, J. A. C. Humphrey, B. Treidler: Drop and swing dispersal behavior of a tropical wandering spider: experiments and numerical model. J. comp. Physiol., A 169 (1991) 313

Barth, F.G., U. Wastl, J.A.C. Humphrey, R. Devarakonda: Dynamics of arthropod filiform hairs. II. Mechanical properties of spider trichobothria (*Cupiennius salei* Keys.) Phil. Trans. R. Soc. Lond. B 340 (1993) 445

Baschwitz, G. G.: Sobre la ultraestructura del endosternito de *Araneus diadematus* Clerck. Rev. Fac. Cienc. Oviedo 14 (1973) 37

Baschwitz, G. G.: Sobre las conexiones musculares de los aracnidos. Rev. Fac. Cienc. Oviedo 15 (1974) 43

Bauer, K.-H.: Funktionsmechanismus der Autotomie bei Spinnen (Araneae) und seine morphologischen Voraussetzungen. Z. Morph. Tiere 72 (1972) 173

Bays, S. M.: A study on the training possibilities of *Araneus diadematus* Cl. Experientia (Basel) 18 (1962) 423

Bell, A. L., D. B. Peakall: Changes in the fine structure during silk protein production in the ampullate gland of the spider *Araneus sericatus*. J. Cell Biol. 42 (1969) 284

Berland, L.: Les Arachnides. In: Encyclopédie Entomologique. Lechevalier, Paris 1932 (p. 79)

Bernheimer, A. W., B. J. Campbell, L. W. Forster: Comparative toxinology of *Loxosceles reclusa* and *Corynebacterium pseudotuberculosis*. Science 228 (1985) 590

Bertkau, P.: Über die Übertragungsorgane und die Spermatozoen der Spinnen. Verh. Naturh. Ver. Preuss. Rheinl. 34 (1877) 28

Bertkau, P.: Über den Verdauungsapparat der Spinnen. Arch. mikr. Anat. 24 (1885) 398

Bihlmayer, S.: Zur Herz- und Kreislauffunktion der Vogelspinne *Eurypelma californicum*. Ph.D. thesis, University of Munich (1991)

Bihlmayer, S., S. Zahler, R. Paul: Zur Struktur und Funktion des Kreislaufsystems der Vogelspinne *Eurypelma californicum*. Verh. dtsch. Zool. Ges. 82 (1989) 223

Bihlmayer, S., S. Zahler, R. Werner, R. Paul: Morphological and physiological studies on the circulatory system of the tarantula *Eurypelma californicum*. Int. Symp.: Mechanisms of Systemic Regulation in Lower Vertebrates (1990) 8

Blanke, R.: Untersuchungen zur Oekophysiologie und Oekethologie von *Cyrtophora citricola* Forskal (Araneae, Araneidae) in Andalusien. Forma et Functio 5 (1972) 125

Blanke, R.: Neue Ergebnisse zum Sexualverhalten von *Araneus cucurbitinus* Cl. (Araneae, Araneidae). Forma et Functio 6 (1973a) 279

Blanke, R.: Nachweis von Pheromonen bei Netzspinnen. Naturwissenschaften 10 (1973b) 481

Blanke, R.: Rolle der Beute beim Kopulationsverhalten von *Meta segmentata* (Cl.) (Araneae, Araneidae). Forma et Functio 7 (1974a) 83

Blanke, R.: Der Zusammenhang zwischen Beuteangebot und Reproduktionsrate bei *Cyrtophora citricola* Forskal (Araneae, Araneidae). Beitr. naturk. Forsch. Südw.-Dtl. 33 (1974b) 223

Blanke, R.: Die Bedeutung der Guanocyten für den physiologischen Farbwechsel bei *Cyrtophora cicatrosa* (Arachnida: Araneidae). Ent. Germ. 2 (1975a)

Blanke, R.: Untersuchungen zum Sexualverhalten von *Cyrtophora cicatrosa* (Stoliczka) (Araneae, Araneidae). Z. Tierpsychol. 37 (1975b) 62

Bleckmann, H.: Die Reaktion der Raubspinne *Dolomedes triton* auf Oberflächenwellen des Wassers. Verh. dtsch. zool. Ges. 1982 (p. 253)

Bleckmann, H., F. G. Barth: Sensory ecology of a semi-aquatic spider (*Dolomedes triton*) II. The release of predatory behavior by surface waves. Behav. Ecol. Sociobiol. 14 (1984) 303

Bleckmann, H., J. S. Rovner: Sensory ecology of a semi-aquatic spider (*Dolomedes triton*). I. Roles of vegetation and wind-generated waves in site selection. Behav. Ecol. Sociobiol. 14 (1984) 297

Blest, A. D.:The fine structure of spider photoreceptors in relation to function. In: Barth, F. G., ed.: Neurobiology of Arachnids. Springer Verlag, Berlin 1985 (p. 89)

Blest, A. D., M. F. Land: The physiological optics of *Dinopis subrufus* L. Koch: a fish-lens in a spider. Proc. Roy. Soc. B 196 (1977) 197

Blest, A. D., G. Pomeroy: The sexual behaviour and genital mechanics of three species of *Mynoglenes* (Araneae: Linyphiidae). J. Zool. Lond. 185 (1978) 319

Blest, A. D., H. H. Taylor: The clypeal glands of *Mynoglenes* and of some other Linyphiid spiders. J. Zool. Lond. 183 (1977) 473

Blumenthal, H.: Untersuchungen über das "Tarsalorgan" der Spinnen. Z. Morph. Oekol. Tiere 29 (1935) 667

Boissin, L.: Etude ultrastructurale de la spermiogénèse de *Meta bourneti* Simon (Arachnides, Aranéides, Metinae). 2ième Réun. Arachnol. Langue Franc., Montpellier 1973 (p. 7)

Bonaric, J. C.: Effects of ecdysterone on the molting mechanisms and duration of the intermolt period in *Pisaura mirabilis*. Gen. comp. Endocr. 30 (1976) 267

Bonaric, J. C.: Contribution a l'étude de la biologie du développement chez l'araignée *Pisaura mirabilis* (Clerck 1758). Approche physiologique des phénomènes de mue et de diapause hivernale. Thèse Etat Universitaire, Montpellier (1980)

Bonaric, J. C., M. de Reggi: Changes in ecdysone levels in the spider *Pisaura mirabilis* nymphs (Araneae, Pisauridae). Experientia 33 (1977) 1664

Bonnet, P.: La mue, l'autotomie et la régéneration chez les Araignées, avec une étude des Dolomèdes d'Europe. Bull. Soc. Hist. Nat. Toulouse 59 (1930) 237

Bösenberg, H.: Beiträge zur Kenntnis der Spermatogenese bei den Arachnoiden. Zool. Jb. Anat. 21 (1905) 505

Boys, C. V.: The influence of a tuning fork on the garden spider. Nature (Lond.) 23 (1881) 149

Brach, V.: The biology of the social spider *Anelosimus eximius* (Araneae: Theridiidae). Bull. 5th Calif. Acad. Sci. 74 (1975) 37

Brach, V.: *Anelosimus eximius* (Araneae: Theridiidae) and the evolution of quasisociality in theridiid spiders. Evolution 31 (1977) 154

Brady, A. R.: The spider genus *Sossipus* in North America, Mexico, and Central America (Araneae, Lycosidae). Psyche 69 (1962) 129

Brandwood, A.: Mechanical properties and factors of safety of spider drag-lines. J. exp. Biol. 116 (1985) 141

Braun, F.: Beiträge zur Biologie und Atmungsphysiologie der *Argyroneta aquatica* Cl. Zool. Jb. Syst. 62 (1931) 175

Braunitzer, C., D. Wolff: Vergleichende chemische Untersuchungen über die Fibroine von *Bombyx mori* und *Nephila madagascariensis*. Z. Naturforsch. 10b (1955) 404

Breene, R. G., M. H. Sweet: Evidence of insemination of multiple females by the black widow spider *Latrodectus mactans* (Araneae, Theridiidae). J. Arachnol. 13 (1985) 331

Brenner, H. R.: Evidence for peripheral inhibition in an arachnid muscle. J. comp. Physiol. 80 (1972) 227

Breymeyer, A.: Preliminary data for estimating the biological production of wandering spiders In: Petrusewicz, K., ed:: Secondary Productivity in Terrestrial Ecosystems, Vol. 2. Paustwowe Wydawnictwo Naukowe, Warsaw 1967 (p. 821)

Breymeyer, A., J. Jozwik: Consumption of wandering spiders (Lycosidae, Araneae) estimated in laboratory conditions. Bull. Acad. Polon. Sci. Sér. Sci. Biol. 23 (1975) 93

Brignoli, P. M.: Über die Gruppe der Haplogynae (Araneae). Proc. 6th Int. Arachnol. Congr. 1974, Free University of Amsterdam 1975 (p. 33)

Brignoli, P. M.: Some remarks on the relationships between the Haplogynae, the semi-Entelegynae and the Cribellatae (Araneae). Symp. Zool. Soc. Lond. 42 (1978) 285

Bristowe, W. S.: The mating habits of spiders, with special reference to the problems surrounding sex dimorphisms. Proc. Zool Soc. Lond. (1929) 309

Bristowe, W. S.: A preliminary note on the spiders of Krakatau. Proc. zool. Soc. Lond. (1931) 1387

Bristowe, W. S.: The liphistiid spiders, with an appendix on their internal anatomy by J. Millot. Proc. Zool. Soc. Lond. 103 (1932) 1015

Bristowe, W. S.: Introductory Notes. In: Reimoser, F., ed.: The Spiders of Krakatau. Proc. zool. Soc. Lond. (1934) 11

Bristowe, W. S.: The classification of spiders. Proc. Zool. Soc. Lond. 108 (1938) 285

Bristowe, W. S.: The Comity of Spiders. Ray Soc. No. 126, London 1939

Bristowe, W. S.: The Comity of Spiders. Ray Soc. No. 128, London 1941

Bristowe, W. S.: The World of Spiders. Collins, London 1958

Bristowe, W. S.: Notes on rare spiders and courtships as a clue to relationships. Proc. 5th London Entomol. Nat. Hist. Soc (1962) 184

Bristowe, W. S.: Eine Familie lebender Fossilien. Endeavour 34 (1975) 115

Bristowe, W. S., G. H. Locket: The courtship of British lycosid spiders and its probable significance. Proc. Zool. Soc. Lond. (1926) 317

Bristowe, W. S., J. Millot: The Liphistiid spiders. Proc. Zool. Soc. Lond. 103 (1933) 1015

Bromhall, C.: Spider heart-rates and locomotion. J. comp. Physiol. B 157 (1987) 451

Brown, K. M.: Foraging ecology and niche partitioning in orb-weaving spiders. Oecologia 50 (1981) 380

Brown, S. C.: Mating behavior of the golden-orb-weaving spider, *Nephila clavipes*. II. Sperm capacitation, sperm competition, and fecundity. J. comp. Psychol. 99 (1985) 167

Browning H. C.: The integument and moult cycle of *Tegenaria atrica* (Araneae). Proc. Roy. Soc. B 131 (1942) 65

Brüssel, A., W. Gnatzy: A somatotopic organization of leg afferents in the spider *Cupiennius salei* Keys. (Araneae, Ctenizidae). Experientia 41 (1985) 468

Buchar, J.: A winter trip among spiders. Ziva 16 (1968) 24

Buche, W.: Beiträge zur Oekologie und Biologie winterreifer Kleinspinnen mit besonderer Berücksichtigung der Linyphiidae *Macrargus rufus rufus* (Wider), *Macrargus rufus carpenteri* (Cambridge) und *Centromerus silvaticus* (Blackwall). Z. Morph. Oekol. Tiere 57 (1966) 329

Bücherl, W.: Novo processo de obtencao de veneno sèco, puro, de *Phoneutria nigriventer* (Keyserling 1891) e titulacao da LD50 em camundongos. Mem. Inst. Butantan 25 (1953) 153

Bücherl, W.: Südamerikanische Spinnen und ihre Gifte. Arzneimittel-Forsch. 6 (1956) 293

Bücherl, W.: Spiders. In: Bücherl, W., E. Buckley, eds: Venomous Animals and their Venoms III. Academic Press, New York 1971 (p. 197)

Buchli, H. H. R.: Hunting behaviour in the Ctenidae. Amer. Zool. 9 (1969) 175

Burch, T. L.: The importance of communal experience to survival for spiderlings of *Araneus diadematus* (Araneae: Araneidae). J. Arachnol. 7 (1979) 1

Burgess, J. W.: Social behavior in groupliving spider species. Symp. zool. Soc. Lond. 42 (1978) 69

Burgess, J. W., G. W. Uetz: Social spacing strategies in spiders. In: Witt, P. N., J. S. Rovner, eds: Spider Communication, Mechanisms and Ecological Significance. Princeton University Press, Princeton, New Jersey 1982 (p. 317)

Bürgis, H.: Eine Spinne, die spuckt. Mikrokosmos 69 (1980) 342
Bürgis, H.: Die Speispinne *Scytodes thoracica* (Araneae: Sicariidae). Ein Beitrag zur Morphologie und Biologie. Mitt. Pollichia 77 (1990) 289
Buskirk, R. E.: Aggressive display and orb defense in a colonial spider, *Metabus gravidus*. Anim. Behav. 23 (1975a) 560
Buskirk, R. E.: Coloniality, activity patterns and feeding in a tropical orb-weaving spider. Ecology 56 (1975b) 1314
Buskirk, R. E.: Sociality in the Arachnida. In: Hermann, H. R., ed.: Social Insects, Vol. II. Academic Press, New York 1981 (p. 281)
Butt, A. G., H. H. Taylor: The function of spider coxal organs: effects of feeding and salt-loading on *Porrhothele antipodiana* (Mygalomorpha: Dipluridae). J. exp. Zool. 158 (1991) 439
Buxton, B. H.: Coxal glands of the arachnids. Zool. Jb. Anat. 37 (1913) 231
Caldwell, R.L.: A unique form of locomotion in a stomatopod—backward somersaulting. Nature (London) 282(1979) 71
Canard, A.: Données sur le développement, la croissance, le cycle biologique et l'évolution démographique de la Mygale (*Atypus affinis* Eichwald 1830) (Atypidae, Mygalomorpha). Mém. Soc. roy. belge Ent. 33 (1986) 47
Canard, A.: Analyse nouvelle du développement postembryonnaire des araignées. Rev. Arachnol. 7 (1987) 91
Canard, A., R. Stockman: Comparative postembryonic develpoment of Arachnids. Proc. 12th Int. Congr. Arachnol, Brisbane. Mem. Queensl. Mus. 33 (1993) 461
Carico, J. E.: Predatory behaviour in *Euryopis funebris* (Hentz) (Araneae: Theridiidae) and the evolutionary significance of web reduction. Symp. Zool. Soc. Lond. 42 (1978) 51
Carico, J. E.: Secondary use of the removed orb web of *Mecogynea lemniscata* (Walckenaer) (Araneae, Araneidae). J. Arachnol. 12 (1984) 357
Carico, J. E.: Web removal patterns in orb weaving spiders. In: Shear, W. A., ed.: Spiders. Webs, Behavior, and Evolution. Stanford University Press, Stanford 1986 (p. 306)
Carrel, J. E.: Behavioral thermoregulation during winter in an orb-weaving spider. Symp. Zool. Soc. Lond. 42 (1978) 41
Carrel, J. E., R. D. Heathcote: Heart rate in spiders: Influence of body size and foraging energetics. Science 193 (1976) 148
Christenson, T. E.: Behaviour of colonial and solitary spiders of the theridiid species *Anelosimus eximius*. Anim. Behav. 18 (1984) 725
Christian, U.: Zur Feinstruktur der Trichobothrien der Winkelspinne *Tegenaria derhami* (Scopoli), (Agelenidae, Araneae). Cytobiologie 4 (1971) 172
Christian, U. H.: Trichobothrien, ein Mechanorezeptor bei Spinnen. Elektronenmikroskopische Befunde bei der Winkelspinne *Tegenaria derhami* (Scopoli) (Agelenidae, Araneae). Verh. dtsch. zool Ges. 66 (1973) 31
Clark, A. W., W. P. Hurlbut, A. Mauro: Changes in the fine structure of the neuromuscular junction of the frog caused by Black Widow spider venom. J. Cell Biol. 52 (1972) 1
Clark, R. J., R. R. Jackson: Self recognition in a jumping spider: *Portia labiata* females discriminate between their own draglines and those of conspecifics. Ethol. Ecol. Evol. 6 (1994) 371
Clarke, J.: The comparative functional morphology of the leg joints and muscles of five spiders. Bull. Brit. Arachnol. Soc. 7 (1986) 37
Clarke, R. D., P.R. Grant: An experimental study of the role of spiders as predators in a forest litter community. Ecology 49 (1968) 1152

Cloudsley-Thompson, J. L.: Studies in diurnal rhythms. V. Nocturnal ecology and water-relations of the British cribellate spiders of the genus Ciniflo Bl. J. Linn. soc. Zool. 43 (1957) 134
Cloudsley-Thompson, J. L.: Adaptive functions of circadian rhythms. Cold Spring Harb. Symp. quant. Biol. 25 (1961) 345
Cloudsley-Thompson, J. L.: The water relations of scorpions and tarantulas from the Sonoran desert. Entomol. Month. Mag. 103 (1968) 217
Cloudsley-Thompson, J. L: Biological clocks in Arachnida. Bull. Brit. Arachnol. Soc. 4 (1978) 184
Cloudsley-Thompson, J. L.: A comparison of rhythmic locomotory activity in tropical forest arthropoda with that in desert species. J. Arid Environments 4 (1981) 327
Coddington: J. A.: Orb webs in "non-orb" weaving ogre-faced spiders (Araneae: Dinopidae): a question of genealogy. Cladistics 2 (1986) 53
Coddington, J. A.: Spinneret silk spigot morphology. Evidence for the monophyly of orb-weaving spiders, Cyrtophorinae (Araneidae), and the group Theridiidae-Nesticidae. J. Arachnol 17 (1989) 71
Coddington, J. A.: Cladistics and spider classification; araneomorph phylogeny and the monophyly of orbweavers (Araneae: Araneaomorphae; Orbiculariae). Acta Zool. Fennica 190 (1990) 75
Coddington, J. A., H. W. Levi: Systematics and evolution of spiders. Ann. Rev. Ecol. Syst. 22 (1991) 565
Coddington, J., C. Sobrevila: Web manipulation and two stereotyped attack behaviors in the ogre-faced spider *Deinopis spinosus* Marx (Araneae, Deinopidae). J. Arachnol. 15 (1987) 213
Cohen, A. C.: Hemolymph chemistry of two species of araneid spiders. Comp Biochem. Physiol., A 66 (1980) 715
Collatz, K.-G., T. Mommsen: Lebensweise und jahreszyklische Veränderungen des Stoffbestandes der Spinne *Tegenaria atrica* C. L. Koch (Agelenidae). J. comp. Physiol. 91 (1974) 91
Collatz, K.-G., T. Mommsen: Veränderung der Körperzusammensetzung und der Stoffwechselintensität der Spinne *Tegenaria atrica* C. L. Koch (Agelenidae) nach kurzem und langem Hunger. J. comp. Physiol. 98 (1975) 205
Colmorgen, M., R. J. Paul: Imaging of physiological functions in transparent animals (*Agonus cataphractus, Daphnia magna, Pholcus phalangioides*) by video microscopy and digital image processing. Comp. biochem. Physiol. 111A (1995) 583
Cooke, J. A. L.: Synopsis of the structure and function of the genitalia in *Dysdera crocata* (Araneae, Dysderidae). Senck Biol. 47 (1966) 35
Cooke, J. A. L.: Spider genitalia and phylogeny. Bull. Mus. Nat. Hist. Natur. 41 (1970) 142
Cooke, J. A. L., V D. Roth, F. H. Miller: The urticating hairs of theraphosid spiders. Amer. Mus. Novitates 2498 (1972) 1
Cooke, J. A. L., F. H. Miller, R. W. Grover, J. L. Duffy: Urticaria caused by tarantula hairs. Amer. J. trop. Med. Hyg. 22 (1973) 130
Coville, R. E.: Spider-hunting sphecid wasps. In: Nentwig, W., ed.: Ecophysiology of Spiders. Springer, Berlin 1987 (p. 309)
Coville, R. E., P. L. Coville: Nesting biology and male behavior of *Trypoxylon (Tripargilum) tenoctitlan* in Costa Rica (Hymenoptera: Sphecidae). Ann. Entomol. Soc. Am. 73 (1980) 110
Coyle, F. A.: Aerial dispersal by mygalomorph spiderlings (Araneae, Mygalomorphae). J. Arachnol. 11 (1983) 283

Coyle, F. A., F. W. Harrison, W. C. McGimsey, J. M. Palmer: Observations on the structure and function of spermathecae in haplogyne spiders. Trans. Amer. microsc. Soc. 102 (1983) 272

Coyle, F. A., M. H. Greenstone, A.-L. Hultsch, C. E. Morgan: Ballooning mygalomorphs: estimates of the masses of *Sphodros* and *Ummidia* ballooners (Araneae: Atypidae, Ctenizidac). J. Arachnol. 13 (1985) 291

Coyle, F.A., T.C. O'Shields, D. G. Perlmutter: Observations on the behavior of the klepto-parasitic spider *Mysmenopsis furtiva* (Araneae, Mysmenidae). J. Arachnol. 19 (1991) 62

Craig, C. L.: Orb-web visibility: the influence of insect flight behaviour and visual physiology on the evolution of web designs within the Araneoidea. Anim. Behav. 34 (1986) 54

Craig, C. L.: Insect perception of spider orb webs in three light habitats. Funct. Ecol. 2 (1988) 277

Craig, C. L.: Effects of background pattern on insect perception of webs spun by orb-weaving spiders. Anim. Behav. 39 (1990) 135

Craig, C. L.: Predator foraging behavior in response to perception and learning by its prey: interactions between orb-spinning spiders and stingless bees. Behav. Ecol. Sociobiol. (1995) in press

Craig, C. L., G. D. Bernard: Insect attraction to ultraviolet-reflecting web decorations. Ecology 71 (1990) 616

Craig, C. L., A. Okubo, V. Andreasen: Effect of spider orb web and insect oscillations on prey interception. J. theor. Biol. 115 (1985) 201

Crane, J: Comparative biology of salticid spiders at Rancho Grande, Venezuela. IV. An analysis of display. Zoologica 34 (1949) 159

Crome, W.: Die Respirations- und Circulationsorgane der *Argyroneta aquatica* Cl. (Araneae). Wiss. Z. Humboldt Univ. Berlin 2, Math.-naturwiss. Reihe 3–4 (1953) 53

Crome, W.: Kokonbau und Eiablage einiger Kreuzspinnenarten des Genus Araneus (Araneae, Araneidae). Dtsch. Ent. Z. (N.F.) 3 (1956a) 28

Crome, W.: Taranteln, Skorpione und Schwarze Witwen. Die Neue Brehm Bücherei. Ziemsen, Wittenberg Lutherstadt 1956b (p. 4)

Crome, W.: Arachnida-Spinnentiere. In: Exkursionsfauna von Deutschland, Bd. 1: Wirbellose, hrsg. von E. Stresemann. VEB Volk und Wissen, Berlin 1957 (p. 289)

Crome, W.: Embryonalentwicklung ohne 'Umrollung' (=Reversion) bei Vogel spinnen (Araneae: Orthognatha). Dtsch. Ent. Z. (N. F.) 10 (1963) 83

Cutler, B., A. G. Richards: On the absence of chitin in the endosternite of chelicerata. Experientia (Basel) 30 (1974) 1393

Czajka, M.: Unknown facts of the biology of the spider *Ero furcata* (Villers). Pol. Pismo Entomol. 33 (1963) 229

Dabelow, S.: Zur Biologie der Leimschleuderspinne *Scytodes thoracica*. Zool. Jb. Syst. 86 (1958) 85

Dahl, F.: Ueber die Hörhaare bei Arachnoideen. Zool. Anz. 6 (1883) 276

Dahlem, B., C. Gack, J. Martens: Balzverhalten von Wolfsspinnen der Gattung *Alopecosa* (Arachnida: Lycosidae). Zool. Beitr. N. F. 31 (1987) 151

D'Andrea, M.: Social behaviour in spiders (Arachnida, Araneae). Ital. J. Zool. New Series Monogr. 3 (1987) 1

Darchen, R.: Ethologie d'une araignée sociale, *Agelena consociata* Denis. Biol. Gabonica 1(1965) 117

Dawydoff, C.: Développement embryonnaire des Arachnides. In: Grassé, P.-P., ed.: Traité de Zoologie VI. Masson, Paris 1949 (p. 320)

Decae, A. E: A theory on the origin of spiders and the primitive function of spider silk. J. Arachnol. 12 (1984) 21

Decae, A. E.: Dispersal: ballooning and other mechanisms. In: Nentwig, W., ed.: Ecophysiology of Spiders. Springer, Berlin 1987 (S. 348)

Deevey, G. B.: The blood cells of the Haitian tarantula and their relation to the moulting cycle. J. Morph. 68 (1941) 457

Deevey, G. B.: The developmental history of *Latrodectus mactans* at different rates of feeding. Amer. Midl. Naturalist 42 (1949) 189

De la Serna de Esteban, C. J., L. Lauria de Cidre: El organo de Millot en *Polybetes pythagoricus* (Araneae, Sparassidae). Physis (Buenos Aires) Secc. C, 43 (1985)

Den Otter, C. J.: Setiform sensilla and prey detection in the Bird-Spider *Sericopelma rubronitens* Ausserer (Araneae, Theraphosidae). Neth. J. Zool. 24 (1974) 219

Denny, M.: The physical properties of spider's silk and their role in the design of orb-webs. J. exp. Biol. 65 (1976) 483

De Voe, R. D.: Ultraviolet and green receptors in principal eyes of jumping spiders. J. gen. Physiol. 66 (1975) 193

Dierkes, S.: Warum können Trichterspinnen (Araneae, Agelenidae) schneller auf ihren Netzen laufen als Insekten? Dipl. Arb. TH Darmstadt 1988

Dierkes, S., F. G. Barth: Mechanism of signal production in the vibratory communication of the wandering spider *Cupiennius getazi* (Arachnida, Araneae). J. comp. Physiol. A 176 (1995) 31

Dijkstra, H.: Comparative research of the courtship behaviour in the genus *Pardosa* (Arachn. Araneae) III. Agonistic behaviour in *Pardosa amentata*. Bull. Mus. Nat. Hist. Natur. 41 (1970) 91

Dijkstra, H.: Searching behaviour and tactochemical orientation in males of the wolfspider *Pardosa amentata* (Cl.) (Araneae, Lycosidae). Proc. kon. ned. Akad. Wer. Ser. C 79 (1976) 235

Dillon, L S.: The myology of the araneid leg. J. Morph. 90 (1952) 467

Dondale, C. D, B M. Hegdekar: The contact sex pheromone of *Pardosa lapidicina* Emerton (Araneida: Lycosidae). Can. J. Zool. 51 (1973) 400

Dong, Z., R. V. Lewis, C. R. Middaugh: Molecular mechanis of spider silk elasticity. Arch. Biochem. Biophys. 284 (1991) 53

Dornfeldt, K.: Die Bedeutung der Haupt- und Nebenaugen für das Heimfindevermögen der Trichterspinne *Agelena labyrinthica* (Clerck) mit Hilfe einer Lichtquelle. Diss. Berlin 1972

Dornfeldt, K.: Die Bedeutung der Haupt und Nebenaugen für die photomenotaktische Orientierung der Trichterspinne *Agelena labyrinthica* (Cl.). Z. Tierpsychol. 38 (1975) 113

Downes, M. F.: A proposal for standardization of the terms used to describe the early development of spiders, based on a study of *Theridion rufipes* Lucas (Araneae: Theridiidae). Bull. Br. arachnol. Soc. 7 (1987) 187

Downes, M. F.: Tolerance, interattraction and co-operation in the behaviour of the social spider *Phryganoporus candidus* (Araneae: Desidae). Bull. Br. arachnol. Soc. 9 (1994) 309

Downes, M. F.: Australasian social spiders: what is meant by "social"? Rec. West. Austral. Mus. Suppl. (1995) 25

Drees, O.: Untersuchungen über die angeborenen Verhaltensweisen bei Springspinnen (Salticidae). Z. Tierpsychol. 9 (1952) 169

Drewes, C. D., R. A. Bernard: Electrophysiological responses of chemosensitive sensilla in the wolf spider. J. exp. Zool. 198 (1976) 423

Duelli, P: Movement detection in the posterolateral eyes of jumping spiders (*Evarcha arcuata*) Salticidae). J. comp. Physiol. 124 (1978) 15

Duffey, E.: Aerial dispersal in a known spider population. J. Anim. Ecol. 25 (1956) 85

Duffey, E.: Spider ecology and habitat structure (Arach., Araneae). Senck. Biol. 47 (1966) 45

Duman, J. C.: Subzero temperature tolerance in spiders: the role of thermal-hysteresis-factors. J. comp. Physiol. 131 (1979) 347

Dumpert, K.: Spider odor receptor: Electrophysiological proof. Experientia (Basel) 34 (1978) 754

Eakin, R. M., J. L. Brandenburger: Fine structure of the eyes of jumping spiders. J. Ultrastruct. Res. 37 (1971) 618

Eberhard, W.: Attack behaviour of diguetid spiders and the origin of prey wrapping in spiders. Psyche 74 (1967) 173

Eberhard, W. G.: The spider *Uloborus diversus* (Marx) (Uloboridae) and its web. Doct. thesis, Harvard University, Cambridge, Massachusetts (1969)

Eberhard, W. G.: The web of *Uloborus diversus* (Araneae, Uloboridae). J. Zool. (Lond.) 166 (1972) 417

Eberhard, W. G.: The 'inverted ladder' orb web of *Scoloderus* sp. and the intermediate orb of *Eustala* (?) sp. Araneae: Araneidae. J. nat. Hist. 9 (1975) 93

Eberhard, W. G.: Photography of orb webs in the field. Bull. Brit. arachnol. Soc. 3 (1976) 200

Eberhard, W. G.: The webs of newly emerged *Uloborus diversus* and of a male *Uloborus* sp. (Araneae: Uloboridae). J. Arachnol. 4 (1977a) 201

Eberhard, W. G.: Aggressive chemical mimicry by a Bolas spider. Science 198 (1977b) 1173

Eberhard, W. G.: The natural history and behavior of the bolas spider *Mastophora dizzydeani* Sp. N. (Araneidae). Psyche 87 (1980) 143

Eberhard, W. G.: The "saw-toothed" orb web of *Eustala* spider (Araneae, Araneidae) with a discussion of ontogenetic changes in spiders' web-building behavior. Psyche 92 (1985) 105

Eberhard, W. G.: How spiders initiate airborne lines. J. Arachnol. 15 (1987) 1

Eberhard, W. G.: Function and phylogeny of spider webs. Ann. Rev. Ecol. Syst. 21 (1990) 341

Eberhard, W. G.: *Chrosiotes tonala* (Araneae, Theridiidae): a web-building spider specializing on termites. Psyche 98 (1991) 7

Eberhard, W. G.: Web construction by *Modisimus* sp. (Araneae, Pholcidae). J. Arachnol. 20 (1992) 25

Eberhard, W., F. Pereira: Ultrastructure of cribellate silk of nine species in eight families and possible taxonomic implications (Araneae: Amaurobiidae, Deinopidae, Desidae, Dictynidae, Filistatidae, Hypochilidae, Stiphidiidae, Tengellidae). J. Arachnol. 21 (1993) 161

Eberhard, W. G., M. Bareto, W. Pfizenmaier: Web robbery by mature male orb-weaving spiders. Bull. Br. arachnol. Soc. 4 (1978) 228

Eckert, M.: Experimentelle Untersuchungen zur Häutungsphysiologie bei Spinnen Zool. Jb. Physiol. 73 (1967) 49

Eckweiler, W., E.-A. Seyfarth: Tactile hairs and the adjustment of body height in wandering spiders: behavior, leg reflexes, and afferent projections in the leg ganglia. J. comp. Physiol. A 162 (1988) 611

Edgar, W. E.: Prey and feeding behaviour of adult females of the wolf spider *Pardosa amentata* (Clerck). Neth. J. Zool. 20 (1970) 487

Edgar, W. E., M. Loenen: Aspects of the overwintering habitat of the wolf spider *Pardosa lugubris*. J. Zool. (Lond.) 172 (1974) 383

Edmonds, D. T., F. Vollrath: The contribution of atmospheric water vapour to the formation and efficiency of a spider's capture web. Proc. R. Soc. Lond. B 248 (1992) 145

Edmunds, M.: On the association between *Myrmarachne* spp. (Salticidae) and ants. Bull. Brit. arachnol. Soc. 4 (1978) 149

Edmunds, M.: Does mimicry of ants reduce predation by wasps on salticid spiders? Mem. Queensland Mus. 33 (1993) 507

Edwards, G. B., R. R. Jackson: The role of experience in the development of predatory behaviour in *Phidippus regius,* a jumping spider (Araneae, Salticidae) from Florida. N. Zeal. J. Zool. 21 (1994) 269

Edwards, G. B., J. F. Carroll, W. H. Whitcomb: *Stoidis aurata* (Araneae: Salticidae), a spider predator of ants. Florida Entomol. 57 (1974) 337

Edwards, J. S.: Derelics of dispersal: arthropod fallout on Pacific Northwest volcanoes. In: Danthanarayana, W., ed.: Insect Flight, Dispersal and Migration. Springer Verlag Berlin 1986 (p. 196)

Ehlers, M.: Untersuchungen über Formen aktiver Lokomotion bei Spinnen. Zool. Jb. Syst. 72 (1939) 373

Ehn, R., H. Tichy: Hygro- and thermoreceptive tarsal organ in the spider *Cupiennius salei.* J. comp. Physiol. A 174 (1994) 345

Eisner T., J. Dean: Ploy and counterploy in predator–prey interactions: orb-weaving spiders versus bombardier beetles. Proc. nat. Acad. Sci. (Wash.) 73 (1976) 1365

Eisner, T., S. Camazine: Spider leg autotomy induced by prey venom injection: an adaptive response to "pain"? Proc. nat. Acad. Sci. (Wash.) 80 (1983) 3382

Eisner, T., S. Nowicki: Spider web protection through visual advertisement: role of the stabilimentum. Science 219 (1983) 185

Eisner, T., R. Alsop, G. Ettershank: Adhesiveness of spider silk. Science 146 (1964) 1058

Elgar, M. A.: Inter-specific associations involving spiders: kleptoparasitism, mimicry and mutualism. Mem. Queensland Mus. 33 (1993) 411

Ellis, C. H.: The mechanism of extension in the legs of spiders. Biol. Bull. 86 (1944) 41

Emerit, M., J. C. Bonaric: Contribution à l'étude du développement de l'appareil mechanorécepteur des Araignées: la trichobothriotaxie de *Pisaura mirabilis* Cl. (Araneae-Pisauridae). Zool. Jb. Anat. 94 (1975) 358

Enders, F.: Vertical stratification in orbweb spiders and a consideration of other methods of coexistence. Ecology 55 (1974) 317

Enders, F.: The influence of hunting manner on prey size, particularly in spiders with long attack distances (Araneidae, Linyphiidae, and Salticidae). Amer. Natur. 109 (1975) 737

Enders, F.: Effects of prey capture, web destruction and habitat physiognomy on web-site tenacity of *Argiope* spiders (Araneidae). J. Arachnol. 3 (1976) 75

Enders, F.: Web-site selection by orb-web spiders, particularly *Argiope aurantia* Lucas. Anim. Behav. 25 (1977) 694

Engelhardt, W.: Die mitteleuropäischen Arten der Gattung *Trochosa* C. L. Koch 1848 (Araneae, Lycosidae). Morphologie, Chemotaxonomie, Biologie, Autoekologie. Z. Morph. Oekol. Tiere 54 (1964) 219

Engelhardt, W.: Gestalt und Lebensweise der "Ameisenspinne" *Synageles venator* (Lucas). Zugleich ein Beitrag zur Ameisenmimikryforschung. Zool. Anz. 185 (1971) 21

Entwistle, I. D., R. A. W. Johnstone, D. Medzihradsky, T. E. May: Isolation of a pure toxic polypeptide from the venom of the spider *Phoneutria nigriventer* and its neurophysiological activity on an insect femur preparation. Toxicon 20 (1982) 1059

Eschrich, J., R. J. Paul: Fluoreszenzspektroskopische In-vivo-Messungen des Zustandes biologischer Redoxsysteme in verschiedenen Geweben bei der Vogelspinne *Eurypelma californicum*. Verh. Dtsch. Zool. Ges. 84 (1991) 407

Eskov, K.: A new fossil spider family from the Jurassic of Transbaikalia (Araneae: Chelicerata). N. Jb. Geol. Paläont. Abh. 11 (1984) 645

Eskov, K.: A new Archaeid spider (Chelicerata: Araneae) from the Jurassic of Kazakhstan, with notes on the so-called "Gondwana" ranges of recent taxa. N. Jb. Geol. Paläont. Abh. 175 (1987) 81

Eskov, K., S. Zonshtein: First mesozoic mygalomorph spiders from the Lower Cretaceous of Siberia and Mongolia, with notes on the system and evolution of the infraorder Mygalomorphae (Chelicerata: Araneae). N. Jb. Geol. Paläont. Abh. 178 (1990) 325

Exline, H., H. W. Levi: American spiders of the genus *Argyrodes* (Araneae, Theridiidae). Bull. Harvard Mus. Comp. Zool. 127 (1962) 75

Fahrenbach, W. H.: The cyanoblast: Hemocyanin formation in *Limulus polyphemus*. J. Cell Biol. 44 (1970) 445

Fahrenbach, W. H.: The brain of the horseshoe crab (*Limulus polyphemus*) III. Cellular and synaptic organization of the corpora pedunculata. Tissue Cell 11 (1979) 163

Ferdinand,W.: Die Lokomotion der Krabbenspinnen (Araneae, Thomisidae) und das Wilsonsche Modell der metachronen Koordination. Zool. Jb. Physiol. 85 (1981) 46

Fincke, O.: Giant damselflies in a tropical forest: reproductive biology of *Megaloprepus coerulatus* with notes on *Mecistogaster* (Zygoptera: Pseudostigmatidae). Advanc. Odonatol. 2 (1984) 13

Fincke, T., R. Paul: Book lung function in arachnids. III. The function and control of the spiracles. J. comp. Physiol. B 159 (1989) 409

Fischel, W.: Wachstum und Häutung der Spinnen. 1. Mitt.: Studien an retitelen Spinnen. Z. wiss. Zool. 133 (1929) 441

Foelix, R. F.: Structure and function of tarsal sensilla in the spider *Araneus diadematus*. J. exp. Zool. 175 (1970a) 99

Foelix, R. F.: Chemosensitive hairs in spiders. J. Morph. 132 (1970b) 313

Foelix, R. F.: Occurrence of synapses in peripheral sensory nerves of arachnids. Nature (London) 254 (1975) 146

Foelix, R. F.: Sensory nerves and peripheral synapses. In: Barth, F. G., ed.: Neurobiology of Arachnids. Springer Verlag Berlin 1985 (p. 189)

Foelix, R. F.: How do crab spiders bite their prey? Rev. Suisse Zool. (1996) in press

Foelix, R. F., A. Choms: Fine structure of a spider joint-receptor and associated synapses. Europ. J. Cell Biol. 13 (1979) 149

Foelix, R. F., I.-W. Chu-Wang: The morphology of spider sensilla. I. Mechanoreceptors. Tissue Cell 5 (1973a) 451

Foelix, R. F., I.-W. Chu-Wang: The morphology of spider sensilla. II. Chemoreceptors. Tissue Cell 5 (1973b) 461

Foelix, R. F., I.-W. Chu-Wang: The structure of scopula hairs in spiders. Proc. 6th Int. Arachnol. Congr. 1974, Free University of Amsterdam 1975 (p.156)

Foelix, R. F., H. Jung: Some anatomical aspects of *Hypochilus thorelli* with special reference to the calamistrum and cribellum. Symp. Zool. Soc. Lond. 42 (1978) 417

Foelix, R. F., G. Müller-Vorholt, H. Jung: Organisation of sensory leg nerves in the spider *Zygiella x-notata* (Araneae Araneidae). Bull. Br. arachnol. Soc. 5 (1980) 20

Foelix, R., R. R Jackson, A. Henksmeyer, S. Hallas: Tarsal hairs specialized for prey capture in the salticid *Portia*. Rev. Arachnol. 5 (1984) 329

Fontana, M. D., O. Vital Brazil: Mode of action of *Phoneutria nigriventer* spider venom at the isolated phrenic nerve-diaphragm of the rat. Braz. J. Med. Biol. Res. 18 (1985) 557

Ford, M. J.: Locomotory activity and the predation strategy of the wolf-spider *Pardosa amentata* (Clerck) (Lycosidae). Anim. Behav. 26 (1978) 31

Formanowicz, D. R.: The antipredator efficacy of spider leg autotomy. Anim. Behav. 40 (1990) 400

Forster, L.: Vision and prey-catching in jumping spiders. Amer. Sci. 70 (1982) 165

Forster, L.: Target discrimination in jumping spiders (Araneae: Salticidae). In: Barth, F. G., ed.: Neurobiology of Arachnids. Springer Verlag Berlin 1985 (S. 249)

Forster, L., S. Kingsford: A preliminary study of development in two *Latrodectus* species (Araneae: Theridiidae). New Zeal. Entomol. 7 (1983) 431

Forster, L. M., F. M. Murphy: Ecology and behaviour in *Portia schultzii*, with notes on related species (Araneae, Salticidae). J. Arachnol. 14 (1986) 29

Forster, R. R.: Evolution of the tarsal organ, the respiratory system and the female genitalia in spiders. Proc. 8th Int. Congr. Arachnol., Vienna, 1980, (p. 269)

Forster, R. R., L. M. Forster: New Zealand Spiders. Collins, Auckland 1973

Forster, R. R., C. L. Wilton: The Spiders of New Zealand. Part IV. Otago Mus. Bull. No.4, Dunedin 1973

Fourtner, C. R.: Chelicerate skeletal neuromuscular systems. Amer. Zool. 13 (1973) 271

Fowler, H. C., J. Diehl: Biology of a Paraguayan colonial orb-weaver, *Eriophora bistriata* (Rengger) (Araneae, Araneidae) Bull. Brit. arachn. Soc. 4 (1978) 241

Fowler, H. C., N. Gobbi: Communication and synchronized molting in a colonial araneid spider, *Eriophora bistriata*. Experientia 44 (1988a) 720

Fowler, H. C., N. Gobbi: Cooperative prey capture by an orb-web spider. Naturwiss. 75 (1988b) 208

Frank, H.: Untersuchungen zur funktionellen Anatomie der lokomotorischen Extremitäten von *Zygiella x-notata*. eincr Radnetzspinne. Zool. Jb. Anat. 76 (1957) 423

Franz, V.: Ueber die Struktur des Herzens und die Entstehung von Blutzellen bei Spinnen. Zool. Anz. 27 (1904) 192

Friedel, T.: Vergleichende Untersuchungen zur Wirkung von Spinnengift auf Insekten. Dipl. Arb. Universität Regensburg (1987)

Friedel, T., W. Nentwig: Immobilizing and lethal effects of spider venoms on the cockroach and the common mealbeetle. Toxicon 27 (1989) 305

Friedrich, V. L., R. M. Langer: Fine structure of cribellate spider silk. Amer. Zool. 9 (1969) 91

von Frisch, K.: The Dance Language and Orientation of Bees. Harvard University Press, Cambridge, Mass., 1967

Fröhlich, A.: Verhaltensphysiologische Untersuchungen zur Lokomotion der Spinnen *Agelena labyrinthica* Cl. und *Sitticus pubescens* F. Diss. Freie Universität Berlin (1978)

Fujii, Y.: Examination of the maternal care of cocoon in *Pardosa astrigera* Koch (Araneae, Lycosidae). Bull. Nippon dent. Univ. Gen. Ed. 3 (1978) 223

Full, R., K. Earls, M. Wong, R. Caldwell: Locomotion like a wheel? Nature (Lond.) 365 (1993) 495

Gabe, M.: Données histologiques sur la neurosécrétion chez les Arachnides. Arch. Anat. micr. Morph. exp. 44 (1955) 351

Gage, P. W., I. Spence: The origin of the muscle fasciculation caused by funnel web spider venom. AJEBAK 55 (1977) 453

Gerhardt, U.: Studien über die Copulation einheimischer Epeiriden. Zool. Jb. Syst. 31 (1911) 643

Gerhardt, U.: Neue Studien zur Sexualbiologie und zur Bedeutung des sexuellen Grössendimorphismus der Spinnen. Z. Morph. Oekol. Tiere 1 (1924) 507

Gerhardt, U.: Weitere Untersuchungen zur Biologie der Spinnen. Z. Morph. Oekol. Tiere 6 (1926) 1

Gerhardt, U.: Neue biologische Untersuchungen an einheimischen und ausländischen Spinnen. Z. Morph.Oekol. Tiere 8 (1927) 96

Gerhardt, U.: Biologische Studien an griechischen, corsischen und deutschen Spinnen. Z. Morph. Oekol. Tiere 10 (1928) 576

Gerhardt, U.: Zur vergleichenden Sexualbiologie primitiver Spinnen, insbesondere der Tetrapneumonen. Z. Morph. Oekol. Tiere 14 (1929) 699

Gerhardt, U.: Biologische Untersuchungen an südfranzösischen Spinnen. Z. Morph Oekol. Tiere 19 (1930) 184

Gerhardt, U.: Neue Untersuchungen zur Sexualbiologie der Spinnen, insbesondere an Arten der Mittelmeerländer und der Tropen. Z. Morph. Oekol. Tiere 27 (1933) 1

Gerhardt, U., A. Kaestner: 8. Ordnung der Arachnida: Araneae = Echte Spinnen = Webspinnen. In: W. Kükenthal, T. Krumbach, eds: Handbuch der Zoologie. De Gruyter, Berlin 1938 (p. 394)

Gertsch, W. J.: American Spiders. Van Nostrand, New York 1949

Gertsch, W. J., S. E. Riechert: The spatial and temporal partitioning of a desert spider community, with descriptions of new species. Novitates (American Museum) 2604 (1976) 1

Gettmann, W. W.: Beutefang bei Wolfspinnen der Gattung *Pirata* (Araneae: Lycosidae). Ent. Germ. 3 (1976) 93

Gettmann, W. W.: Untersuchungen zum Nahrungsspektrum von Wolfsspinnen (Lycosidae) der Gattung *Pirata*. Mitt. dtsch. Ges. allg angew. Ent. 1 (1978) 63

Gillespie, R. G.: Impaled prey. Nature (Lond.) 355 (1992) 212

Glatz, L.: Der Spinnapparat haplogyner Spinnen. Z. Morph. Tiere 72 (1972) 1

Glatz, L.: Der Spinnapparat der Orthognatha (Arachnida, Araneae). Z. Morph. Tiere 75 (1973) 1

Glick, P. A.: The distribution of insects, spiders and mites in the air. USDA Tech. Bull. No. 673, US Department of Agriculture, Washington DC 1939

Gonzalez-Fernandez, F., R. G. Sherman: Cardioregulatory nerves in the spider *Eurypelma marxi* Simon. J. exp. Zool. 231 (1984) 27

Gorb, S. N., F. G. Barth: Locomotor behavior during prey-capture of a fishing spider, *Dolomedes plantarius* (Araneae: Araneidae): Galloping and stopping. J. Arachnol. 22 (1994) 89

Görner, P.: Die optische und kinästhetische Orientierung der Trichterspinne *Agelena labyrinthica* (Cl.). Z. vergl. Physiol. 41 (1958) 111

Görner, P.: Orientierung der Trichterspinne nach polarisiertem Licht. Z. vergl. Physiol. 45 (1962) 307

Görner, P.: Mehrfach innervierte Mechanorezeptoren bei Spinnen. Naturwissenschaften 52 (1965) 437

Görner, P.: Über die Koppelung der optischen und kinaesthetischen Orientierung der Trichterspinne *Agelena labyrinthica* (Clerck) und *Agelena gracilens* C. L. Koch. Z. vergl. Physiol. 53 (1966) 253

Görner, P.: Resultant positioning between optical and kinesthetic orientation in the spider *Agelena labyrinthica* Clerck. In: Wehner, R., ed.: Information Processing in the Visual Systems of Arthropods. Springer Verlag, Berlin 1972 (p. 269)

Görner, P.: Beispiele einer Orientierung ohne richtende Aussenreize. Fortschr. Zool. 21 (1973) 20

Görner, P.: Homing behavior of funnel web spiders (Agelenidae) by means of web-related cues. Naturwiss. 75 (1988) 209

Görner, P., P. Andrews: Trichobothrien, ein Ferntastsinnesorgan bei Webspinnen (Araneen). Z. vergl. Physiol. 64 (1969) 301

Görner, P., B. Claas: Homing behavior and orientation in the funnel-web spider *Agelena labyrinthica* Clerck. In: Barth, F. G., ed.: Neurobiology of Arachnids. Springer Verlag, Berlin 1985 (p. 275)

Görner, P., I. Zeppenfeld: The runs of *Pardosa amentata* (Araneae, Lycosidae) after removing its cocoon. Proc. 8th Int. Arach. Kongr. Wien 1980 (p. 243)

Gosline, J. M., M. W. Denny, M. E. DeMont: Spider silk as rubber. Nature 309 (1984) 551

Gosline, J. M., M. E. DeMont, M. W. Denny: The structure and properties of spider silk. Endeavour 10 (1986) 37

Grasshoff, M.: Die Kreuzspinne *Araneus pallidus*—ihr Netzbau und ihre Paarungsbiologie. Natur u. Museum 94 (1964) 305

Grasshoff, M.: Morphologische Kriterien als Ausdruck von Artgrenzen bei Radnetzspinnen der Subfamilie Araneinae (Arachnida: Araneae: Araneidae). Abh. Senckenberg naturforsch. Ges. 516 (1968) 1

Grasshoff, M.: Konstruktions und Funktionsanalyse an Kopulationsorganen einiger Radnetzspinnen. Aufs. Red. Senckenberg naturforsch. Ges. 24 (1973) 129

Grasshoff, M.: Reconstruction of an evolutionary transformation—the copulatory organs of *Mangora* (Arachnida. Araneae, Araneidae). Proc. 6th Int. Arachnol. Congr. 1974, Free University Amsterdam 1975 (p. 12)

Grasshoff, M.: A model of the evolution of the main chelicerate groups. Symp. Zool. Soc. Lond. 42 (1978) 273

Gray, M. R.: Getting to know funnel webs. Austral. Nat. Hist. 20 (1981) 265

Greene, E., L. J. Orsak, D. W. Whitman: A tephritid fly mimics the territorial displays of its jumping spider predators. Science 236 (1987) 310

Greenquist, E. A., J. S. Rovner: Lycosid spiders on artificial foliage: stratum choice, orientarion preferences, and prey-wrapping. Psyche 83 (1976) 196

Greenstone, M. H., C. E. Morgan, A.-L. Hultsch, R. A. Farrow, J. E. Dowse: Ballooning spiders in Missouri, USA, and New South Wales, Australia: family and mass distributions. J. Arachnol. 15 (1987) 163

Grégoire, C.: Sur la coagulation du sang des Araignées. Arch. int. Physiol. 60 (1952) 100

Griffiths, D. J., G. T. Smyth Jr.: Action of Black Widow spider venom at insect neuromuscular junctions. Toxicon 11 (1973) 369

Griswold, C. E.: Investigations into the phylogeny of the lycosoid spiders and their kin (Arachnida: Araneae: Lycosoidea). Smith. Contr. Zool. 539 (1993) 1

Groh, G., M. Lemieux, A. Saint-Jean: Bildgewordenes Verhalten: Das Spinnennetz. Image Roche 18 (1966) 3

Gronenberg, W.: Anatomical and physiological observations on the organization of mechanoreceptors and local interneurons in the central nervous system of the wandering spider *Cupiennius salei*. Cell Tissue Res. 258 (1989) 163

Gronenberg, W.: The organization of plurisegmental mechanosensitive interneurons in the central nervous system of the wandering spider *Cupiennius salei*. Cell Tissue Res. 260 (1990) 260

Gruber, J.: Fatherless spiders. Br. Arachnol. Soc. Newsl. 58 (1990) 3

Grünbaum, A. A: Über das Verhalten der Spinne *Epeira diademata*, besonders gegenüber vibratorischen Reizen. Psychol. Forsch. 9 (1927) 275

Gundermann, J.-L.: Etude des interactions mère–jeunes chez l'araignée *Coelotes terrestris* (WIDER) (Araneae, Agelenidae). Comportements 6 (1986) 31

Gundermann, J.-L., A. Horel, B. Krafft: Maternal food-supply activity and its regulation in *Coelotes terrestris* (Araneae, Agelenidae). Behaviour 107 (1988) 278

Gundermann, J.-L., A. Horel, C. Roland: Mother–offspring food transfer in *Coelotes terrestris* (Araneae, Agelenidae). J. Arachnol. 19 (1991) 97

Gunnarsson, B., A. Andersson: Chromosome variation in the embryos of a solitary spider, *Pityhyophantes phrygianus* with skewed sex ratio. Hereditas 117 (1992) 85

Gwinner-Hanke, H.: Zum Verhalten zweier stridulierender Spinnen *Steatoda bipunctata* Linné und *Teutana grossa* Koch (Theridiidae, Araneae) unter besonderer Berücksichtigung des Fortpflanzungsverhaltens. Z. Tierpsychol. 27 (1970) 649

Gwynne, D.T.: Nesting biology of the spider wasps (Hymenoptera, Pompilidae) which prey on burrowing wolf spiders (Araneae: Lycosidae, Geolycosa). J. Nat. Hist. 13 (1979) 681

Habermehl, G.: Vergiftungen durch die Dornfingerspinne. Naturwissenschaften 61 (1974) 368

Habermehl, G.: Die biologische Bedeutung tierischer Gifte. Naturwissenschaften 62 (1975) 15

Hackmann, W.: Chromosomenstudien an Araneen mit besonderer Berücksichtigung der Geschlechtschromosomen. Acta. zool. Fenn. 54 (1948) 1

Hågvar, S.: Ecological studies on a winteractive spider *Bolyphantes index* (Thorell) (Araneida, Linyphiidae). Nor. Entomol. Tidsskr. 20 (1973) 309

Hansell, M.: Secondhand silk. Nat. Hist. 102 (1993) 40

Hanström, B.: Über die Histologie und vergleichende Anatomie der Sehganglien und Globuli der Araneen. Kungl. Sv. Vet. Akad. Handlingar 61 (1921) 2

Hanström, B.: Vergleichende Anatomie des Nervensystems der wirbellosen Tiere. Berlin 1928 (p. 628)

Hanström, B.: Fortgesetzte Untersuchungen über das Araneengehirn. Zool. Jb. Anat. 59 (1935) 455

Harkness, R. D.: The relation between an ant, *Cataglyphis bicolor* (F.) (Hym., Formicidae) and a spider, *Zodarium frenatum* (SIMON) (Araneae, Zodariidae). Entomol. Month. Mag. 111 (1975) 141

Harkness, R. D.: Further observations on the relation between an ant *Cataglyphis bicolor* (F.) (Hym. Formicidae) and a spider, *Zodarium frenatum* (SIMON) (Araneae, Zodariidae). Entomol. Month. Mag. 112 (1976) 111

Harkness, R., R. Wehner: *Cataglyphis*. Endeavour (N. S.) 1 (1977) 115

Harm, M.: Beiträge zur Kenntnis des Baues, der Funktion und der Entwicklung des akzessorischen Kopulationsorgans von *Segestria bavarica* C. L Koch. Z Morph. Ökol. Tiere 22 (1931) 629

Harris, D. J.: Hair regeneration during moulting in the spider *Ciniflo similis* (Araneae, Dictynidae). Zoomorphologie 88 (1977) 37

Harris, D. J., P. J. Mill: The ultrastructure of chemoreceptor sensilla in *Ciniflo* (Araneida, Arachnida). Tissue Cell 5 (1973) 679

Harris, D. J., P. J. Mill: Observations on the leg receptors of *Ciniflo* (Araneida: Dictynidae). I. External mechanoreceptors. J. comp. Physiol. 119 (1977a) 37

Harris, D. J., P. J. Mill: Observations on the leg receptors of *Ciniflo* (Araneida: Dictynidae). II. Chemoreceptors. J. comp. Physiol. 119 (1977b) 55

Harrison, J. B.: Acoustic behavior of a wolf spider, *Lycosa gulosa*. Anim. Behav. 17 (1969) 14

Haupt, J.: Preliminary report on the mating behaviour of the primitive spider *Heptathela kimurai* (Kishida) (Araneae, Liphistiomorphae). Z. Naturforsch. 32c (1977) 312

Haupt, J.: Lebensweise und Sexualverhalten der mesothelen Spinne *Heptathela nishihirai* n. sp. (Araneae, Liphistiidae). Zool. Anz. Jena 202 (1979) 348

Heil, K. H.: Beiträge zur Physiologie und Psychologie der Springspinnen. Z. vergl. Physiol. 23 (1936) 1

Heinzberger, R.: Verhaltensphysiologische Untersuchungen an *Argyroneta aquatica* Cl. Diss. Universität Bonn (1974)

Heller, G.: Zur Biologie der ameisenfressenden Spinne *Callilepis nocturna* L. 1758 (Aranea, Drassodidae). Diss. Universität Mainz (1974)

Heller, G.: Zum Beutefangverhalten der ameisenfressenden Spinne *Callilepis nocturna* (Arachnida: Araneae: Drassodidae). Ent. Germ. 3 (1976) 100

von Helversen, O.: Gedanken zur Evolution der Paarungsstellungen bei den Spinnen (Arachnida: Araneae). Ent. Germ. 3 (1976) 13

Henschel, J. R.: Spiders wheel to escape. South Afr. J. Sci. 86 (1990a) 151

Henschel, J. R.: The biology of *Leucorchestris arenicola* (Araneae: Heteropodidae), a burrowing spider of the Namib Dunes. In: Seely, M. K., ed.: Namib Ecology: 25 Years of Namib Research. Transvaal Mus. Monogr. 7, Pretoria (1990b) 115

Henschel, J. R.: Is solitary life an alternative for the social spider *Stegodyphus dumicola*? Namibia Sci. Soc. Windhoek 43 (1992) 71

Henschel, J. R., R. Jocqué: Bauble spiders: a new species of *Achaearanea* (Araneae, Theridiidae) with ingenious spiral retreats. J. Nat. Hist. 28 (1994) 1287

Henton, W. W., F. T. Crawford: The discrimination of polarized light by the tarantula. J. comp. Physiol. 52 (1966) 26

Hergenröder, R., F. G. Barth: The release of attack behavior by vibratory stimuli in a wandering spider (*Cupiennius salei* Keys.) J. comp. Physiol. 152 (1983a) 347

Hergenröder, R., F. G. Barth: Vibratory signals and spider behavior: how do the sensory inputs from the eight legs interact in orientation? J. comp. Physiol. 152 (1983b) 361

Hieber, C. S.: Orb-web orientation and modification by the spiders *Araneus diadematus* and *Araneus gemmoides* (Araneae: Araneidae) in response to wind and light. Z. Tierpsychol. 65 (1984) 250

Hieber, C. S.: The "insulation" layer in the cocoons of *Argiope aurantia* (Araneae: Araneidae). J. therm. Biol. 10 (1985) 171

Higashi, G. A., J. S. Rovner: Post-emergent behaviour of juvenile lycosid spiders. Bull. Brit. Arach. Soc. 3 (1975) 113

Higgins, L.E., R. E. Buskirk: A trap-building predator exhibiting different tactics for different aspects of foraging behaviour. Anim. Behav. 44 (1992) 485

Hill, D. E.: The pretarsus of salticid spiders. Zool. J. Linn. Soc. 60 (1977a) 319

Hill, D. E.: Some observations on the physiology of living *Lyssomanes viridis* which should apply to the Araneae in general. Peckhamia 1 (1977b) 41

Hill, D. E.: Orientation by jumping spiders of the genus *Phidippus* (Araneae: Salticidae) during pursuit of prey. Behav. Ecol. Sociobiol. 5 (1979) 301

Hiller, U.: Untersuchungen zum Feinbau und zur Funktion der Haftborsten von Reptilien. Z. Morph. Tiere 62 (1968) 307

Hingston, R. W. G.: A Naturalist in Himalaya. Willoby, London 1920

Hingston, R. W. G.: Field observations on spider mimics. Proc. Zool. Soc. Lond. 2 (1927) 841

Hinman, M.D., R. V. Lewis: Isolation of a clone encoding a second dragline fibroin. J. Biol. Chem. 267 (1992) 19320

Hinton, H. E., R. S. Wilson: Stridulatory organs in spiny orb-weaver spiders. J. Zool. (Lond.) 162 (1970) 481

Hirschberg, D.: Beiträge zur Biologie, insbesondere zur Brutpflege einiger Theridiiden. Z. wiss. Zool. 179 (1969) 189

Hoffmaster, D.K.: Aggression in tropical orb-weaving spiders: a quest for food? Ethology 72 (1986) 265

Hogstad, O.: Variation in numbers, territoriality and flock size of a gold crest *Regulus regulus* population in winter. Ibis 126 (1984) 296

Holl, A.: Reifefärbung und Hypodermispigmente männlicher *Micrommata virescens* (Arachnida, Araneidae, Eusparrasidae). Verh. naturwiss. Ver. Hamburg, (NF) 29 (1987a) 181

Holl, A.: Coloration and chromes. In: Nentwig, W., ed.: Ecophysiology of Spiders. Springer Verlag, Berlin 1987b (p. 16)

Hollander, J., H. Dijkstra: *Pardosa vliimi* sp. nov. a new ethospecies sibling *Pardosa procima* (C. L. Koch, 1848), from France, with description of courtship display (Araneae, Lycosidae). Beaufortia 289 (1974) 57

Holm, A.: Studien über die Entwicklung und Entwicklungsbiologie der Spinnen. Zool. Bidr. (Uppsala) 19 (1940) 1

Holm, A.: Experimentelle Unrersuchungen über die Entwicklung und Entwicklungsphysiologie der Spinnenembryos. Zool. Bidr. (Uppsala) 29 (1952)

Holzapfel, M.: Die Bedeutung der Netzstarrheit für die Orientierung der Trichterspinne *Agelena labyrinthica* (Cl.). Rev. Suisse Zool. 40 (1933) 247

Holzapfel, M.: Die nicht-optische Orientierung der Trichterspinne *Agelena labyrinthica* (Cl.) vergl. Physiol. 20 (1934) 55

Homann, H.: Beiträge zur Physiologie der Spinnenaugen. I. Untersuchungsmethoden. II. Das Sehvermögen. Z. vergl. Physiol. 7 (1928) 201

Homann, H.: Beiträge zur Physiologie der Spinnenaugen. III. Das Sehvermögen der Lycosiden. Z. vergl. Physiol. 14 (1931) 40

Homann, H.: Beiträge zur Physiologie der Spinnenaugen. IV. Das Sehvermögen der Thomisiden. Z. vergl. Physiol. 20 (1934) 420

Homann, H.: Die Funktion des männlichen Spinnentasters im Versuch. Zool. Anz. 109 (1935) 73

Homann, H: Beiträge zur Physiologie der Spinnenaugen. V. Der Lichtsinn von *Aranea sexpunctata* (Argiopidae). Biol. Zbl. 66 (1947) 251

Homann, H.: Über das Wachstum und die mechanischen Vorgange bei der Häutung von *Tegenaria agrestis* (Araneae). Z. vergl. Physiol. 31 (1949) 413

Homann, H.: Die Nebenaugen der Araneen. Zool. Jb. Anat. 71 (1950) 1

Homann, H.: Die Nebenaugen der Araneen. Zool. Jb. Anat. 72 (1952) 345

Homann, H.: Haften Spinnen an einer Wasserhaut? Naturwissenschaften 44 (1957) 318

Homann, H.: Die Augen der Araneae. Anatomie, Ontogenie und Bedeutung für die Systematik (Chelicerata, Arachnida). Z. Morph. Tiere 69 (1971) 201

Homann, H: Die Stellung der Thomisidae und der Philodromidae im System der Araneae (Chelicerata, Arachnida). Z.. Morph. Tiere 80 (1975) 181

Homann, H.: Die Cheliceren der Araneae, Amblypygi und Uropygi mit den Skleriten, den Plagulae (Chelicerata, Arachnomorpha). Zoomorph. 105 (1985) 69

Horstmann, K.: Oekologische Untersuchungen über die Ichneumoniden (Hymenoptera) der Nordseeküste Schleswig Holsteins, Oecologia 4 (1970) 29

Huber, B. A.: Genital mechanics and sexual selection in the spider *Nesticus cellulanus* (Araneae: Nesticidae). Can. J. Zool. 71 (1993) 2437

Huber, B.A.: Genital morphology, copulatory mechanism and reproductive biology in *Psilochorus simoni* (Berland, 1911) (Pholcidae; Araneae). Netherl. J. Zool. 44 (1994) 85

Humphreys, W. F.: Behavioral thermoregulation in a wolf spider. Nature (Lond.) 251 (1974) 502

Humphreys, W. F.: The influence of burrowing and thermoregulatory behaviour on the water relations of *Geolycosa godeffroyi* (Araneae: Lycosidae), an australian wolf spider. Oecologia 21 (1975) 291

Humphreys, W. F.: The thermal biology of the wolf spider *Lycosa tarentula* (Araneae: Lycosidae) in Northern Greece. Bull. Brit. arachnol. Soc.7 (1987) 117

Humphreys, W. F.: Stabilimenta as parasols: shade construction by *Neogea* sp. (Araneae: Araneidae, Argiopinae) and its thermal behaviour. Bull. Br. arachnol. Soc. 9 (1992) 47

Husby, J. A., K. E. Zachariassen: Antifreeze agents in the body fluid of winteractive insects and spiders. Experientia 36 (1980) 963

Irving, W. G., E. H. Hinman: The blue mud-dauber as a predator of the Black Widow spider. Science 82 (1935) 395

Jackson, H., T. N. Parks: Spider toxin: recent applications in neurobiology. Ann. Rev. Neurosci. 12 (1989) 405

Jackson, R. R.: Fine structure of the thread connections in the orb web of *Araneus diadematus*. Psyche 78 (1971) 12

Jackson, R. R.: Courtship versatility in the jumping spider *Phidippus johnsoni* (Araneae: Salticidae). Anim. Behav. 25 (1977) 953

Jackson, R. R.: Comparative studies of *Dictyna* and *Mallos* (Araneae, Dictynidae). I. Social organization and web characteristics. Rev. Arachnol. 1 (1978) 133

Jackson, R. R.: Nests of *Phidippus johnsoni* (Araneae, Salticidae): characteristics, patterns of occupation, and function. J. Arachnol. 7 (1979) 47

Jackson, R. R.: Nest mediated sexual discrimination by a jumping spider (*Phidippus johnsoni*). J. Arachnol. 9 (1981) 87

Jackson, R. R.: The behavior of communicating in jumping spiders (Salticidae). In: Rovner, J., P. Witt, eds: Communication in spiders. Princeton University Press, Princeton 1982 (p. 213)

Jackson, R. R.: The biology of ant-like jumping spiders (Araneae, Salticidae): prey and predatory behavior of *Myrmarachne* with particular attention to *M. lupara* from Queensland. Zool. J. Linn. Soc. 88 (1986a) 179

Jackson, R. R.: Cohabitation of males and juvenile females: a prevalent mating tactic of spiders. J. nat. Hist. 20 (1986b) 1193

Jackson, R R: Comparative study of releaser pheromones associated with the silk of jumping spiders (Araneae, Salticidae). N. Zeal. J. Zool. 14 (1987) 1

Jackson, R. R.: Predator–prey interactions between jumping spiders (Araneae, Salticidae) and *Pholcus phalangioides* (Araneae, Pholcidae). J. Zool. Lond. 220 (1990) 553

Jackson, R. R.: Eight-legged tricksters. Spiders that specialize in catching other spiders. BioScience 42 (1992) 590

Jackson, R. R., A. D. Blest: The biology of *Portia fimbriata*, a web-building jumping spider (Araneae, Salticidae) from Queensland: utilization of webs and predatory versatility. J. Zool. Lond. 196 (1982) 255

Jackson, R. R., R. J. Brassington: The biology of *Pholcus phalangioides* (Araneae, Pholcidae): predatory versatility, araneophagy and aggressive mimicry. J. Zool. Lond. 211 (1987) 327

Jackson, R. R., K. J. Cooper: The influence of body size and prior residency on the outcome of male–male interactions of *Marpissa marina*, a New Zealand jumping spider (Araneae, Salticidae). Ethol. Ecol. Evol. 3 (1991) 79

Jackson, R. R., S. E. A. Hallas: Comparative biology of *Portia africana, P. albimana, P. fimbriata, P. labiata,* and *P. shultzi*, araneophagic, web-building jumping spiders (Araneae: Salticidae): utilisation of webs, predatory versatility, and intraspecific interactions. N. Zeal. J. Zool. 13 (1986) 423

Jackson, R. R., A. Van Olphen: Prey capture techniques and prey preferences of *Corythalia canosa* and *Pystira orbiculata*, ant-eating jumping spiders (Araneae, Salticidae). J. Zool. Lond. 223 (1991) 577

Jackson, R. R., M. S. Tarsitano: Responses of jumping spiders to motionless prey. Bull. Brit. arachnol. Soc. 9 (1993) 105

Jackson, R. R., M. E. A. Whitehouse: The biology of New Zealand and Queensland pirate spiders (Araneae: Mimetidae): aggressive mimicry, araneophagy and prey specialization. J. Zool. Lond. (A) 210 (1986) 279

Jackson, R. R., M. B. Willey: The comparative study of the behaviour of *Myrmarachne*, ant-like jumping spiders (Araneae: Salticidae). Zool. J. Linn. Soc. 110 (1994) 77

Jackson, R. R., R. J. Brassington, R. J. Rowe: Anti-predator defences of *Pholcus phalangioides* (Araneae, Pholcidae), a web-building and web-invading spider. J. Zool. Lond. 220 (1990) 543

Jacobi-Kleemann, M.: Über die Lokomotion der Kreuzspinne *Aranea diadema* beim Netzbau (Nach Filmanalysen). Z. vergl. Physiol. 34 (1953) 606

Järlsfors, U., D. S. Smith, F. E. Russell: Innervation of the venom-secreting cells in the Black Widow Spider (*Latrodectus mactans*). In: A. De Vries, E. Kochva, eds: Toxins of Animal and Plant Origin. Vol. 1. Gordon & Breach, London 1971 (p. 159)

Job, W.: Das Röhrengewebe von *Aulonia albimanus* Walckenaer (Araneida: Lycosidae) und seine systematische Bedeutung. Zool. Anz. 180 (1968) 403

Job, W.: Beiträge zur Biologie der fangnetzbauenden Wolfsspinne *Aulonia albimana* Walckenaer 1805) (Arachnida, Araneae, Lycosidae, Hipassinae). Zool. Jb. Syst. 101 (1974) 560

Jocqué, R., J. Billen: The femoral organ of the Zodariinae (Araneae, Zodariidae). Rev. Zool. Afr. 101 (1987) 165

Juberthie, C.: Neurosecretory systems and neurohemal organs of terrestrial Chelicerata (Arachnida). In Gupta, A. P., ed.: Neurohemal Organs of Arthropods. Thomas, Springfield, Illinois 1983 (p. 149)

Juperthie-Jupeau, L., A. Lopez, T. Kronestedt: Structure et ultrastructure de la glande tibiale chez le mâle *d'Alopecosa cuneata* (Clerck) (Araneae, Lycosidae). Rev. Arachnol. 9 (1990) 63

Kaestner, A.: Bau und Funktion der Fächertracheen einiger Spinnen. Z. Morph. Oekol. Tiere 13 (1929) 463

Kaestner, A.: Reaktionen der Hüpfspinnen (Salticidae) auf unbewegte farblose und farbige Gesichtsreize. Zool. Beitr. 1 (1950) 12

Kaestner, A.: Die Mundwerkzeuge der Spinnen, ihr Bau, ihre Funktion und ihre Bedeutung für das System. 1. Teil. Orthognatha, Palaeocribellata. Zool. Jb. Anat. 72 (1952)101

Kaestner, A.: Lehrbuch der Speziellen Zoologie, 3. Aufl., Bd. I: Wirbellose. Fischer, Stuttgart 1969

Kaiser, E., W. Raab: Collagenolytic activity of snake and spider venoms. Toxicon 4 (1967) 251

Kaston, B. J.: The senses involved in the courtship of some vagabond spiders. Ent. Amer. 16 (1936) 97

Kaston, B. J.: The evolution of spider webs. Amer. Zool. 4 (1964) 191

Kaston, B. J.: Comparative biology of American Black Widow spiders. Transact. San Diego Soc. Nat. Hist. 16 (1970) 33

Kaston, B. J.: How to Know the Spiders. Browne, Dubuque/Iowa 1972

Kavanagh, E. J., E. K. Tillinghast: Fibrous and adhesive components of the orb webs of *Araneus trifolium* and *Argiope trifasciata*. J. Morphol. 160 (1979) 17

Kawai, N., A. Niwa, A. Abe: Spider venom contains specific receptor blocker of glutaminergic synapses. Brain Res. 247 (1982) 169

Kessler, A.: Relation between egg production and food consumption in species of the genus *Pardosa* (Lycosidae, Araneae) under experimental conditions of food-abundance and food-shortage. Oecologia 8 (1971a) 93

Kessler, A.: Hymenopterous egg parasites in *Pardosa* cocoons. 5th Int. Archnol. Congr. Brno 1971b (p. 145)

Kessler, A.: A comparative study of the production of eggs in eight *Pardosa* species in the field (Araneae, Lycosidae). T. Ent. 116 (1973) 23

Kessler, A., A. Fokkinga: Hymenopterous parasites in egg sacs of spiders of the genus *Pardosa* (Araneida, Lycosidae). T. Ent. 116 (1973) 43

Kirchner, W.: Bisher Bekanntes über die forstliche Bedeutung der Spinnen—Versuch einer Literaturanalyse. Waldhygiene 5 (1964) 161

Kirchner, W.: Ecological aspects of cold resistance in spiders (a comparative study). In: Wieser, W., ed.: Effects of Temperature on Ectothermic Organism. Springer Verlag, Berlin 1973 (p. 271)

Kirchner, W.: Das Netz der Zitterspinne (*Pholcus phalangioides* Fuesslin) (Araneae: Pholcidae). Zool. Anz. 216 (1986) 151

Kirchner, W.: Behavioral and physiological adaptations to cold. In: Nentwig, W., ed.: Ecophysiology of Spiders. Springer Verlag, Berlin 1987 (p. 66)

Kirchner, W., P. Kestler: Untersuchungen zur Kälteresistenz der Schilfradspinne *Araneus cornutus* (Araneidae). J. Insect Physiol. 15 (1969) 41

Kirchner, W., E. Kullmann: Überwinterung und Kälteresistenz der Haubennetzspinnenarten *Theridion impressum* (L: Koch) und *Theridion sisyphium* (Clerck)(Araneae, Theridiidae). Decheniana 127 (1975) 241

Klärner, D., F. G. Barth: Vibratory signals and prey capture in web spiders (*Zygiella x-notata, Nephila clavipes*). J. comp. Physiol. 148 (1982) 445

Klingel, H.: Beobachtungen an *Lipistius battuensis* Abr. (Araneae, Mesothelae). Verh. dtsch. zool. Ges. (Gött.) Suppl. 30 (1966) 246

Knost, S. J., J. S. Rovner: Scavenging by wolf spiders (Araneae: Lycosidae). Amer. Midl. Natur. 93 (1975) 239

Kolb, A.: Nahrung und Nahrungsaufnahme bei Fledermäusen. Z. Säugetierkunde, Berlin 73 (1958) 84

Komiya, M., S. Yamashita, H. Tateda: Turning reactions to real and apparent motion stimuli in the posterolateral eyes of jumping spiders. J. comp. Physiol. A 163 (1988) 585

Kosok, G., E.-A. Seyfarth: Vascularization and hemolymph supply of the spider CNS. Proc. 22th Göttingen Neurobiol. Conf. (1994) in press

Kovoor, J.: Etude histochimique et cytologique des glandes séricigènes de quelques Argiopidae. Ann. Sci. Nat. Zool. 14 (1972) 1

Kovoor, J.: Comparative structure and histochemistry of silk producing organs in Arachnids. In: Nentwig, W., ed.: Ecophysiology of Spiders. Springer Verlag, Berlin, 1987 (p. 160)

Kovoor, J., H. M. Peters: The spinning apparatus of *Polenecia producta* (Araneae, Uloboridae) : Structure and histochemistry. Zoomorphology 108 (1988) 47

Kovoor, J., L. Zylberberg: Histologie et infrastructure de la glande chélicerienne de *Scytodes delicatula* Sim. (Araneida, Scytodidae) Ann. Sci. Nat. Zool. 14 (1972) 333

Kovoor, J., L. Zylberberg: Fine structural aspects of silk secretion in a spider (*Araneus diadematus*). 1. Elaboration in the pyriform glands. Tissue Cell 12 (1980) 547

Krafft, B.: La société *d'Agelena consociata* Denis, araignée sociale du Gabon. 96 Congr. Nat. Soc. Savantes, Toulouse 1971

Krafft, B.: La tolerance réciproque chez l'araignée sociale *Agelena consociata* Denis. Proc. 6th Arachnol. Congr. 1974. Free University Amsterdam 1975 (p. 107)

Krafft, B.: The recording of vibratory signals performed by spiders during courtship. Symp. Zool. Soc. Lond. 42 (1978) 59

Krafft, B.: Les araignées sociales. La Recherche 16 (1985) 884

Krakauer, T.: Thermal responses of the orbweaving spider *Nephila clavipes* (Araneae: Argiopidae). Amer. Midl. Natur. 88 (1972) 245

Kraus, O.: *Liphistius* and the evolution of spider genitalia. Symp. Zool. Soc. Lond. 42 (1978) 235

Kraus, O., M. P. Kraus: The genus *Stegodyphus* (Arachnida, Araneae). Sibling species, species groups, and parallel origin of social living. Verh. Naturwiss. Ver. Hamburg (NF) 30 (1988) 151

Kraus, O., M. Kraus: Divergent transformation of chelicerae and original arrangement of eyes in spiders (Arachnida, Araneae). Mem. Queensland Mus. 33 (1993) 579

Krishnakumaran, A., H. A. Schneiderman: Chemical control of moulting in arthropods. Nature (Lond.) 220 (1968) 601

Kronestedt, T.: Study of a stridulatory apparatus in *Pardosa fulvipes* (Collett) (Araneae, Lycosidae) by scanning electron microscopy. Zool. Scripta 2 (1973) 43

Kronestedt, T.: A presumptive pheromone-emitting structure in wolf spiders (Araneae, Lycosidae). Psyche 93 (1986) 127

Kronestedt, T.: Separation of two species standing as *Alopecosa aculeata* (Clerck) by morphological, behavioural and ecological characters, with remarks on related species in the *pulverulenta* group (Araneae, Lycosidae). Zool. Scripta 19 (1990) 203

Kuenzler, E. J.: Niche relations of three species of lycosid spiders. Ecology 39 (1958) 494

Kühne, H.: Die neurosekretorischen Zellen und der retrocerebrale neuroendokrine Komplex von Spinnen (Araneae, Labidognatha) unter Berücksichtigung einiger histologisch erkennbarer Veränderungen während des postembryonalen Lebens. Zool. Jb. Anat. 77 (1959) 527

Kullmann, E.: Beobachtungen und Betrachtungen zum Verhalten der Theridiide *Conopista argyrodes* Walckenaer (Araneae). Mitt. Zool. Mus. 35 (1959) 275

Kullmann, E.: Der Eierkokonbau von *Cyrtophora citricola* Forskal (Araneae, Araneidae). Zool. Jb. Syst. 89 (1961) 369 0

Kullmann, E.: Spinnorgan mit 40000 "Düsen." Umschau Wiss. Techn. 3 (1969) 82

Kullmann, E.: The convergent development of orb-webs in cribellate and ecribellate spiders. Amer. Zool. 12 (1972a) 395

Kullmann, E.: Evolution of social behaviour in spiders (Araneae; Eresidae and Theridiidae). Amer. Zool. 12 (1972b) 419

Kullmann, E.: *Theridion impressum* (Theridiidae) Brutfürsorge und periodisch-soziales Verhalten. Encycl. Cinemat. E 1864, Göttingen 1974

Kullmann, E., W. Kloft: Traceruntersuchungen zur Regurgitationsfütterung bei Spinnen (Araneae, Theritiidae). Zool. Anz., Suppl. 32 (1969) 487

Kullmann, E., H. Stern: Leben am seidenen Faden. Bertelsmann München 1975 (p. 32)

Kullmann, E., W. Zimmermann: Ein neuer Beitrag zum Cribellaten-Ecribellaten-Problem: Beschreibung von *Uroecobius ecribellatus* n. gen. n. spec. und Diskussion seiner phylogenetischen Stellung (Arachnida: Araneae: Oecobiidae). Ent. Germ. 3 (1976) 29

Kullmann, E., H. Sittertz, W. Zimmermann: Erster Nachweis von Regurgitationsfütterungen bei einer cribellaten Spinne (*Stegodyphus lineatus* Latreille, 1817, Eresidae). Bonn. Zool. Beitr. 22 (1971) 175

Kullmann, E., S. Nawabi, W. Zimmermann: Neue Ergebnisse zur Brutbiologie cribellater Spinnen aus Afghanistan und der Serengeti. Z. d. Kölner Zoo 14 (1971/1972) 87

Kurpiewski, G., L. J. Forster, T. J. Barrett, B. J. Campbell: Platelet aggregation and sphingomyelinase A activity of a purified toxin of *Loxosceles reclusa*. Biochim. Biophys. Acta 678 (1981) 467

Lamoral, B. H.: On the ecology and habitat adaptation of two intertidal spiders *Desis formidabilis* and *Amaurobioides africanus* at "The Island" (Kommetjie, Cape Peninsula). Ann. Nat. Mus. 20 (1968) 151

Lamy, E.: Recherches anatomiques sur les trachées des Araignées. Ann. Sci. Nat. Zool. 15 (1902) 149

Land, M. F.: Structure of the retinae of the principal eyes of jumping spiders (Salticidae: Dendryphantinae) in relation to visual optics. J. exp. Biol. 51 (1969a) 443

Land, M. F.: Movements of the retinae of jumping spiders (Salticidae: Dendryphantinae) in response to visual stimuli. J. exp. Biol. 51 (1969b) 471

Land, M. F.: Orientation by jumping spiders in the absence of visual feedback. J. exp. Biol. 54 (1971) 119

Land, M. F.: Stepping movements made by jumping spiders during turns mediated by the lateral eyes. J. exp. Biol. 57 (1972a)

Land, M. F.: Mechanism of orientation and pattern recognition by jumping spiders. In: Wehner, R., ed.: Information Processing in the Visual Systems of Arthropods. Springer Verlag, Berlin 1972b (p. 231)

Land, M. F.: The morphology and optics of spider eyes. In: Barth, F. G., ed.: Neurobiology of Arachnids. Springer Verlag, Berlin 1985 (p. 53)

Larcher, S. F., D. H. Wise: Experimental studies of the interactions between a web-invading spider and two host species. J. Arachnol. 13 (1985) 43

Le Berre, M.: Modifications temporelles de l'activité chez les jeunes Argiopidés. Rev. Arachnol. 2 (1979) 193

Lee, C. K., T. K. Chan, B. C. Ward, D.E. Howell, G. V. Odell: Characterization of a neurotoxin from tarantula, *Dugesiella hentzi* (Girard) venom. Arch. Biochem. Biophys. 164 (1974) 341

Lee, R. C. P., M. Nyffeler, E. Krelina, B. W. Pennycook: Acoustic communication in two species of the genus *Steatoda* (Araneae, Theridiidae). Mitteilg Schweiz. Entomol. Ges. 59 (1986) 337

Legendre, R.: Le système sympathique stomatogastrique (organe de Schneider) des araignées du genre *Tegenaria*. C. R. Acad. Sci. 237 (1953) 1283

Legendre, R.: Localisation d'un organe olfactif non encore décrit ches les Aranéides. C. R. Acad. Sci. (Paris) 143 (1956) 1237

Legendre, R.: Contributions a l'étude du système nerveux des Aranéides. Ann. Biol. 34 (1958) 13

Legendre, R.: Le méchanism de la prise de nourriture chez les Araignées C. R. Acad. Sci. (Paris) 252 (1961a) 321

Legendre, R.: Sur deux particularités du système nerveux central de la mygale *Scodra calceata* Fabr. Ann. Sci. Nat. Zool. 3 (1961b) 767

Legendre, R.: L'audition et l'émission de sons chez les aranéides. Ann. Biol. 2 (1963) 371
Legendre, R.: Morphologie et développement des chélicerates. Embryologie, développement et anatomie des aranéides. Fortschr. Zool. 17 (1965) 237
Legendre, R.: Sur la présence d'un nerf cardiaque chez les Araignées orthognathes. C. R. Acad. Sci. (Paris) 267 (1968) 84
Legendre, R.: Un organ stridulent nouveau chez les Archeidae (Araneida). Bull. Soc. Zool. France 95 (1970) 1
Legendre, R.: Les Arachnides de Madagascar. In: Battistini,R., G. Richard-Vindard, eds: Biogeography and Ecology in Madagascar. Junk, Den Haag 1972 (p. 427)
Legendre, R.: The stomatogastric nervous system and neurosecretion. In: Barth, F. G., ed.: Neurobiology of Arachnids. Springer Verlag, Berlin 1985 (p. 38)
LeGuelte, L.: Learning in spiders. Amer. Zool. 9 (1969) 145
Lehmensick, R., E. Kullmann: Ueber den Feinbau der Fäden einiger Spinnen. (Vergleich des Aufbaues der Fangfäden cribellater und ecribellater Spinnen). Zool. Anz. 19, Suppl. (1956) 123
Lehtinen, P. T.: Classification of the cribellate spiders and some allied families. Ann. Zool. Fennici 4 (1967) 199
Lehtinen, P. T.: Notes on the phylogenetic classification of Araneae. Proc. 6th Int. Arachnol. Congr. 1974, Free University Amsterdam 1975 (p. 26)
Lehtinen, P. T.: Trichobothrial patterns in high-level taxonomy of spiders. Proc. 8th Int. Arachnol. Congr. 1980, Vienna 1980 (p. 466)
Levi, H. W. Adaptations of respiratory systems of spiders. Evolution 21 (1967) 571
Levi, H. W.: Orb-webs: primitive or specialized. Proc. 8th Int. Congr. Arachnol. Wien 1980 (p. 367)
Liesenfeld, F.: Untersuchungen am Netz und über den Erschütterungsinn von *Zygiella x-notata* (Cl.) (Araneidae). Z. vergl. Physiol. 38 (1956) 563
Liesenfeld, F.: Über Leistung und Sitz des Erschütterungssinnes von Netzspinnen. Biol. Zbl. 80 (1961) 465
Linskens, H.F., F. Ciampolini, M. Cresti: Spider webs as pollen traps. Proc. Kon. Ned. Akad. Wetensch. 96 (1993) 415
Linzen, B., P. Gallowitz: Enzyme activity pattern in muscles of the lycosid spider, *Cupiennius salei*. J. comp. Physiol. 96 (1975) 101
Linzen, B., W. Schartau, H. J. Schneider: Primary structure of arthropod hemocyanins. In:Lamy, J., J. P. Truchot, R. Gilles, eds: Respiratory Pigments in Animals. Springer Verlag, Berlin 1985 (p. 59)
Loewe, R., B. Linzen, W. v. Stackelberg: Die gelösten Stoffe in der Haemolymphe einer Spinne, *Cupiennius salei* Keyserling. Z. vergl. Physiol. 66 (1970) 27
Lombardi, S. J., D. L. Kaplan: The *Nephila clavipes* major ampullate gland silk protein: amino acid composition and the detection of silk gene-related nucleic acids in the genome. Acta Zool. Fennica 190 (1990) 243
Lopez, A.P., M. Emerit: New data on the epigastric apparatus of male spiders. Bull. Brit. arachnol. Soc. 7 (1988) 220
Lopez, A., J. C. Bonaric, L. Boissin: Etude ultrastructurale de la spermiogénèse chez l'araignée *Pisaura mirabilis* (Clerck, 1758) (Pisauridae). Rev. Arachnol. 5 (1983) 55
Lubin, Y. D.: Web structure and function: the non-adhesive orb-web of *Cyrtophora mollucensis* (Doleschall) (Araneae: Araneidae). Forma Functio 6 (1973) 337
Lubin, Y. D.: The predatory behavior of *Cyrtophora* (Araneae: Araneidae). J. Arachnol. 8 (1980a) 159
Lubin, Y. D.: Population studies of two colonial orb-weaving spiders. Zool. J. Linn. Doc. 70 (1980b) 265

Lubin, Y. D.: Web building and prey capture in the Uloboridae. In: Shear, W. A., ed.: Spiders: Webs, Behavior, and Evolution. Stanford University Press, Stanford 1986 (p. 132)

Lubin, Y. D.: Patterns of variation in female-biased colony sex ratios in a social spider. Biol. J. Linn. Soc. 43 (1991) 297

Lubin, Y. D., J. R. Henschel: Foraging at the thermal limit: burrowing spiders (*Seothyra,* Eresidae) in the Namib desert dunes. Oecologia 84 (1990) 461

Lubin, Y. D., W. G. Eberhard, G. G. Montgomery: Webs of *Miagrammopes* (Araneae: Araneidae) in the neotropics. Psyche 85 (1978) 1

Lubin, Y. D., B. D. Opell, W. C. Eberhard, H. W. Levi: Orb plus cone-webs in Uloboridae (Araneae), with a description of a new genus and four new species. Psyche 89 (1982) 29

Lucas, F.: Spiders and their silks. Discovery 25 (1964) 1

Lucas, S.: Spiders in Brazil. Toxicon 26 (1988) 759

Luczak, J.: The distribution of wandering spiders in different layers of the environment as a result of interspecies competition. Ekol. Polska A 14 (1966) 233

Ludwig, M., G. Alberti: Mineral congregations, "spherites," in the midgut gland of *Coelotes terrestris* (Araneae): structure, composition and function. Protoplasma 143 (1988) 43

Lüters, H.: Der Lebenszyklus von *Agroeca brunnea* Blackwall (Araneae, Clubionidae) unter besonderer Berücksichtigung des Kokonbau- und des Häutungsverhaltens. Diss. Universität Göttingen (1966)

Maddison, W. P.: XXXY Sex chromosomes in males of the jumping spider genus *Pellenes* (Araneae: Salticidae). Chromosoma (Berlin) 85 (1982) 23

Magni, F.: Analysis of polarised light in wolf spiders. In: Bernhard, C. G., ed.: The Functional Organisation of the Compound Eye. Pergamon Press, Oxford 1966 (p. 201)

Maier, L., T. M. Root, E.-A. Seyfarth: Heterogeneity of spider leg muscle: histochemistry and electrophysiology of identified fibers in the claw levator. J. comp. Physiol. B 157 (1987) 285

Main, B. Y.: Spiders. Collins, Sydney 1976

Main, B. Y.: Adaptations to arid habitats by mygalomorph spiders. In: Barker, W. R., P. J. M. Greensdale, eds: Evolution of the Flora and Fauna of Arid Australia. S.A Peacock Publ., Frewville, 1982 (p. 273)

Main, B. Y.: Social behaviour of a nonsnare-building spider (Thomisidae). Actas X Congr. Int. Aracnol. Jaca/Espana 1986 (p. 134)

Main, B. Y.: The biology of a social thomisid spider. Austral. Arachnol. (Austr. ent. Soc. misc. Publ. No. 5, Brisbane) (1988) 55

Mansour, F., W. H. Whitcomb: The spiders of a citrus grove in Israel and their role as biocontrol agents of *Ceroplastes floridensis* (Homoptera: Coccidae). Entomophaga 31 (1986) 269

Mansour, F., M. Wysoki, W. H. Whitcomb: Spiders inhabiting avocado orchards and their role as natural enemies of *Boarmia selenaria* Schiff Lepidoptera: Geometridae larvae in Israel. Acta Oecologia/Oecol. Applic. 6 (1985) 315

Marc, P.: Nycthemeral activity rhythm of adult *Clubiona corticalis* (Walckenaer, 1802) (Araneae, Clubionidae). Acta Zool. Fennica 190 (1990) 279

Maretić, Z.: Toxicity of "non-venomous" spiders. Proc. 5th Int. Arachnol. Congr. Brno 1971 (p. 201)

Maretić, Z.: European Araneism. Bull. Brit. arachnol. Soc. 3 (1975) 126

Maretić, Z.: Electrocardiographic changes following bites and stings of venomous animals. Arh. Hig. Rada. 33 (1982) 325

Maretić, E: Latrodectism: variations in clinical manifestations provoked by *Latrodectus* species of spiders. Toxicon 21 (1983) 457

Maretić, Z.: Spider venoms and their effect. In: Nentwig, W., ed.: Ecophysiology of Spiders. Springer Verlag, Berlin 1987 (p 142)

Maretić, Z., S. Habermehl: Latrodektismus bei Menschen und Tieren sowie grosse Tierseuchen des vorigen Jahrhunderts, hervorgerufen durch *Latrodektus*. Dtsch. tierärztl. Wschr. 92 (1985) 245

Markl, H.: Borstenfelder an den Gelenken als Schweresinnesorgane bei Ameisen und anderen Hymenopteren. Z. vergl. Physiol. 45 (1962) 475

Markl, H: Die Verständigung durch Stridulationssignale bei Blattschneiderameisen. I. Die biologische Bedeutung der Stridulation. Z. vergl. Physiol. 57 (1967) 299

Markl, H.: Vom Eigennutz des Uneigennützigen. Die Evolution hochentwickelter Sozialsysteme im Tierreich. Naturw. Rdsch. 24 (1971) 281

Maroli, M., S. Bettini, B. Panfili: Toxicity of *Latrodectus mactans tredecimguttatus* venom on frog and birds. Toxicon 11 (1973) 203

Marples, B. J.: The spinnerets and epiandrous glands of spiders. J. Linn. Soc. Zool. 46 (1967) 209

Martin, A. W.: Circulation in invertebrates. Ann. Rev. Physiol. 36 (1974) 171

Martin, D.: Zum Radnetzbau der Gattung *Pachygnatha* Sund. (Araneae: Tetragnathidae). Mitteilg. Zool. Mus. Berlin 54 (1978) 83

Masters, W. M.: Vibrations in the orbwebs of *Nuctenea sclopetaria* (Araneidae). I. Transmission through the web. Behav. Ecol. Sociobiol. 15 (1984a) 207

Masters, W. M.: Vibrations in the orbwebs of *Nuctenea sclopetaria* (Araneidae). II. Prey and wind signals and the spider's response threshold. Behav. Ecol. Sociobiol. 15 (1984b) 217

Masters, W. M., H. Markl: Vibration signal transmission in spider orb-webs. Science 213 (1981) 363

Masters, W. M., A. J. M. Moffat: A functional explanation of top–bottom asymmetry in vertical orbwebs. Anim. Behav. 31 (1983) 1043

Masters, W. M., H. S. Markl, A. J. Moffat: Transmission of vibration in a spider's web. In: Shear, W. A., ed.: Spiders. Webs, Behavior, and Evolution. Stanford University Press, Stanford 1986 (p. 49)

Masumoto, T.: The effect of the copulatory plug in the funnel-web spider, *Agelena limbata* (Araneae, Agelenidae). J. Arachnol. 21 (1993) 55

Mather, M. H., B. D. Roitberg: A sheep in wolf's clothing: tephritid flies mimic spider predators. Science 236 (1987) 308

Mathew, A. P.: Observations on the habits of two spider mimics of the red ant *Oecophylla smaragdina* (Fabr.). J. Bombay Nat. Hist. Soc. 52 (1954) 249

McCormick, S. J., G. A. Polis: Arthropods that prey on vertebrates. Biol. Rev. 57 (1982) 29

McCrone, J. D., R. J. Hatla: Isolation and characterization of a lethal compound from the venom of *Latrodectus mactans mactans*. In: Russell, F. E., P. R. Saunders, eds: Animal Toxins. Pergamon Press, New York 1967 (p. 29)

McCrone, J. D., H. W. Levi: North American widow spiders of the *Latrodectus curacaviensis* group (Araneae: Theridiidae). Psyche 71 (1964) 12

McCrone, J. D., M. L. Netzloff: An immunological and electrophoretical comparison of the venoms of the North American *Latrodectus* spiders. Toxicon 3 (1965) 107

McQueen, D. J.: Interactions between the pompilid wasp *Anoplius relativus* (Fox) and the burrowing wolf spider *Geolycosa domifex* (Hancock). Can. J. Zool. 57 (1979) 542

Meier, F.: Beiträge zur Kenntnis der postembryonalen Entwicklung der Spinnen, Araneida, Labidognatha, unter besonderer Berücksichtigung der Histogenese des Zentralnervensystems. Rev. Suisse Zool. 74 (1967) 127

Melchers, M.: Zur Biologie und zum Verhalten von *Cupiennius salei* (Keyserling), einer amerikanischen Ctenide. Zool. Jb. Syst. 91 (1963) 1

Melchers, M.: Zur Biologie der Vogelspinnen (Fam. Aviculariidae). Z. Morph. Oekol. Tiere 53 (1964) 517

Melchers, M.: Der Beutefang von *Cupiennius salei* (Keyserling) (Ctenidae). Z. Morph. Oekol. Tiere 58 (1967) 321

Meldolesi, J., H. Scher, L. Madeddu, E. Wanke: Mechanism of action of α-latrotoxin: the presynaptic stimulatory toxin of the black widow spider venom. Trans. Pharmacol. Sci. 7 (1986) 151

Meyer, E: Neue sinnesbiologische Beobachtungen an Spinnen. Z. Morph. Oekol. Tiere 12 (1928) 1

Meyer, W., C. Schlesinger, H. M. Poehling, W. Ruge: Comparative quantitative aspects of putative neurotransmitters in the central nervous system of spiders (Arachnida: Araneida). Comp. Biochem. Physiol. 78c (1984) 357

Michaelis, E. K., N. Galton, S. L. Early: Spider venoms inhibit L-glutamate binding to brain synaptic membrane receptors. Proc. nat. Acad. Sci. 81 (1984) 5571

Midttun, B.: Ultrastructure of cardiac muscle of *Trochosa terricola* Thor., *Pardosa amentata* Clerck, *P. pullata* Cl., and *Pisaura mirabilis* Cl. (Araneae: Lycosidae, Pisauridae). Cell Tiss. Res 181 (1977) 299

Milde, J. J., E.-A. Seyfarth: Tactile hairs and leg reflexes in wandering spiders: physiological and anatomical correlates of reflex activity in the leg ganglia. J. comp. Physiol. A 162 (1988) 623

Mill, P. J., D. J. Harris: Observations on the leg receptors of *Ciniflo* (Araneida: Dictynidae) III. Proprioreceptors. J. comp. Physiol. 119 (1977) 63

Millot, J.: Contribution à l'histophysiologie des Aranéides. Bull. biol. Franç. Belg. Suppl. 8 (1926) 1

Millot, J.: Glandes venimeuses et glandes séricigènes chez les Sicariides. Bull. Soc. Zool. Franç. 55 (1930) 150

Millot, J.: Le sens du goût chez les Araignées. Bull. Soc. Zool. Franç. 61 (1936) 27

Millot, J.: Sens chimiques et sens visuel chez les Araignées. Ann. Biol. 22 (1946) 1

Millot, J.: Ordre des Aranéides (Araneae). In: Grassé, P.-P., ed.: Traité de Zoologie, vol. VI. Masson, Paris 1949 (p. 589)

Minch, E. W.: The molting sequence in *Aphonopelma chalcodes* Chamberlin. J. Arachnol. 5 (1977) 133

Minch, E. W.: Daily activity patterns in the tarantula *Aphonopelma chalcodes* Chamberlin. Bull. Brit. arachnol. Soc. 4 (1978) 231

Mittal, O. P.: Karyological studies on the Indian spiders. II. An analysis of the chromosomes during spermatogenesis in five species belonging to the family Salticidae. Res. Bull. (N. S.) Panjab Univ. 15 (1964) 315

Mittal, O. P.: Sex mechanism in spiders. Proc. 53rd Ind. Sci. Congr. 1966. Abstract (p. 320)

Miyashita, K.: Growth and development of *Lycosa T-insignata* Boes. et Str. (Araneae: Lycosidae) under different feeding conditions. Appl. entomol. Zool. 3 (1968) 81

Moller, P.: Die systematischen Abweichungen bei der optischen Richtungsorientierung der Trichterspinne *Agelena labyrinthica*. Z. vergl. Physiol. 66 (1970) 78

Monterosso, B.: Note araneologiche.- Su la biologia degli Scitotidi e la ghiandola glutinifera di essi. Arch. Zool. Ital. 12 (1928) 63

Montgomery, T. H.: On the maturation mitosis and fertilization of the egg of *Theridium*. Zool. Jb. Anat. 25 (1908) 237

Mullen, G. R.: Morphology and histology of the silk glands in *Araneus sericatus* Cl. Trans. Amer. microsc. Soc. 88 (1969) 232

Müller, H.-G.: Ein Mermithide als Parasitoid von *Coelotes inermis* (L. Koch 1855) (Arachnida: Araneae: Agelenidae). Entomol. Zeitschr. 93 (1983) 358

Müller, M. C., W. Westheide: Comparative morphology of the sexually dimorphic orb-weaving spider *Argiope bruennichi* (Araneae: Araneidae). Mem. Queensland Mus. Brisbane 33 (1993) 615

Nawabi, S.: Histologische Untersuchungen an der Mitteldarmdrüse von *Stegodyphus pacificus* (Pocock 1900) (Araneae, Eresidae). Diss. Universität Bonn 1974

Neet, C. R.: Function and structural variability of the stabilimenta of *Cyclosa insulana* (Costa) (Araneae, Araneidae). Bull. Brit. arachnol. Soc. 8 (1990) 161

Nemenz, H.: Der Wasserhaushalt einiger Spinnen. Oest. Zool. Z. 5 (1954) 123

Nemenz, H.: Ueber den Bau der Kutikula und dessen Einfluss auf die Wasserabgabe bei Spinnen. S.-B. Ost. Akad. Wiss., math.-naturw. Kl., Abt. I, 164 (1955) 65

Nentwig, W.: The selective prey of linyphiid-like spiders and of their webs. Oecologia (1980) 236

Nentwig, W.: Social spiders catch larger prey: a study of *Anelosimus eximius* (Araneae: Theridiidae). Behav. Ecol. Sociobiol. 17 (1985) 79

Nentwig, W., S. Heimer: Ecological aspects of spider webs. In: Nentwig, W., ed.: Ecophysiology of Spiders. Springer Verlag, Berlin 1987 (p. 211)

Nentwig, W., H. Rogg: The cross-stabilimentum of *Argiope argentata* (Araneae: Araneidae). Nonfunctional or a nonspcific stress reaction? Zool. Anz. 221 (1988) 248

Nentwig, W., C. Wissel: A comparison of prey length among spiders. Oecologia 68 (1986) 595

Nielsen, E.: Contributions to the life history of pimpline spider parasites. Ent. Med. 14 (1923) 137

Nielsen, E.: The Biology of Spiders. With Especial Reference to the Danish Fauna. Lewin & Munksgaard, Kopenhagen 1932 (p. 218)

Nitzsche, R.: Beutefang und "Brautgeschenk" bei der Raubspinne *Pisaura mirabilis* (Cl.) (Araneae: Pisauridae). Diss. Universität Kaiserslautern 1981

Nørgaard, E.: On the ecology of two Lycosid spiders (*Pirata piraticus* and *Lycosa pullata*) from a Danish Sphagnum bog. Oikos 3 (1951) 1

Nørgaard, E.: Environment and behaviour of *Theridion saxatile*. Oikos 7 (1956) 159

Nossek, M. E., J. S. Rovner: Agonistic behavior in female wolf spiders (Araneae, Lycosidae). J. Arachnol. 11 (1984) 407

Nyffeler, M.: Die ökologische Bedeutung der Spinnen in Forst-Ökosystemen, eine Literaturzusammenstellung. Anz. Schädlingskunde, Pflanzenschutz Umweltschutz 55 (1982) 134

Nyffeler, M., G. Benz: Zur ökologischen Bedeutung der Spinnen der Vegetationsschicht von Getreide- und Rapsfeldern bei Zürich (Schweiz). Z. angew. Entomol. 87 (1979) 348

Nyffeler, M., G. Benz: Kleptoparasitismus von juvenilen Kreuzspinnen und Skorpionsfliegen in den Netzen adulter Spinnen. Rev. Suisse Zool. 87 (1980) 907

Nyffeler, M., G. Benz: Spiders in natural pest control: a review. Z. angew. Entomol. 103 (1987) 321

Okuma, C., R. Kisimoto: Airborne spiders collected over the East China Sea. Jap. J. appl. Entomol. Zool. 25 (1981) 296

Olive, C.W.: Behavioral responses to a sit-and-wait predator to spatial variation in foraging gain. Ecology 63 (1982) 912

Oliveira, P.S., I. Sazima: The adaptive bases of ant-mimicry in a neotropical aphantochilid spider (Araneae: Aphantochilidae). Biol. J. Linn. Soc. 33 (1984) 145

Opell, B. D.: Eggsac differences in the spider family Uloboridae (Arachnida: Araneae). Trans. Amer. microsc. Soc. 103 (1984) 122

Opell, B. D.: The influence of web monitoring tactics on the tracheal systems of spiders in the family Uloboridae (Arachnida, Araneida). Zoomorph. 107 (1987) 255

Oppenheim, S.: Regeneration und Autotomie bei Spinnen. Zool. Anzeiger 33 (1908) 56

Osaki, H.: Electron microscope study on the spermatozoon of the Liphistiid spider *Heptathela*. Acta Arachnol. 22 (1969) 1

Osaki, H.: Electron microscope studies on developing oocytes of the spider *Plexippus paykulli*. Annot. Zool. Jap. 45 (1972) 187

Oxford, G.S., A. Plowman: Do large house spiders *Tegenaria gigantea* and *T. saeva* (Araneae, Agelenidae) hybridise in the wild ? A multivariate approach. Bull. Brit. arachnol. Soc. 8 (1991) 293

Oxford, G. S., C. J. Smith: The distribution of *Tegenaria gigantea* Chamberlin & Ivie, 1935 and *T. saeva* Blackwall, 1844 (Araneae, Agelenidae) in Yorkshire. Bull. Brit. arachnol. Soc. 7 (1987) 123

Pacala, S., J. Roughgarden: Control of arthropod abundance by *Anolis* lizards on St. Eustatius (Neth. Antilles). Oecologia 64 (1984) 160

Pack, M.: Morphologische Untersuchungen am peripheren Blutgefäßsystem der Spinnen *Zygiella* sp. und *Tegenaria* sp. Staatsexamensarbeit Ruhr Universität Bochum (1983)

Painter, T. S.: Spermatogenesis in spiders. Zool. Jb. Anat. 38 (1914) 509

Palmer, J. M.: The silk and silk production system of the funnel-web mygalomorph spider Euagrus (Araneae, Dipluridae). J. Morph. 186 (1985) 195

Palmer, J. M., F. A. Coyle, F. W. Harrison: Structure and cytochemistry of the silk glands of the mygalomorph spider *Antrodiaetus unicolor* (Araneae, Antrodiaetidae). J. Morph. 174 (1982) 269

Palmgren, P.: On the muscular anatomy of spiders. Acta Zool. Fennica 155 (1978) 1

Palmgren, P.: The mechanism of the extrinsic coxal muscles of spiders. Ann. Zool. Fennici 18 (1981) 203

Papi, F.: Astronomische Orientierung bei der Wolfspinne *Arctosa perita* (Latr.). Z. vergl. Physiol. 37 (1955) 230

Parry, D. A.: The small leg-nerve of spiders and a probable mechanoreceptor. Quart. J. microsc. Sci. 101 (1960) 1

Parry, D. A., R. H. J. Brown: The hydraulic mechanism of the spider leg. J. exp. Biol. 36 (1959a) 423

Parry, D. A., R. H. J. Brown: The jumping mechanism of salticid spiders. J. exp. Biol. 36 (1959b) 654

Pasquet, A.: Proies capturées et stratégies prédatrices chez deux espèces d'araignées orbitèles: *Argiope bruennichi* et *Araneus marmoreus*. Entomol. exp. appl. 36 (1984) 177

Paul, J., T. Fincke, B. Linzen: Respiration in the tarantula *Eurypelma californicum*: evidence for diffusion lungs. J. comp. Physiol. B 157 (1987) 209

Paul, R.: Gas exchange and gas transport in the tarantula *Eurypelma californicum*—an overview. In: Linzen, B., ed.: Invertebrate Oxygen Carriers. Springer Verlag, Berlin 1986 (p. 321)

Paul, R.J.: La respiration des arachnides. La Recherche 226 (1990) 1338

Paul, R.J.: Das blaue Blut der Vogelspinnen. Naturwiss. Rundschau 45 (1992a) 216

Paul, R. J.: Gas exchange, circulation, and energy metabolism in arachnids. In: Wood, S. C., R. E. Weber, A. R. Hargens, R. W. Millard, eds: Physiological Adaptations in Invertebrates. Respiration, Circulation, and Metabolism. Dekker, New York 1992b (p. 169)

Paul, R. J., S. Bihlmayer: Circulatory physiology of a tarantula (*Eurypelma californicum*). Zoology 98 (1995) 69

Paul, R. J., K. Tiling, P. Focke, B. Linzen: Heart and circulatory functions in a spider (*Eurypelma caifornicum*): the effects of hydraulic force generation. J. comp. Physiol. B 158 (1989) 673

Paul, R. J., S. Zahler, R. Werner: Adaptation of an open circulatory system to the oxidative capacity of different muscle cell types. Naturwiss. 78 (1991) 134

Paul, R. J., S. Bihlmayer, M. Colmorgen, S. Zahler: The open circulatory system of spiders (*Eurypelma californicum, Pholcus phalangioides*): a survey of functional morphology and physiology. Physiol. Zool. 67 (1994) 1360

Peakall, D. B.: Synthesis of silk, mechanism and location. Amer. Zool. 9 (1969) 71

Peakall, D. B.: Conservation of web proteins in the spider *Araneus diadematus*. J. exp. Zool. 176 (1971) 257

Peaslee, A. G., G. Wilson: Spectral sensitivity in jumping spiders (Araneae, Salticidae). J. comp. Physiol. A 164 (1989) 359

Perret, B. A.: Biologie, Gift und Giftigkeit der orthognathen Spinne *Pterinochilus* spec. (Fam. Theraphosidae). Diss. Universität Basel (1973)

Peters, H. M.: Die Fanghandlung der Kreuzspinne (*Epeira diademata* L.). Experimentelle Analysen des Verhaltens. Z. vergl. Physiol. 15 (1931) 693

Peters, H. M.: Experimente über die Orientierung der Kreuzspinne *Epeira diademata* Cl. im Netz. Zool. Jb. Zool. 51 (1932) 239

Peters, H. M.: Weitere Untersuchungen über die Fanghandlung der Kreuzspinne (*Epeira diademata* Cl.). Z. vergl. Physiol. 19 (1933) 47

Peters, H. M.: Über das Kreuzspinnennetz und seine Probleme. Naturwissenschaften 47 (1939a) 777

Peters, H. M.: Probleme des Kreuzspinnennetzes. Z. Morph. Oekol. Tiere 36 (1939b) 179

Peters, H. M.: Orientierungsvermögen und Gedächtnis der Kreuzspinne (*Aranea diadema*). Erläuterung zum Hochschulfilm C 595/1950, Institut fur Film und Bild in Wissenschaft und Unterricht, Göttingen 1951

Peters, H. M: Über den Spinnapparat von *Nephila madagascariensis*. Z. Naturforsch. 10b (1955) 395

Peters, H. M.: Maturing and coordination of webbuilding activity. Amer. Zool. 9 (1969) 223

Peters, H. M.: Wie Spinnen der Familie Uloboridae ihre Beute einspinnen und verzehren. Verh. naturwiss. Ver. Hamburg 25 (1982) 147

Peters, H. M.: The spinning apparatus of Uloboridae in relation to the structure and construction of capture threads (Arachnida, Araneida). Zoomorph. 104 (1984) 96

Peters, H. M.: Fine structure and function of capture threads. In: Nentwig, W., ed.: Ecophysiology of Spiders. Springer Verlag, Berlin 1987 (p. 187)

Peters, H. M.: Über Struktur und Herstellung von Fangfäden cribellater Spinnen der Familie Eresidae (Arachnida, Araneae). Verh. naturwiss. Ver. Hamburg 33 (1992) 213

Peters, H. M.: Über das Problem der Stabilimente in Spinnennetzen. Zool. Jb. Physiol. 97 (1993a) 245

Peters, H. M.: Functional organization of the spinning apparatus of *Cyrtophora citricola* with regard to the evolution of the web. Zoomorphol. 113 (1993b) 153

Peters, H. M.: Ultrastructure of orb spider's gluey capture threads. Naturwiss. 82 (1995) 380

Peters, H. M., J. Kovoor: Un complément à l'appareil séricigène des Uloboridae (Araneae): le paracribellum et ses glandes. Zoomorph. 96 (1980) 91

Peters, H. M., J. Kovoor: Die Herstellung der Eierkokons bei der Spinne *Polenecia producta* (SIMON 1873) in Beziehung zu den Leistungen des Spinnapparates. Zool. Jb. Physiol. 93 (1989) 125

Peters, H. M., J. Kovoor: The silk-producing system of *Linyphia triangularis* (Araneae, Linyphiidae) and some comparisons with Araneidae. Structure, histochemistry and function. Zoomorph. 111 (1991) 1

Peters, H. M., P. N. Witt, D. Wolff: Die Beeinflussung des Netzbaues der Spinnen durch neurotrope Substanzen. Z. vergl. Physiol. 32 (1950) 29

Peters, P. J.: Orb-web construction: interaction of spider (*Araneus diadematus* Cl.) and thread configuration. Behaviour 18 (1970) 478

Peters, R.: Vergleichende Untersuchungen über Bau und Funktion der Spinnwarzen und Spinnwarzenmuskulatur einiger Araneen. Zool. Beitr. 13 (1967) 29

Peters, W., C. Pfreundt: Die Verteilung von Trichobothrien und lyraförmigen Organen an den Laufbeinen von Spinnen mit unterschiedlicher Lebensweise. Zool. Beitr. N. F. 29 (1986) 209

Petersen, B.: The relation between size of mother and number of eggs and young in some spiders and its significance for the evolution of size. Experientia (Basel) 6 (1950) 96

Petrunkevitch, A.: Contributions to our knowledge of the anatomy and relationships of spiders. Ann. Ent. Soc. Amer. 2 (1909) 11

Petrunkevitch, A.: Ueber die Circulationsorgane von *Lycosa carolinensis*. Zool. Jb. Anat. 31 (1910) 161

Petrunkevitch, A.: Tarantula versus tarantulahawk: a study in instinct. J. exp. Zool. 45 (1926) 367

Petrunkevitch, A.: An inquiry into the natural classification of spiders, based on a study of their internal anatomy. Trans. Connect. Acad. Arts Sci. 31 (1933) 299

Petrunkevitch, A.: A study of amber spiders. Trans. Connect. Acad. Arts Sci. 34 (1942) 119

Petrunkevitch, A.: The spider and the wasp. Sci. Amer. 187 (1952) 20

Petrunkevitch, A.: Arachnida. In: Moore, R.C., ed.: Treatise on Invertebrate Paleontology. Arthropoda 2. University Press of Kansas, Lawrence/K 1955 (p. 42)

Piéron, H.: Autotomie et "autospasie". C. R. Séanc. Soc. Biol. 63 (1907) 427

Pinkston, K. N., J. H. Frick: Determinations of osmolarity on the fresh unfixed hemolymph of four species of spiders. Ann. Ent. Soc. Amer. 66 (1973) 696

Plateau, F.: Recherches sur la structure de l'appareil digestif et sur les phénomènes de la digestion chez les aranéides dipneumones. Bull. Acad. roy. Belg. 44 (1877) 129

Platnick, N. I.: The evolution of courtship behaviour in spiders. Bull. Brit. arachnol. Soc. 2 (1971) 40

Platnick, N.I.: On the validity of Haplogynae as a taxonomic grouping in spiders. Proc. 6th Int. Arachnol. Congr. Amsterdam 1975 (p. 30)

Platnick, N. I.: The hypochiloid spiders: a cladistic analysis, with notes on the Atypoidea (Arachnida Aranaea). Amer. Mus. Novitates No. 2627 (1977) 1

Platnick, N. I., W. J. Gertsch: The suborders of spiders: a cladistic analysis (Arachnida, Araneae). Amer. Mus. Novitates No. 2607 (1976) 1

Platnick, N. I., J. A. Coddington, R. R. Forster, C. E. Griswold: Spinneret morphology and the phylogeny of haplogyne spiders (Araneae, Araneomorphae). Amer. Mus. Novitates 3016 (1991) 1

Poinar, G. O.: Nematode parasites of spiders. In: Nentwig, W., ed.: Ecophysiology of Spiders. Springer Verlag, Berlin 1987 (p. 299)

Pollard, S. D.: Little murders. Nat. Hist. 102 (1993) 58

Pollard, S. D.: Consequences of sexual selection on feeding in male spiders (Araneae: Salticidae). J. Zool. Lond. 234 (1994) 203

Pollard, S. D., R. R. Jackson: The biology of *Clubiona cambridgei* (Araneae, Clubionidae): interspecific interactions. N. Zeal. J. Ecol. 5 (1982) 44

Pollard, S. D., A. M. Macnab, R. R. Jackson: Communication with chemicals: pheromones and spiders. In: Nentwig, W., ed.: Ecophysiology of Spiders. Springer Verlag, Berlin 1987 (p. 133)

Pötzsch, J.: Von der Brutfürsorge heimischer Spinnen. Neue Brehm Bücherei. Ziemsen, Wittenberg Lutherstadt 1963

Prestwich, K. N.: Anaerobic metabolism in spiders. Physiol. Zool. 56 (1983a) 112

Prestwich, K. N.: The roles of aerobioc and anaerobic metabolism in active spiders. Physiol. Zool. (1983b) 122

Prestwich, K. N.: The constraints on maximal activity in spiders. I. Evidence against the fluid insufficiency hypothesis. J. comp. Physiol. B 158 (1988a) 437

Prestwich, K. N.: The constraints of maximal activity in spiders. II. Limitations imposed by phosphagen depletion and anaerobic metabolism. J. comp. Physiol. B 158 (1988b) 449

Prestwich, K. N., N. H. Ing: The activities of enzymes associated with anaerobic pathways, glycolysis, and the Krebs cycle in spiders. Comp. Biochem. Physiol. 72 B (1982) 295

Pringle, J. W. S.: Proprioception in insects. III. The function of the hair sensilla at the joints. J. exp. Biol. 15 (1938) 467

Pulz, R.: Temperature-related behavior and temperature perception in spiders: review of present knowledge. In: Nachtigall, W., H. Laudien, eds: Thermal Relations in Animals and Man. Biona-Report 4, Akad. Wiss. Mainz. Fischer, Stuttgart 1986 (p. 41)

Purcell, E. F.: Development and origin of the respiratory organs of Araneae. Quart. J. microsc. Sci. 54 (1909) 1

Purcell, W. F.: The phylogeny of trachea in Araneae. Quart. J. microsc. Sci. 54 (1910) 519

Ramousse, R.: Organisation temporelle du comportement constructeur chez huit espèces d'Argiopides. C. R. 5. Colloq. Arachnol. IX Barcelona 1980 (p. 203)

Ramousse, R., F. Davis: Web-building time in a spider: preliminary application of ultrasonic detection. Physiol. Behav. 17 (1976) 997

Ramousse, R., L. LeGuelte: Relations spatio-temporelles dans le comportement constructeur d'*Araneus diadematus*. Rev. Arachnol. 2 (1979) 183

Randall, J. B.: Regeneration and autotomy exhibited by the black widow spider, *Latrodectus variolus* Walckenaer. Wilhelm Roux Arch. dev. Biol. 190 (1981) 230

Rathmayer, W.: Die Innervation der Beinmuskeln einer Spinne, *Eurypelma hentzi* Chamb. (Orthognatha, Aviculariidae). Verh. dtsch. zool. Ges. (Jena) (1965a) 505

Rathmayer, W.: Polyneurale Innervation bei Spinnen. Naturwissenschaften 52 (1965b) 114

Rathmayer, W.: Neuromuscular transmission in a spider and the effect of calcium. Comp. biochem. Physiol. 14 (1965c) 673

Rathmayer, W.: The effect of the poison of spider and diggerwasps on their prey (Hymenoptera: Pompilidae, Sphecidae). Mem. Inst. Butantan, Simp. Internac. 33 (1966) 651

Rathmayer, W.: Elektrophysiologische Untersuchungen an Propriorezeptoren im Bein einer Vogelspinne (*Eurypelma hentzi* Chamb.). Z. vergl. Physiol 54 (1967) 438

Rathmayer, W.: Venoms of Sphecidae, Pompilidae, Mutillidae and Bethylidae In: Bettini, S., ed.: Handbook of Experimental Pharmacology, Vol. XLVIII. Springer Verlag, Berlin 1978 (p. 661)

Rathmayer, W., J. Koopmann: Die Verteilung der Propriorezeptoren im Spinnenbein. Z. Morph. Tiere 66 (1970) 212

Raven, R. J.: The spider infraorder Mygalomorphae: cladistics and systematics. Bull. Amer. Mus. Nat. Hist. 182 (1985) 1

Reed, C. F., P. N. Witt, R. L. Jones: The measuring function of the first legs of *Araneus diadematus* Cl. Behaviour 25 (1965) 98

Reed, C. F., P. N. Witt, M. B. Scarboro, D. B. Peakall: Experience and the orb web. Devel. Psychobiology 3 (1970) 251

Reger, J. F.: Spermiogenesis in the spider, *Pisaurina* sp.: A fine structure study. J. Morph. 130 (1970) 421

Rehnberg, B. G.: Selection of spider prey by *Trypoxylon politum* (Say) (Hymenoptera: Sphecidae). Can. Entomol. 119 (1987) 189

Reisinger, P. W., M., P. Focke, B. Linzen: Lung morphology of the tarantula, *Eurypelma californicum* Ausserer 1871 (Araneae: Theraphosidae). Bull. Brit. arachnol. Soc. 8 (1990) 165

Reisinger, P. W., I. Tutter, U. Welsch: Fine structure of the horseshoe crabs *Limulus polyphemus* and *Tachypleus tridentatus* and of the book lungs of the spider *Eurypelma californicum*. Zool. Jb. Anat. 121 (1991) 331

Reiskind, J.: Multiple mimetic forms in an antmimicking clubionid spider. Science 169 (1970) 587

Reiskind, J.: Ant-mimicry in Panamian clubionid and salticid spiders (Araneae: Clubionidae, Salticidae). Biotropica 9 (1977) 1

Reissland, A., P. Görner: Trichobothria. In: Barth, F. G., ed. : Neurobiology of Arachnids. Springer Verlag, Berlin 1985 (p. 138)

Richardson, C. E. W.: Specificity of the spider web pattern and its efficiency in trapping prey. Amer. Zool. 13 (1973) 59

Richman, D. B., R. R. Jackson: A review of the ethology of jumping spiders (Araneae, Salticidae). Bull. Brit. arachnol. Soc. 9 (1992) 33

Richter, C. J. J.: Morphology and function of the spinning apparatus of the wolf spider *Pardosa amentata* (Cl) (Araneae, Lycosidae). Z. Morph. Tiere 68 (1970) 37

Richter, C. J. J.: Some aspects of the serial dispersal in different populations of wolf spiders, with particular reference to *Pardosa amentata* (Araneae, Lycosidae). Misc. Pap. Landb. Hogesch Wageningen 8 (1971) 77

Riechert, S. E.: Thoughts on the ecological significance of spiders. Bio. Sci. 24 (1974) 352

Riechert, S. E.: Web-site selection in the desert spider *Agelenopsis aperta*. Oikos 27 (1976) 311

Riechert, S. E.: Games spiders play: behavioral variability in territorial disputes. Behav. Ecol. Sociobiol. 3 (1978) 135

Riechert, S. E.: Spider interaction strategies: communication vs. coercion. In: Witt P. N., J. S. Rovner, eds : Spider Communication: Mechanisms and Ecological Significance. Princeton University Press, Princeton, New Jersey 1982 (p. 281)

Riechert, S. E., A. B. Cady: Patterns of resource use and tests for competitive release in a spider community. Ecology 64 (1983) 899

Riechert, S. E., T. Lockley: Spiders as biological control agents. Ann. Rev. Entomol. 29 (1984) 299
Robinson, B. C., M. H. Robinson: The biology of some *Argiope* species from New Guinea: predatory behaviour and stabilimentum construction (Araneae: Araneidae). Zool. J. Linn. Soc. 54 (1974) 145
Robinson, M. H.: The evolution of predatory behaviour in araneid spiders. In: Baerends, G., C. Beer, A. Manning, eds: Function and Evolution in Behaviour, Clarendon Press, Oxford 1975 (p. 292)
Robinson, M. H.: Ecology and behaviour of tropical spicers. Proc. 8th Int. Arachnol. Congr. Wien 1980 (p. 13)
Robinson, M. H., H. Mirick: The predatory behaviour of the golden-web spider *Nephila clavipes* (Araneae: Araneidae). Psyche 78 (1971) 123
Robinson, M. H., B. Robinson: The predatory behaviour of the ogre-faced spider *Dinopis longipes* F. Cambridge. Amer. Midl. Natl. 85 (1971) 85
Robinson, M. H., B. Robinson: The structure, possible function and origin of the remarkable ladder-web built by a New Guinea orb-web spider. J. Nat. Hist. 6 (1972) 687
Robinson, M. H., B. Robinson: The stabilimenta of *Nephila clavipes* and the origins of stabilimentum-building in Araneids. Psyche 80 (1973) 277
Robinson, M. H., B. C. Robinson: Adaptive complexity: the thermoregulatory postures of the Golden-web spider, *Nephila clavipes*, at low latitudes. Amer. Midl. Natur. 92 (1974) 368
Robinson, M. H., B. Robinson: Evolution beyond the orb web: the web of the araneid spider *Pasilobus* sp., its structure, operation and construction. Zool. J. Linn. Soc. 56 (1975) 301
Robinson, M. H., B. Robinson: Associations between flies and spiders: bibiocommensalism and dipsoparasitism. Psyche 84 (1977) 150
Robinson, M. H., B. Robinson: The evolution of courtship systems in tropical araneid spiders. Symp. Zool. Soc. Lond. 42 (1978) 17
Robinson, M. H., B. Robinson: Comparative studies of the courtship and mating behaviour of tropical araneid spiders. Pacific Insects Monogr. 36 (1980) 1
Robinson, M. H., C. L. Valerio: Attacks on large or heavily defended prey by tropical salticid spiders. Psyche 84 (1977) 1
Roland, C., J. S. Rovner: Chemical and vibratory communication in the aquatic pisaurid spider *Dolomedes triton*. J. Arachnol. 11 (1983) 77
Roscoe, D. T., G. Walker: The adhesion of spiders to smooth surfaces. Bull. Brit. arachnol Soc. 8 (1991) 224
Ross, K., R. L. Smith: Aspects of the courtship behavior of the black widow spider *Latrodectus hesperus* (Araneae: Theridiidae), with evidence for the existence of a contact pheromone. J. Arachnol. 7 (1979) 69
Roth, V. D.: A new genus of spider (Agelenidae) from California exhibiting a third type of leg autospasy. Bull. Amer. Mus. Nat. Hist. 170 (1981) 137
Roth, V. D., B. M. Roth: A review of appendotomy in spiders and other arachnids. Bull. Brit. arachnol. Soc. 6 (1984) 137
Rovner, J. S.: Courtship in spiders without prior sperm induction. Science 152 (1966) 543
Rovner, J. S.: Copulation and sperm induction by normal and palpless male Linyphiid spiders. Science 157 (1967) 835
Rovner, J. S.: An analysis of display in the lycosid spider *Lycosa rabida* Walckenaer. Anim. Behav. 16 (1968a) 358
Rovner, J. S.: Territoriality in the sheet-web spider *Linyphia triangularis* (Clerck) (Araneae, Linyphiidae). Z. Tierpsychol. 25 (1968b) 232

Rovner, J. S.: Sound production by nearctic wolf spiders: a substratum-coupled stridulatory mechanism. Science 190 (1975) 1309

Rovner, J. S.: Detritus stabilimenta on the webs of *Cyclosa turbinata* (Araneae, Araneidae). J. Arachnol. 4 (1976) 215

Rovner, J. S.: Adhesive hairs in spiders: behavioral functions and hydraulically mediated movement. Symp. Zool. Soc. Lond. 42 (1978) 99

Rovner, J. S.: Morphological and ethological adaptations for prey capture in wolf spiders (Araneae, Lycosidae). J. Arachnol. 8 (1980a) 201

Rovner, J. S.: Vibration in *Heteropoda venatoria* (Sparassidae): a third method of sound production in spiders. J. Arachnol. 8 (1980b) 193

Rovner, J. S.: Morphological and ethological adaptations for prey capture in wolf spiders (Araneae, Lycosidae). J. Arachnol. 8 (1980c) 201

Rovner, J. S.: Spider hairiness: air stores and low activity enhance flooding survival in inland terrestrial species. Actas X Congr. Int. Arachnol. Jaca (España) 1986 (Vol. 1, p. 123)

Rovner, J. S.: Nests of terrestrial spiders maintain a physical gill: flooding and the evolution of silk constructions. J. Arachnol. 14 (1987) 327

Rovner, J. S.: Wolf spiders lack mirror-image responsiveness seen in jumping spiders. Anim. Behav. 38 (1989) 526

Rovner, J. S., F. G. Barth: Vibratory communication through living plants by a tropical wandering spider. Science 214 (1981) 464

Rovner, J., S. J. Knost: Post-immobilization wrapping of prey by lycosid spiders of the herbaceous stratum. Psyche 81 (1974) 398

Rovner, J. S., G. A. Higashi, R. F. Foelix: Maternal behaviour in wolf spiders: the role of abdominal hairs. Science 182 (1973) 1153

Rowell, D. M.: Complex sex-linked fusion heterozygosity in the Australian huntsman spider *Delena cancerides* (Araneae: Sparassidae). Chromosoma 93 (1985) 169

Rowell, D., L. Avilés: Sociality in a bark-dwelling huntsman spider from Australia, *Delena cancerides* Walckenaer (Aranea: Sparassidae). Ins. Soc. 42 (1995) 287

Ruhland, M.: Die neuromuskuläre Organisation normaler und regenerierter Laufbeine einer Vogelspinne (*Dugesiella hentzi* Ch., Orthognatha, Aviculariidae). Diss. Universität Konstanz 1976

Ruhland, M., W. Rathmayer: Die Beinmuskulatur und ihre Innervation bei der Vogelspinne *Dugesiella hentzi* (Ch.) (Araneae: Aviculariidae). Zoomorphologie 89 (1978) 33

Rypstra, A. L.: The effect of kleptoparasitis on prey consumption and web relocation in a Peruvian population of the spider *Nephila clavipes* (L.) (Araneae: Araneidae). Oikos 37 (1981) 179

Rypstra, A. L.: Aggregations of *Nephila clavipes* (L.) (Araneae, Araneidae) in relation to prey availability. J. Arachnol. 13 (1985) 71

Rypstra, A. L.: Prey capture and feeding efficiency of social and solitary spiders: a comparison. Acta Zool. Fennica 1540 (1990) 339

Sacher, P.: Rudimentäre Radnetze bei adulten Wespenspinnen (*Argiope bruennichi*). Veröff. Naturhist. Mus. Schleusingen 6 (1991) 30

Sadana, G. L.: A note on the entry of eggs into the oviduct of *Lycosa chaperi* Simon (Lycosidae: Araneida). Science Culture 36 (1970) 111

Saint Rémy, G.: Contribution à l'étude du cerveau chez les arthropodes trachéates. Arch. zool. exp. 5, Suppl. (1887) 1

Samu, F., F. Vollrath: Spider orb web as bioassay for pesticide side effects. Entomol. exp. appl. 62 (1992) 117

Sato, Y.: A spider, *Thomisius labefactus*, changing colour by reflected UV rays. The Insectarium Tokyo 24 (1987) 10
Savel-Niemann, A.: Tarantula (*Eurypelma californicum*) venom, a multicomponent system. Biol. Hoppe-Seyler 370 (1989) 485
Savory, T. H.: Daddy longlegs. Sci. Amer. 207 (1962) 119
Savory, T. H.: Evolution in the Arachnida. Merrow, Watford 1971
Schaefer, M.: Ökologische Isolation und die Bedeutung des Konkurrenzfaktors am Beispiel des Verteilungsmusters der Lycosiden einer Küstenlandschaft. Oecologia 9 (1972) 171
Schaefer, M.: Experimentelle Untersuchungen zur Bedeutung der interspezifischen Konkurrenz bei 3 Wolfspinnen-Arten (Araneida Lycosidae) einer Salzwiese. Zool. Jb. Syst. 101 (1974) 213
Schaefer, M.: Experimentelle Untersuchungen zum Jahreszyklus und zur Überwinterung von Spinnen (Araneida). Zool. Jb. Syst. 103 (1976a) 127
Schaefer, M.: An analysis of diapause and resistance in the egg stage of *Floronia bucculenta* (Araneida: Linyphiidae). Oecologia 25 (1976b) 155
Schaefer, M.: Winter ecology of spiders (Araneida). Z. ang. Ent. 83 (1977) 113
Schaible, U., C. Gack, H. F. Paulus: Zur Morphologie, Histologie und biologischen Bedeutung der Kopfstrukturen männlicher Zwergspinnen (Linyphiidae: Erigoninae). Zool. Jb. Syst. 113 (1986) 389
Schanbacher, F. L., C. K. Lee, J. E. Hall, I. D. Wilson, D. E. Howell, G. V. Odell: Composition and properties of tarantula *Dugesiella hentzi* (Girard) venom. Toxicon 11 (1973) 21
Schartau, W., T. Leidescher: Composition of the hemolymph of the tarantula *Eurypelma californicum*. J. comp. Physiol. 152 (1983) 73
Schenberg, S., F. A. Pereira Lima: *Phoneutria nigriventer*. Venom pharmacology and biochemistry of its components. In: Bucherl, W., E. Buckley, eds.: Venomous Animals and their Venoms III. Academic Press, New York 1971 (p. 279)
Scheuring, L.: Die Augen der Arachnoideen. Zool. Jb. Anat. 37 (1914) 369
Schildknecht, H., P. Kunzelmann, D. Krauss, C.Kuhn: Über die Chemie der Spinnwebe, 1. Arthropodenabwehrstoffe. LVII. Naturwissenschaften 59 (1972) 98
Schimkewitsch, W.: Etude sur l'anatomie de l'Epeire. Ann. Sci. Nat. 17 (1884) 7
Schlinger, E. I.: The biology of Acroceridae (Diptera): true endoparasites of spiders. In: Nentwig, W., ed.: Ecophysiology of Spiders. Springer Verlag, Berlin 1987 (p. 319)
Schmid, A., M. Duncker: Histamine immunoreactivity in the central nervous system of the spider *Cupiennius salei*. Cell Tissue Res. 273 (1993) 533
Schmid, A., G. Sperk, H. Reither: Quantitative determination of neuroactive substances in the CNS of the spider *Cupiennius salei* Keys. Comp. Biochem. Physiol. 102 (1992) 447
Schmidt, G: Giftspinnen—auch ein Problem des Ferntourismus. Münch. med. Wschr. 115 (1973) 2237
Schmitt, A., M. Schuster, F. G. Barth: Daily locomotor activity patterns in three species of *Cupiennius* (Araneae: Ctenidae): the males are the wandering spiders. J. Arachnol. 18 (1990) 249
Schmitt, A., Schuster, M., F. G. Barth: Male competetion in a wandering spider (*Cupiennius getazi*, Ctenidae). Ethology 90 (1992) 293
Schneider, P.: Elastic properties of the viscid silk of orb-weaving spiders (Araneidae). Naturwiss. 82 (1995) 144
Schoener, T. W., D. A. Spiller: Effect of lizards on spider populations: manipulative reconstruction of a natural experiment. Science 236 (1987) 949

Schoener, T. W., C. A. Toft: Spider populations: extraordinarily high densities on islands without top predators. Science 219 (1983) 1353

Schroer, W.-D.: Zum Mechanismus der Analyse polarisierten Lichtes bei *Agelena gracilens* C. L. Koch (Araneae, Agelenidae). Z. Morph. Tiere 79 (1974) 215

Schroer, W.-D.: Polarisationsempfindlichkeit rhabdomerialer Systeme in den Hauptaugen der Trichterspinne *Agelena gracilens* (Arachnida: Araneae: Agelenidae). Ent. Germ. 3 (1976) 88

Schüch, W., F. G. Barth: Temporal patterns in the vibratory courtship signals of the wandering spider *Cupiennius salei* Keys. Behav. Ecol. Sociobiol. 16 (1985) 263

Schulz, S., S. Toft: Identification of a sex pheromone from a spider. Science 260 (1993) 1635

Seibt, U., W. Wickler: Interspecific tolerance in social *Stegodyphus* spiders (Eresidae, Araneae). J. Arachnol. 16 (1988a) 35

Seibt, U., W. Wickler: Bionomics and social structure of "family spiders" of the genus *Stegodyphus*, with special reference to the African species *S. dumicola* and *S. mimosarum* (Araneida, Eresidae). Verh. naturwiss. Ver. Hamburg (NF) 30 (1988b) 255

Seibt, U., W. Wickler: Why do "family spiders" *Stegodyphus* (Eresidae) live in colonies? J. Arachnol. 16 (1988c) 193

Seitz, K.-A.: Normale Entwicklung des Arachniden-Embryos *Cupiennius salei* Keyserling und seine Regenerationsfähigkeit nach Röntgenbestrahlung. Zool. Jb. Anat. 83 (1966) 327

Seitz, K.-A.: Embryonale Defekt- und Doppelbildungen im Ei der Spinne *Cupiennius salei* (Ctenidae) als Folgen röntgeninduzierter Koagulationsarbeiten. Zool. Jb. Anat. 87 (1970) 588

Seitz, K.-A.: Licht- und elektronenmikroskopische Untersuchungen zur Ovarentwicklung und Oogenese bei *Cupiennius salei* Keys. (Araneae, Ctenidae). Z. Morph. Tiere 69 (1971) 283

Seitz, K-A.: Zur Histologie und Feinstruktur des Herzens und der Hämocyten von *Cupiennius salei* Keys. (Araneae, Ctenidae). I. Herzwandung. Bildung und Differenzierung der Hämocyten. Zool. Jb. Anat. 89 (1972a) 351

Seitz, K.-A.: Zur Histologie und Feinstruktur des Herzens und der Hämocyten von *Cupiennius salei* Keys (Araneae, Ctenidae). II. Zur Funktionsmorphologie der Phagocyten. Zool. Jb. Anat. 89 (1972b) 385

Seitz, K.-A.: Elektronenmikroskopische Untersuchungen an den Guaninspeicherzellen von *Araneus diadematus* Clerck (Araneae, Araneidae). Z. Morph. Tiere 72 (1972c) 245

Seitz, K.-A.: Licht- und elektronenmikroskopische Untersuchungen an den Malpighischen Gefässen der Spinne *Cupiennius salei* Keys. (Ctenidae, Araneae). Zool. Jb. Anat. 94 (1975) 413

Seitz, K.-A.: Zur Feinstruktur der Häutungshämocyten von *Cupiennius salei* Keys. (Araneae, Ctenidae). Zool. Jb. Anat. 96 (1976) 280

Selden, P. A.: Orb weaving spiders in the early Cretaceous. Nature (Lond.) 340 (1989) 711

Selden, P. A.: Lower cretaceous spiders from Sierra de Montsech, northeast Spain. Paleontology 33 (1990) 257

Selden, P. A., W. A. Shear, P. M. Bonamo: A spider and other arachnids from the Devonian of New York, and reinterpretations of Devonian Araneae. Paleontology 34 (1991) 241

Seligy, V. L.: Ommochrome pigments of spiders. Comp. Biochem. Physiol. 42 A (1972) 699 Seyfarth, E.-A.: Daily patterns of locomotor activity in a wandering spider. Physiol. Entomol. 5 (1980) 199

Seyfarth, E.-A.: Spider proprioception: receptors, reflexes, and control of locomotion. In: Barth, F. G., ed. : Neurobiology of Arachnids. Springer Verlag, Berlin 1985 (p. 230)

Seyfarth, E.-A., F. G. Barth: Compound slit sense organs on the spider leg: mechanoreceptors involved in kinesthetic orientation. J. comp. Physiol. 78 (1972) 176

Seyfarth, E.-A., J. Bohnenberger: Compensated walking of tarantula spiders and the effects of lyriform slit sense organ ablation. Proc. 8th Int. Arachnol. Congr. Wien 1980 (p. 249)

Seyfarth, E.-A., R. Hergenröder, H. Ebbes, F. G. Barth: Idiothetic orientation of a wandering spider: compensation of detours and estimates of goal distance. Behav. Ecol. Sociobiol. 11 (1982) 139

Seyfarth, E.-A., W. Eckweiler, K. Hammer: Proprioreceptors and sensory nerves in the legs of spider, *Cupiennius salei* (Arachnida, Araneida). Zoomorph. 105 (1985) 190

Seyfarth, E.-A., W. Gnatzy, K. Hammer: Coxal hair plates in spiders: physiology, fine structure, and specific central projections. J. comp. Physiol., A 166 (1990) 633

Seyfarth, E.-A., K. Hammer, U. Spörhase-Eichmann, M. Hörner, H. G. B. Vullings: Octopamine immunoreactive neurons in the fused central nervous system of spiders. Brain Res. 611 (1993) 197

Shear, W. A.: The evolution of social phenomena in spiders. Bull. Brit. arachnol. Soc. 1 (1970) 65

Shear, W. A., J. Kukalova-Peck: The ecology of paleozoic terrestrial arthropods: the fossil evidence. Can. J. Zool. 68 (1990) 1807

Shear, W. A., J. M. Palmer, J. A. Coddington, P. M. Bonamo: A Devonian spinneret: early evidence of spiders and silk use. Science 246 (1989) 479

Sherman, P. M.: The orb-web: an energetic and behavioural estimator of a spider's dynamic foraging and reproductive strategies. Anim. Behav. 48 (1994) 19

Sherman, R. G.: Ultrastructural features of cardiac muscle cells in a tarantula spider. J. Morph. 140 (1973a) 215

Sherman, R. G.: Ultrastructurally different hemocytes in a spider. Can. J. Zool. 51 (1973b) 1155

Sherman, R. G.: Chelicerates. In: Ratcliffe, N. A., A. F. Rowley, eds: Invertebrate Blood Cells, Vol. II. Academic Press, London 1981 (p. 355)

Sherman, R. G., A. R. Luff: Structural features of the tarsal claw muscles of the spider *Eurypelma marxi* Simon. Can. J. Zool. 49 (1971) 1549

Sherman, R. G., R. A. Pax: The heartbeat of the spider *Geolycosa missouriensis*. Comp. biochem. Physiol. 26 (1968) 529

Sherman, R. G., R. A. Pax: The spider heart. In: Kerkut, G.A., ed.: Experiments in Physiology and Biochemistry. Academic Press, London 1970a (p. 351)

Sherman, R. G., R. A. Pax: Spider cardiac physiology. II. Responses of a tarantula heart to cholinergic compounds. Comp. Gen. Pharmac. 1 (1970b) 171

Sherman, R. G., R. A. Pax: Spider cardiac physiology. III. Responses of a tarantula heart to certain catecholamines, aminoacids, and 5-hydroxytryptamine. Comp. Gen. Pharmac. 1 (1970c) 185

Sherman, R. G., C. R. Bursey, C. R. Fourtner, R. A. Pax: Cardiac ganglia in spiders (Arachnida, Araneae). Experientia (Basel) 25 (1969) 438

Sheumack, D. D., R. Claassens, N. M. Whitley, M. E. H. Howden: Complete amino acid sequence of a new type of lethal neurotoxin from the venom of the funnel-web spider *Atrax robustus*. FEBS Lett. 181 (1985) 15

Shiel, C. B., C. M. McAney, J. S. Fairley: Analysis of the diet of Natterer's bat *Myotis nattereri* and the common longeared bat *Plecotus auritus* in the west of Ireland. J. Zool. Lond. 223 (1991) 299

Shultz, J. W.: Walking and surface film locomotion in terrestrial and semi-aquatic spiders. J. exp. Biol. 128 (1987a) 427

Shultz, J. W.: The origin of the spinning apparatus in spiders. Biol. Rev. 62 (1987b) 89

Shultz, J. W.: Evolutionary morphology and phylogeny of Arachnida. Cladistics 6 (1990) 1

Shultz, J. W.: Evolution of locomotion in Arachnida: The hydraulic pressure pump of the giant whipscorpion, *Mastigoproctus giganteus* (Uropygi). J. Morph. 210 (1991) 13

Simon, E.: Histoire Naturelle des Araignées, Vol. 1 (p. 1084), Vol. 2 (p. 1030) Paris 1892–1903

Simon, U.: Die Spinnenzönosen (Arachn.: Araneae) der Kiefernrinde (*Pinus sylvestris* L.). Verh. Ges. Ökol. Freiburg 2 (1991) 107

Smith, D. S., U. Järlfors, F. E. Russell: The fine structure of muscle attachments in a spider (*Latrodectus mactans*, Fabr.). Tissue Cell 1 (1969) 673

Smith, R. B., T. P. Mommsen: Pollen feeding in an orb-weaving spider. Science 226 (1984) 1330

Sotelo, J. R., O. Trujillo-Cenóz: Electron microscope study of the vitellin body of some spider oocytes J. biophysic. biochem. Cytol. 3 (1957) 301

Spence, I., D. J. Adams, P. W. Gage: Funnel web spider venom produces spontaneous action potentials in nerve. Life Sci. 20 (1977) 243

Stewart, D. M., A. W. Martin: Blood and fluid balance of the common tarantula, *Dugesiella hentzi*. Z. vergl. Physiol. 70 (1970) 223

Stewart, P. M., A. W. Martin: Blood pressure in the tarantula, *Dugesiella hentzi*. J. comp. Physiol. 88 (1974) 141

Størmer, L.: Chelicerata. In: Moore, R. C., ed.: Treatise on Invertebrate Palaeontology, Arthropoda 2. University Press of Kansas, Lawrence, KS 1955 (p. 1)

Stowe, M. K.: Observations of two nocturnal orbweavers that build specialized webs: *Scoloderus cordatus* and *Wixia ectypa* (Araneae: Araneidae). J. Arachnol. 6 (1978) 141

Stowe, M. K.: Prey specialization in the Araneidae. In: Shear, W. A., ed.: Spiders. Webs, Behavior, and Evolution. Stanford University Press, Stanford 1986 (p. 101)

Stratton, C. E.: Behavioral studies of wolf spiders: a review of recent research. Rev. Arachnol. 6 (1985) 57

Stratton, G. E., G. W. Uetz: Acoustic communication and reproductive isolation in two species of wolf spiders. Science 214 (1981) 575

Stratton, G. E., G. W. Uetz: Communication via substratum-coupled stridulation and reproductive isolation in wolf spiders (Araneae: Lycosidae). Anim. Behav. 31 (1983) 164

Stratton, G. E., G. W. Uetz: The inheritance of courtship behavior and its role as a reproductive isolating mechanism in two species of *Schizocosa* wolf spiders (Araneae; Lycosidae). Evolution 40 (1986) 129

Strausfeld, N. J., F. G. Barth: Two visual systems in one brain: neuropils serving the secondary eyes of the spider *Cupiennius salei*. J. comp. Neurol. 328 (1993) 43

Strausfeld, N. J., P. Weltzien, F. G. Barth: Two visual systems in one brain: neuropils serving the principal eyes of the spider *Cupiennius salei*. J. comp. Neurol. 328 (1993) 63

Strazny, F., S. F. Perry: Morphometric diffusing capacity and functional anatomy of the book lungs in the spider *Tegenaria* spp. (Agelenidae). J. Morph. 182 (1984) 339

Streble, H.: Untersuchungen über das hormonale System der Spinnentiere (Chelicerata) unter besonderer Berücksichtigung des "endokrinen Gewebes" der Spinnen (Araneae). Zool. Jb. Physiol. 72 (1966) 157

Suter, R. B., A. J. Hirscheimer: Multiple web-borne pheromones in a spider *Frontinella pyramitela* (Araneae: Linyphiidae). Anim. Behav. 34 (1986) 748

Suzuki, H., A. Kondo: The second maturation division and fertilization in the spider *Achaearanea japonica* (Bös. et Str.). Zool. Sci. 11 (1994a) 433

Suzuki, H., A. Kondo: Changes at the egg surface during the first maturation division in the spider *Achaearanea japonica* (Bös. et Str.). Zool. Sci. 11 (1994b) 693

Szlep, R.: Developmental changes in the web spinning instinct of Uloboridae: construction of the primary-type web. Behaviour 17 (1961) 60

Tahiri, A., A. Horel, B. Krafft: Etude préliminaire sur les interactions mère–jeunes et jeunes–jeunes chez deux espèces d'*Amaurobius* (Araneae, Amaurobiidae) Rev. Arachnol. 8 (1989) 115

Tarsitano, M.S., R. R. Jackson: Influence of prey movements on the performance of simple detours by jumping spiders. Behaviour 123 (1992) 106

Thornhill, R.: Scorpionflies as kleptoparasites of web-building spiders. Nature (Lond.) 258 (1975) 709

Thornton, I. W. B., T. R. New: Krakatau invertebrates: the 1980s fauna in the context of a century of recolonization. Phil. Trans. Ser. B 322 (1988) 493

Thorp, R. W., W. D. Woodson: Black Widow, America's most poisonous spider. University North Carolina Press, Chapel Hill 1945 (p. 193)

Tietjen, W. J.: Dragline-following by male lycosid spiders. Psyche 84 (1977) 165

Tietjen, W. J.: Tests for olfactory communication in four species of wolf spiders (Araneae, Lycosidae). J. Arachnol. 6 (1979a) 197

Tietjen, W. J.: Is the sex pheromone of *Lycosa rabida* (Araneae, Lycosidae) deposited on a substratum? J. Arachnol. 6 (1979b) 207

Tietjen, W. J., J. S. Rovner: Trail following in two species of wolf spiders: sensory and etho-ecological concomitants. Anim. Behav. 28 (1980) 735

Tietjen, W. J., J. S. Rovner: Chemical communication in Lycosida and other spiders. In: Witt, P. N., J. S. Rovner, ed. : Spider Communication. Mechanisms and Ecological Significance. Princeton University Press, Princeton, New Jersey 1982 (p. 249)

Tillinghast, E. K.: Selective removal of glycoproteins from the adhesive spiral of the spider's orb web. Naturwiss. 68 (1981) 526

Tillinghast, E. K.: The chemical fractionation of the orb web of *Argiope* spiders. Insect Biochem. 14 (1984) 115

Tillinghast, E. K., M. Townley: Chemistry, physical properties, and synthesis of Araneidae orb webs. In: Nentwig, W., ed.: Ecophysiology of Spiders. Springer Verlag, Berlin 1987 (p. 203)

Tillinghast, E. K., S. F. Chase, M. A. Townley: Water extraction by the major ampullate duct during silk formation in the spider *Argiope aurantia* Lucas. J. Insect Physiol. 30 (1984) 591

Toft, S.: Life-histories of spiders in a Danish beech-wood. Nat. Jutl. 19 (1976) 5

Toft, S.: Phenology of some Danish beech-wood spiders. Nat. Jutl. 20 (1978) 285

Toft, S.: Microhabitat identity in two species of sheet-web spiders: field experimental demonstration. Oecologia 72 (1987) 216

Toft, S.: Mate guarding in two *Linyphia* species (Araneae: Linyphiidae). Bull. Brit. arachnol. Soc. 8 (1989) 33

Toft, S.: Interaction among two coexisting *Linyphia* spiders. Acta Zool. Fennica 190 (1990) 367

Tolbert, W.W.: Aerial dispersal behavior of two orb weaving spiders. Psyche 84 (1977) 13

Tretzel, E.: Reife- und Fortpflanzungszeit bei Spinnen. Z. Morph. Ökol. Tiere 42 (1954) 634

Tretzel, E.: Intragenerische Isolation und interspezifische Konkurrenz bei Spinnen. Z. Morph. Ökol. Tiere 44 (1955) 43

Tretzel, E.: Zum Begegnungsverhalten von Spinnen. Zool. Anz. 163 (1959) 194

Tretzel, E.: Biologie, Ökologie und Brutpflege von *Coelotes terrestris* (Wider) (Araneae, Agelenidae), T. I: Biologie und Ökologie. Z. Morph. Ökol. Tiere 49 (1961a) 658

Tretzel, E.: Biologie. Ökologie und Brutpflege von *Coelotes terrestris* (Wider) (Araneae, Agelenidae), T. II: Brutpflege Z. Morph. kol. Tiere 50 (1961b) 375

Trujillo-Cenóz, O.: Some aspects of the structural organization of the arthropod eye. Cold Spring Harb. Symp. quant. Biol. 30 (1965) 371

Turnbull, A. L.: The prey of the spider *Linyphia triangularis* (Clerck) (Araneae, Linyphiidae). Can. J. Zool. 38 (1960) 859

Turnbull, A. L.: Quantitative studies of the food of *Linyphia triangularis* Clerck (Araneae: Linyphiidae). Can. Entomol. 94 (1962) 1233

Turnbull, A. L.: The search for prey by a web-building spider *Achaearanea tepidariorum* (C. L. Koch) (Araneae, Theridiidae). Can. Entomol. 96 (1964) 568

Turnbull, A. L.: Ecology of the true spiders (Araneomorphae). Ann. Rev. Entomol. 18 (1973) 305

Tzeng, M. C., P. Siekevitz: The effect of the purified major protein factor (α-latrotoxin) of black widow spider venom on the release of acetylcholine and norepinephrine from mouse cerebral cortex. Brain Res. 139 (1978) 190

Ude, J., K. Richter: The submicroscopic morphology of the heart ganglion of the spider *Tegenaria atrica* (C. L. Koch) and its neuroendocrine relations to the myocard. Comp. biochem. Physiol. 48A (1974) 301

Uetz, G. W.: Web building and prey capture in communal orb weavers. In: Shear, W. A., ed.: Spiders. Webs, Behavior, and Evolution. Stanford University Press, Stanford 1986 (p. 207)

Uetz, G. W., G. Denterlein: Courtship behavior, habitat and reproductive isolation in *Schizocosa rovneri* Uetz u. Dondale (Araneae; Lycosidae). J. Arachnol. 7 (1979) 121

Uetz, G. W., G. E. Stratton: Acoustic communication and reproductive isolation in spiders. In: Witt, P. N., J. S. Rovner, eds: Spider Communication. Mechanisms and Ecological Significance. Princeton University Press, Princeton, New Jersey 1982 (p. 123)

Uetz, G. W., A. D. Johnson, D. W. Schemske: Web placement, web structure, and prey capture in orb-weaving spiders. Bull. Brit. arachnol. Soc. 4 (1978) 141

Uhl, G.: Sperm storage and repeated egg production in female *Pholcus phalangioides* Fuesslin (Araneae). Bull. Soc. neuchâtel. Sci. nat. 116 (1993) 245

Uhl, G.: Genital morphology and sperm storage in *Pholcus phalangioides* (Fuesslin, 1775) (Pholcidae; Araneae). Acta Zool. 75 (1994) 1

Uhl, G., B. A. Huber, W. Rose: Male pedipalp morphology and copulatory mechanism in *Pholcus phalangioides* (Fuesslin, 1775) (Araneae, Pholcidae). Bull. Brit. arachnol. Soc. 10 (1995) 1

Vachon, M.: Contribution à l'étude du développement postembryonnaire des araignées. Première note. Généralités et nomenclature des stades. Bull. Soc. Zool. France 82 (1957) 337

Van Wingerden, W. K. R. E., H. F. Vugts: Factors influencing aeronautic behaviour of spiders. Bull. Brit. arachnol. Soc. 3 (1974) 6

Vogel, B. R.: Individual interactions of *Pardosa*. Armadillo Pap. 5 (1971) 1

Vogel, B. R.: Apparent niche sharing of two *Pardosa* species (Araneidae, Lycosidae). Armadillo Pap. 7 (1972) 1

Vollrath, F.: Konkurrenzvermeidung bei tropischen kleptoparasitischen Haubennetzspinnen der Gattung *Argyrodes* (Arachnida: Araneae: Theridiidae). Ent. Germ. 3 (1976) 104

Vollrath, F.: A close relationship between two spiders (Arachnida; Araneidae): *Curimagua bayano* synecious on a Diplura species. Psyche 85 (1978) 347

Vollrath, F.: Vibrations: their signal function for a spider kleptoparasite. Science 205 (1979) 1149

Vollrath, F.: Male body size and fitness in the web-building spider *Nephila clavipes*. Z. Tierpsychol. 53 (1980) 61

Vollrath, F.: Gravity as an orientation guide during web construction in the orb spider *Araneus diadematus* (Araneae, Araneidae). J. comp. Physiol. A 159 (1986a) 275

Vollrath, F.: Eusociality and extraordinary sex ratios in the spider *Anelosimus eximius* (Araneae: Theridiidac). Behav. Ecol. Sociobiol. 18 (1986b) 283

Vollrath, F.: Spiders with regenerated legs can build normal webs. Nature (Lond.) 328 (1987a) 247

Vollrath, F.: Kleptobiosis in spiders. In: Nentwig, W., ed.: Ecophysiology of Spiders. Springer Verlag, Berlin 1987b (p. 274)

Vollrath, F.: Spiral orientation of *Araneus diadematus* orb webs built during vertical rotation. J. comp. Physiol. A 162 (1988) 413

Vollrath, F.: Leg regeneration in web spiders and its implications for orb weaver phylogeny. Bull. Brit. arachnol. Soc. 8 (1990) 177

Vollrath, F.: Spider webs and silk. Sci. Amer. 266 (1992) 52

Vollrath, F.: Spinnenseide—Superwerkstoff der Natur. In: Kunststoffe im Automobilbau. VDI-Verlag, Düsseldorf 1993 (p. 1–18)

Vollrath, F., D. T. Edmonds: Modulation of the mechanical properties of spider silk by coating with water. Nature (Lond.) 340 (1989) 305

Vollrath, F., W. Mohren: Spiral geometry in the garden spider's orb web. Naturwiss. 72 (1985) 666

Vollrath, F., E. K. Tillinghast: Glycoprotein glue beneth a spider web's aqueous coat. Naturwiss. 78 (1991) 557

Vollrath, F., W. J. Fairbrother, R. J. P. Williams, E. K. Tillinghast, D.T. Bernstein, K. S. Gallagher, M.A. Townley: Compounds in the droplets of the orb spider's viscid spiral. Nature (Lond.) 345 (1990) 526

Vugts, H., W. K. R. E. van Wingerden: Meteorological aspects of aeronautic behaviour of spiders. Oikos 27 (1976) 433

Walcott, C., W. G. van der Kloot: The physiology of the spider vibration receptor. J. exp. Zool. 141 (1959) 191

Walker, R. J.: Current trends in invertebrate neuropharmacology. Verh. dt. Zool. Ges. (1982) 31

Wallstabe, P.: Beiträge zur Kenntnis der Entwicklungsgeschichte der Araneinen. Die Entwicklung der äusseren Form und Segmentierung. Zool. Jb. Anat. 26 (1908) 683

Wanke, E., A. Ferroni, P. Gattanini, J. Meldolesi: α-latrotoxin of the black widow spider venom opens a small, non-closing cation channel. Biochem. Biophys. Res. Comm. 134 (1986) 320

Ward, D., J. R. Henschel: Experimental evidence that a desert parasitoid keeps its host cool. Ethology 92 (1992) 135

Ward, D., Y. Lubin: Habitat selection and the life history of a desert spider, *Stegodyphus lineatus* (Eresidae). J. Anim. Ecol. 62 (1993) 353

Ward, P. I., M. M. Enders: Conflict and cooperation in the group feeding of the social spider *Stegodyphus mimosarum*. Behaviour 94 (1985) 167

Watson, P. J.: Transmission of a female sex pheromone thwarted by males in the spider *Linyphia litigosa* (Linyphiidae). Science 233 (1986) 219

Watson, P. J.: Multiple paternity and first mate sperm precedence in the sierra dome spider, *Linyphia litigiosa* Keyserling (Linyphiidae). Anim. Behav. 41 (1991) 135

Wehner, R.: Polarized-light navigation by insects. Sci. Amer. 235 (1976) 106

Wehner, R.: Spatial vision in arthropods. In: Autrum, H., ed.: Handbook of Sensory Physiology. Springer Verlag, Berlin 1981 (p. 287)

Weigel, G.: Färbung und Farbwechsel der Krabbenspinne *Misumena vatia.* Z. vergl. Physiol. 29 (1941) 195

Weiss, I.: Konstruktions- und Funktionsanalyse der Kopulationsorgane von *Zodarion aurorae* n. sp. aus Rumänien (Arachnida, Araneae, Nesticidae). Reichenbachia 20 (1982) 77

Weltzien, P., F. G. Barth: Volumetric measurements do not demonstrate that the spider brain "central body" has a special role in web building. J. Morph. 208 (1991) 91

Weygoldt, P.: Communication in crustaceans and arachnids. In: Sebeok, T. A., ed.: How Animals Communicate. Indiana University Press, Bloomington, IN 1977 (p. 303)

Weygoldt, P., H. F. Paulus: Untersuchungen zur Morphologie, Taxonomie, und Phylogenie der Cheliceraten. II. Cladogramme und die Entfaltung der Chelicerata. Z. zool. Syst. Evol. Forsch. 17 (1979) 177

Whitehead, W. F., J. G. Rempel: A study of the musculature of the Black Widow Spider, *Latrodectus mactans* (Fabr.). Can. J. Zool. 37 (1959) 831

Whitehouse, M. E. A.: The foraging behaviours of *Argyrodes antipodiana* (Theridiidae), a kleptoparasitic spider from New Zealand. N. Zeal. J. Zool. 13 (1986) 151

Wickler, W., U. Seibt: Aerial dispersal by ballooning in adult *Stegodyphus mimosarum.* Naturwiss. 73 (1986) 628

Wiehle, H.: Beiträge zur Kenntnis des Radnetzbaues der Epeiriden, Tetragnathiden und Uloboriden. Z. Morph. Ökol. Tiere 8 (1927) 468

Wiehle, H.: Beiträge zur Biologie der Araneen, insbesondere zur Kenntnis des Radnetzbaues. Z. Morph. Ökol. Tiere 11 (1928) 115

Wiehle, H.: Weitere Beiträge zur Biologie der Araneen, insbesondere zur Kenntnis des Radnetzbaues. Z. Morph. Ökol. Tiere 15 (1929) 262

Wiehle, H.: Neue Beiträge zur Kenntnis des Fanggewebes der Spinnen aus den Familien Argiopidae, Uloboridae und Theridiidae. Z. Morph. Ökol. Tiere 22 (1931) 349

Wiehle, H.: Vom Fanggewebe einheimischer Spinnen. Neue Brehm-Bücherei. Akademische Verlagsgesellschaft Geest & Portig, Leipzig 1949

Wiehle, H.: Der Embolus des männlichen Spinnentasters. Zool. Anz. Suppl. 74 (1961) 457

Wiehle, H.: *Meta*—eine semientelegyne Gattung der Araneae (Arach.). Senck. biol. 48 (1967) 183

Willem, V.: Observations sur la circulation sanguine et la respiration pulmonaire chez les araignées. Arch. néerl. sci. 1 (1918) 226

Willey, M. B., R. R. Jackson: Olfactory cues from conspecifics inhibit the web-invasion behavior of *Portia,* web-invading araneophagic jumping spiders (Araneae: Salticidae). Can. J. Zool. 71 (1993) 1415

Williams, D. S.: The feeding behavior of New Zealand *Dolomedes* species (Araneae: Pisauridae). N. Zeal. J. Zool. 6 (1979) 95

Williams, D. S., P. McIntyre: The principal eyes of a jumping spider have a telephoto component. Nature (Lond.) 288 (1980) 578

Wilson, D. M.: Stepping patterns in tarantula spiders. J. exp. Biol. 47 (1967) 133

Wilson, E. O.: The Insect Societies. Belknap Press of Harvard University Press, Cambridge, Massachusetts 1971 (p. 130)

Wilson, R. S.: The structure of the dragline control valves in the garden spider. Quart. J. microsc. Sci. 103 (1962a) 549

Wilson, R. S.: The control of the dragline spinning in the garden spider. Quart. J. microsc. Sci. 104 (1962b) 557

Wilson, R. S.: The heartbeat of the spider *Heteropoda venatoria*. J. Insect Physiol. 13 (1967) 1309

Wilson, R. S.: Some comments on the hydrostatic system of spiders (Chelicerata, Araneae). Z. Morph. Tiere 68 (1970) 308

Wilson, R. S., J. Bullock: The hydraulic interaction between prosoma and opisthosoma in *Amaurobius ferox* (Chelicerata, Arancac). Z. Morph. Tiere 74 (1973) 221

Wirth, E., F. G. Barth: Forces in the spider orb web. J. comp. Physiol. A 171 (1992) 359

Wise, D. H.: Food limitation of the spider *Linyphia marginata*: experimental food studies. Ecology 56 (1975) 637

Wise, D. H.: Predation by a commensal spider, *Argyrodes trigonum*, upon its host: an experimental study. J. Arachnol. 10 (1982) 111

Wise, D. H.: Spiders in Ecological Webs. Cambridge University Press, Cambridge 1993

Wise, D. H., L. L. Barata: Prey of two syntopic spiders with different web structures. J. Arachnol. 11 (1983) 271

Witt, P. N.: Die Wirkung von Substanzen auf den Netzbau der Spinne als biologischer Test. Springer Verlag, Berlin 1956

Witt, P. N.: Behavioral consequences of laser lesions in the central nervous system of *Araneus diadematus* Cl. Amer. Zool. 9 (1969) 121

Witt, P. N.: Drugs alter web-building of spiders: a review and evaluation. Behav. Sci. 16 (1971) 98

Witt, P. N.: The web as a means of communication. Biosci. Comm. 1 (1975) 7

Witt, P. N., C. F. Reed: Spider web building. Measurement of web geometry identifies components in a complex invertebrate behaviour pattern. Science 149 (1965) 1190

Witt, P. N., C. F. Reed, D. B. Peakall: A spider's web. Springer Verlag, Berlin 1968

Witt, P. N., J. O. Rawlings, C. F. Reed: Ontogeny of web-building behaviour in two orb-weaving spiders. Amer. Zool. 12 (1972) 445

Witt, P. N., M. B. Scarboro, D. B. Peakall, R. Gause: Spider web-building in outer space: evaluation of records from the skylab spider experiment. Amer. J. Arachnol. 4 (1977) 115

von Wittich, G. H.: Observationes quaedam de Aranearum ex ovo evolutione. Inaug.-Diss. Halle 1845

Wolf, A.: *Cheiracanthium punctorium*—Portait einer berüchtigten Spinne. Natur. u. Mus. 118 (1988) 210

Wolz, I.: Zur Oekologie der Bechsteinfledermaus *Myotis bechsteini* (Kuhl 1818). (Mammalia: Chiroptera). Diss. Universität Erlangen-Nürnberg (1992)

Wood, F. D.: Autotomy in Arachnida. J. Morphol. 42 (1926) 143

Work, R. W.: Dimensions, birefringences, and force elongation behavior of major and minor ampullate silk fibers from orbweb-spinning spiders. The effects of wetting on these properties. Textile Res. J. 47 (1977) 650

Work, R. W., N. Morosoff: A physicochemical study of the super contraction of spider major ampullate silk fibers. Textile Res. J. 52 (1982) 349

Work, R. W., C. T. Young: The amino acid compositions of major and minor ampullate silks of certain orb-web building spiders (Araneae, Araneidae). J. Arachnol. 15 (1987) 65

Wunderlich, J.: Mitteleuropäische Spinnen der Baumrinde (Araneae). Z. angew. Entom. 94 (1982) 9

Wunderlich, J.: Spinnenfauna gestern und heute. Fossile Spinnen in Bernstein und ihre heute lebenden Verwandten. Bauer Verlag, Wiesbaden 1986

Wunderlich, J.: Die fossilen Spinnen im Dominikanischen Bernstein. Wunderlich, Straubenhardt 1988

Wurdak, E., R. Ramousse: Organisation sensorielle de la larve et de la première nymphe chez l'araignée *Araneus suspicax* (O. Pickard-Cambridge). Rev. Arachnol. 5 (1984) 287

Xu, M., R.V. Lewis: Structure of a protein superfiber: spider dragline silk. Proc. natl. Acad. Sci. 87 (1990) 7120

Yamashita, S., H. Tateda: Spectral sensitives of jumping spider eyes. J. comp. Physiol. 105 (1976) 29

Yoshikura, M.: Embryological studies on the Liphistiid spider *Heptathela kimurai*. Kumamoto J. Sci. B 2 (1955) 607

Yoshikura, M.: Comparative embryology and phylogeny of Arachnida. J. Sci. Biol. 12 (1975) 71

Young, A. M.: Foraging for insects by a tropical hummingbird. Condor 73 (1971) 36

Young, M. R., F. Wanlass: Observations on fluorescence and function of spider's eyes. J. Zool. Proc. Zool. Soc. Lond. 151 (1967) 1

Zachariassen, K. E.: Physiology of cold tolerance in insects. Amer. Physiol. Soc. 65 (1985) 799

Zahl, P. A.: What's so special about spiders? Nat. Geogr. 140 (1971) 190

Zahler, S., S. Bihlmayer, J. Markl, R. Paul: Feinanatomie des Gefäßsystems und Herzphysiologie bei der Vogelspinne *Eurypelma*. Verh. Dtsch. Zool. Ges. 83 (1990) 564

Zebe, E., W. Rathmayer: Elektronenmikroskopische Untersuchungen an Spinnenmuskeln. Z. Zellforsch. 92 (1968) 377

Zill, S. N., M. A. Underwood, Carter Rowley III, Jr., D. T. Moran: A somatotopic organization of groups of afferents in insect peripheral nerves. Brain Res. 198 (1980) 253

Zimmermann, E. W.: Untersuchungen über den Bau des Mundhöhlendaches der Gewebespinnen. Rev. Suisse Zool. 41 (1934) 149

Zschokke, S.: The influence of the auxiliary spiral on the capture spiral in *Araneus diadematus* Clerck (Araneidae). Bull. Brit. arachnol. Soc. 9 (1993) 169

Zschokke, S., F. Vollrath: Unfreezing spider behaviour: orb web geometry and construction behaviour. Ethology (1995) in press

Index